Palgrave Studies in Natural Resource Management

Series Editor
Justin Taberham
London, UK

This series is dedicated to the rapidly growing field of Natural Resource Management (NRM). It aims to bring together academics and professionals from across the sector to debate the future of NRM on a global scale. Contributions from applied, interdisciplinary and cross-sectoral approaches are welcome, including aquatic ecology, natural resources planning and climate change impacts to endangered species, forestry or policy and regulation. The series focuses on the management aspects of NRM, including global approaches and principles, good and less good practice, case study material and cutting edge work in the area.

More information about this series at
http://www.palgrave.com/gp/series/15182

Meg Parsons • Karen Fisher
Roa Petra Crease

Decolonising Blue Spaces in the Anthropocene

Freshwater management in Aotearoa New Zealand

palgrave
macmillan

Meg Parsons
School of Environment
University of Auckland
Auckland, New Zealand

Karen Fisher
School of Environment
University of Auckland
Auckland, New Zealand

Roa Petra Crease
School of Environment
University of Auckland
Auckland, New Zealand

Palgrave Studies in Natural Resource Management
ISBN 978-3-030-61070-8   ISBN 978-3-030-61071-5   (eBook)
https://doi.org/10.1007/978-3-030-61071-5

To Ali and Den Parsons for all their aroha

Acknowledgments

We wish to thank various people for their contribution to this book. First, we need to acknowledge the tremendous work undertaken by Leane Makey throughout the research collection phase of this project. She helped us collect interviews, drove numerous research team members (ourselves, research assistants and students) around the Waipā River catchment on various site visits, and used for superior GIS skills to create maps. Accordingly, we offer our special thanks to Leane, ehara koe i a ia! Second, we would like to express our gratitude to our three other research assistants on the project: Esther McGill, Bradley Jones, and Jack Barrett. Their collective efforts to collect and organise our data sets were an invaluable contribution to our capacities to analyse and write up our results, so a big tēnā koutou. Our third thank you, tēnā kōrua, needs to be given to Melanie Mayall-Nahi and Heather Paterson-Shallard, who occupied numerous roles during our research project (Honours and Masters' students, research assistants, friends), and provided us with different perspectives, new tangents to journey down, and practical help at key stages.

We also recognise that the book and research for the book was supported (directly and indirectly) by numerous people within The University of Auckland. The long journey leading to this book began in 2009 with attendance at hui and wānanga about the Waipā River settlement, and which then led to securing a research grant from the Marsden Fund in 2015 to explore the environmental geographies of the Waipā River. We apologise in advance if we forget to acknowledge some of the people who have supported and guided us along the way. Our special thanks to our colleagues from the School of the Environment and the wider Faculty of Science at The University of Auckland including: Mrs Anna-Marie Simcock; Professor Richard Le Heron; Associate Professor Nick Lewis; Professor Paul Kench; Associate Professor JR Rowland; Dr David Hayward; Dr Gretel Boswijk, and Ms Mel Wall. They all provided us with advice, empathic words of support, and practical actions that helped us undertake our research when we (the author team) faced incredibly difficult personal circumstances that necessitated us adopting a different and highly flexible way of working. In addition, the mahi of other University of Auckland staff was critical to the research being undertaken and the book written. Most notably Wendy Rhodes, Amy Weir and Kathryn Howard supported us with contract variations, sub-contracting and personnel requests, reporting requirements, budget inquiries, and helped to ensure the book could be made open access. Ngā mihi nui ki a koutou mo to tautoko.

This book and research for the book was supported by the Marsden Fund Council from Government funding, managed by Royal Society Te Apārangi.

We also express our gratitude to the following organisations for their assistance with the collection of our historical data: Alexander Turnbull Library; Archives of New Zealand; Auckland Libraries; Hamilton City Libraries. Our special thanks are extended to the staff of Māniapoto Māori Trust Board, Waikato River Authority, Pūniu River Care, and Waikato Regional Council as well as others we encountered in completing the research for this book. We are particularly grateful to the assistance provided by Ngahuia Herangi at various stages of this journey. We also express our very great appreciation to all the research participants who were interviewed as part of our research project and whose words are quoted in this book, a huge tēnā koutou to you all.

Our personal acknowledgements are as follows:

Meg Parsons would like to express her tremendous thanks to her parents Ali and Dennis Parsons for their endless support, compassion, and practical problem-solving efforts that enable Meg to undertake researching and writing the book in between hospital stays, medical treatments, and the lockdown. She would also like to offer her special thanks to Dr Anthony Jordan for his professional expertise that helped ensure that Meg was able to keep working despite her health-related challenges.

Roa Crease would like to acknowledge her parents, Marivic and Luke, and partner, Kelvin, for their encouragement and support. She would like to thank her mum, in particular, for her unwavering support, positivity and kindness.

Karen Fisher would like to acknowledge her parents, Carol (Ngāti Maniapoto, Waikato-Tainui) and Graeme, both of whom were proud children of the King Country and both of whom would have been delighted to see this book come to fruition. To her Ngāti Maniapoto whanau—ngā mihi ki a koutou katoa.

Contents

Abbreviations

Crown	New Zealand Crown
EJ	Environmental Justice
ETS	The New Zealand Emissions Trading Scheme
GEC	The Guardians Establishment Committee
GHG	Greenhouse Gas
IEJ	Indigenous Environmental Justice
IMP	Iwi Management Plan
JMA	Joint Management Agreement
MMTB	Maniapoto Māori Trust Board
NPSFM	The National Policy Statement on Freshwater Management
RMC	Te Nehenehenui Regional Management Committee
NIWA	National Institute of Water and Atmospheric Research
ODC	Otorohanga District Council
OTS	Office of Treaty Settlements
PAC	The Pollution Advisory Council
RMA	Resource Management Act 1991
SHMAK	Stream Health Monitoring Assessment kit
UNDRIP	The United Nations Declaration on the Rights of Indigenous Peoples
UNFCCC	The United Nations Framework on Climate Change Convention
V&S	Vision and strategy
WDC	Waikato District Council
WRA	Waikato Regional Authority
WRC	Waikato Regional Council
WRCuT	The Waikato River Clean-up Trust

List of Figures

List of Tables

1

Introduction

Freshwater is essential to the health and wellbeing of both human and ecological communities. Around the world, including within Aotearoa New Zealand (henceforth Aotearoa), there are increasing imbalances between freshwater demand and supplies, with notable declines in the quality and quantity of water, and issues to do with who can and should be able to access and use freshwater and freshwater biota (Bradford et al. 2016; de Leeuw 2017; Deitz and Meehan 2019; Julian et al. 2017; Larned et al. 2016; Mohai 2018; National Institute of Water and Atmospheric Research Ltd 2014; Wutich 2009). Many commentators term these issues, contestations and conflicts as a water crisis, with the abilities of populations to access and use freshwater (both now and in the future) viewed by some as one of the most ubiquitous social, political, cultural, economic and ecological issues that humanity faces in the twenty-first century (Moggridge 2018; Paerregaard and Andersen 2019). In Aotearoa, which provides the case study for this book, the freshwater systems are affected by ongoing degradation directly connected to human activities over the last two centuries (Knight 2016, 2019; Larned et al. 2016). Scientists suggest that the most pressing problems include heavy metals, nutrient contamination (such as e coli. and nitrogen from

© The Author(s) 2021
M. Parsons et al., *Decolonising Blue Spaces in the Anthropocene*, Palgrave Studies in Natural Resource Management, https://doi.org/10.1007/978-3-030-61071-5_1

effluent), loss of biodiversity, over-extraction (connected to the expansion of irrigated agriculture), flooding, and invasive species (Ballantine and Davies-Colley 2014; Biggs et al. 1990; Bollen 2015; Boulton et al. 1997; Duncan 2017; Hughes and Quinn 2014; McDowell et al. 2016; Wilcock et al. 1999). Such problems present ongoing threats to human health and wellbeing and those of other life forms within Aotearoa. Recently scholars have begun to question the efficacy of established management approaches, the extent to which current land-use practices are to blame and whether the continued environmental decline in our waterways is inevitable (Knight 2016; Salmond et al. 2014; Te Aho 2015). The continued degradation of freshwater systems under conventional management approaches, moreover, necessitates a rethinking of how freshwater systems are governed, managed, and restored.

In this book, we explore the origins of the freshwater crisis (a manifestation of multiple environmental injustices) within a single freshwater system: the Waipā River (Te Waipā Awa). The headwaters of the Waipā River are located in the Rangitoto Range (the present-day the township Te Kuiti lies to the east) within Te Rohe Pōtae district (also known as the King Country) (Cunningham 2014). The river flows through hill country where it is joined by various tributaries (Otamaroa, Okurawhango, Tunawaea and Waimahora Streams). The Mangapu, Mangawhero, and Mangaokewa streams join the Waipā River at Otorohanga. The Waipa River's downstream journey passes the mountain (maunga) of Pirongia, where it is joined by tributaries including Maunguika, Ngakoaohia, and Turitea. The largest tributary Puniu River meets the Waipā River soon after, just south of the township of Te Awamutu; followed on by the streams Mangapiko and Mangaotama, which drain from two lakes (Lake Ngāroto and Lake Mangakawere). The journey of the Waipā River ends at the township of Ngāruawāhia where the Waipā River joins with the Waikato River (Fig. 1.1).

Te Waipā o Awa is now, in 2020, one of the most degraded freshwater systems in Aotearoa. Yet, as we argue through this book, the waters of the Waipā were not always unhealthy, nor was its degradation an inevitable consequence of the establishment of Aotearoa as a developed democratic (settler-colonial) nation-state. Instead, we demonstrate how the deterioration of the Te Waipā o Awa was a product of incremental actions taken

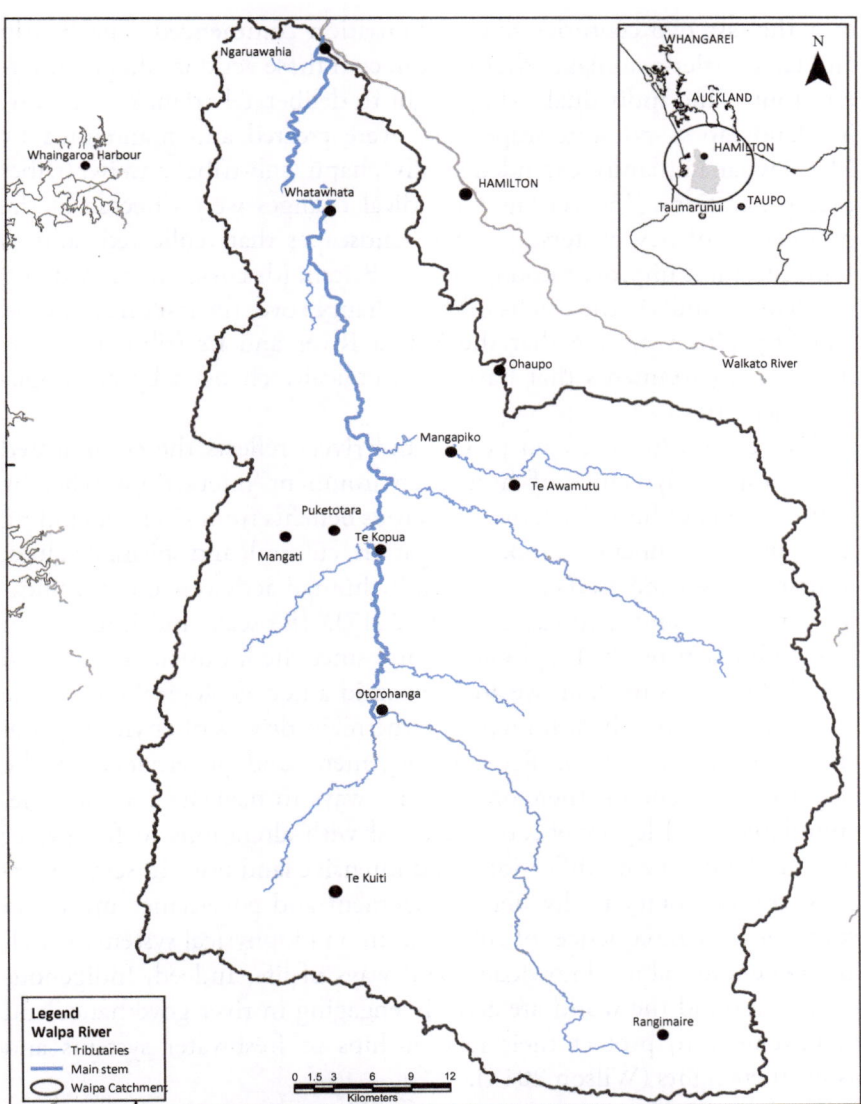

Fig. 1.1 Map showing location of the Waipā River

over the last two centuries since colonisation commenced. The British and later settler-colonial authorities directed these actions, aligned interest groups, and individuals who sought to deliberately remake the existing landscapes and waterscapes that were created and maintained by Māori whanau (family, extended family), hapū (sub-tribe) and iwi (tribe) over generations. The actions for radical changes were directed at the production of new waterscapes and landscapes that replicated (and in some instances improved upon) those of Britain (discussed in more depth in Chaps. 3 and 4). The results of these changes over the last one hundred and fifty plus years are that the Waipā River and its tributaries flow through environments that have been radically changed by anthropogenic activities.

The relationship between people and rivers reflects the complicated and complex dynamics of human-environment interactions whereby humans and nonhumans derive numerous benefits from rivers (including life-supporting functions, food, navigation, cultural, and spiritual values, to name a few) and rivers are affected by human activities and modifications to rivers and landscapes (Kelly 2017). The scale and intensity of human impacts on the biophysical world since the Industrial Revolution have led scholars to claim we are now amid a new geological epoch, the Anthropocene, in which humans are the main drivers of environmental change (Crutzen 2002). River management and governance in the Anthropocene entails, therefore, finding ways to navigate the multiple, cumulative, and legacy effects associated with alterations to freshwater systems, landscape modifications, and intensive land uses. In settler societies, contemporary freshwater management and governance must also confront the consequences of colonisation on biophysical systems as well as Indigenous values, knowledges and ways of life. Indeed, Indigenous peoples around the world are actively engaging in river governance and management to protect their relationships to freshwater systems and assert their rights (Wilson 2014).

The term "Anthropocene" was invented by an atmospheric chemist in 2000 and defined a new geological epoch that followed the Holocene wherein human beings acted as geological agents that altered the global Earth systems (Steffen et al. 2011b, 2015). The whole purpose of identifying the Anthropocene is to define the end of one geological epoch and the start of a new one. The Anthropocene also indicates a period of time now or in the near future where radical climate destabilisation causes drastic environmental changes, and there is no capacity to return to past conditions (as research into tipping points emphasises) (Steffen et al. 2011b). The Working Group on the Anthropocene, which met in August 2016, recommended the mid-twentieth century as the preferred boundary as there were so many measurable changes from this time onwards. Steffen et al. (2015) and many other researchers term this period as the "great acceleration" wherein changes became globally observable and written into the geologic strata (geological markers include mass extinctions, carbon dioxide levels, and abundant use of petrochemicals including plastics) (Steffen et al. 2015, p. 81). Scientists are continuing to debate what indicators should be used to mark out and measure the Anthropocene using scientific language, levels of greenhouse gas emissions (GHG), and other indicators of change within the geological records. Some scholars propose different start dates and measurements of radical environmental changes (Castree 2014; Steffen et al. 2011a, b). Countless environmental issues are tied to the Anthropocene, including biodiversity loss, pollution of land and water bodies, toxic waste and contamination, and the ongoing and worsening impacts of climate change.

While geologists are still debating the 'ifs' and 'when' the Anthropocene began (indeed some dispute the Anthropocene as an epoch), the term is widely embraced outside geological discussions. The Anthropocene—the reverberating concept that is promoting a wealth of new transdisciplinary research—is proving to be a particular intellectual meeting point for scholars from across disciplinary divides (Bashford 2013, p. 346). Scholars from across the traditional disciplinary divides of the humanities, social sciences and natural sciences are all wrestling with the implications of the Anthropocene. Historian Alison Bashford writes of her puzzlement over why humanities and social science scholars continue to largely cede "the Anthropocene to climate scientists" and approach the concept as "as if it were not their territory" (Bashford 2013, p. 346). She argues that the Anthropocene does not refer to the distant past or even prehistory (to so-called "deep time"), but the last two hundred or so years (Hulme 2011; Mahony and Hulme 2018). It is the modern era (intractably linked to the Industrial Revolution, global capitalism, European colonialism).

Thus, the Anthropocene is a perfectly recognisable topic of study for historians and social scientists to investigate. It is neither an issue of the ancient past or the distant future; it is a problem in and of modernity. While climate change involves biophysical phenomena, it is not a matter of geological time itself. The Anthropocene, therefore, lies firmly within the comfort zone of scholars whose investigations focus on settler-colonialism, Indigenous studies, and environmental justice.

We are not scientists and therefore cannot evaluate the question of start dates in terms of scientific accuracy or merit. However, following on from the works of scholars including Davis and Todd (2017), we suggest that the dating of the Anthropocene to mid-twentieth century overlooks the critical value of examining the concept and expanding it beyond its "current Eurocentric framing" (Davis and Todd 2017, p. 763). Scholars, both Indigenous (Todd and Whyte) and non-Indigenous (Lewis, Maslin and Davis), argue that the beginning of the Anthropocene is inextricably bound to the commencement of European colonialism (first the Americas and later elsewhere) and its compatriot, settler-colonialism (Davis and Todd 2017, p. 766). Geographers Lewis and Maslin were the first to propose the start date of 1610 AD as it was the period where large exchanges of plants, animals, and infectious diseases (Columbian Exchange) between Europe and the Americas (Lewis and Maslin 2015). These ecological exchanges radically re-shaped the landscapes (and waterscapes) of both landmasses. Such changes can be found in the biomass accumulated in the geomorphological and geological layers on both continents. The second reason for the 1610 start date, as Davis and Todd aptly write, "which is a much more chilling indictment against the horrifying realities of colonialism, is the drop in carbon dioxide levels that correspond to the genocide of the peoples of the Americas and the subsequent re-growth of forests and other plants" (Davis and Todd 2017, p. 766). In 1492, the Americas was home to between 54 to 61 million Indigenous peoples, Lewis and Maslin observe, by 1650 this number was reduced to just 6 million. In making this argument, Lewis and Maslin acknowledge that the consequences move beyond the strict confines of geology and stratigraphic measurables into the social and political concerns, particularly:

unequal power relationships different groups of people, economic growth, the impacts of globalised trade, and our current reliance on fossil fuels. The onward effects of the arrival of Europeans in the Americas also highlight the long-term and large-scale example of human actions unleashing processes that are difficult to predict or manage. (Lewis and Maslin 2015, p. 177)

Here, they acknowledge that such a framing of the Anthropocene affects how human activities on the environment are conceptualised, and more broadly human-ecological-geological relationships; this framing also explicitly recognises that the Anthropocene unleashes processes (often characterised by highly differential and unequitable power relations) that are difficult to predict and manage.

Such evidence, which records the worst offences of Euro-Western colonialism against Indigenous bodies, nations, biota, and landscapes, are only one type of knowledge of radical human-induced environmental changes (on tracer of the Anthropocene). As the works of Métis sociologist Zoe Todd and Anishinaabek philosopher Kyle Powys Whyte remind us, Indigenous stories, histories, and knowledge(s) also provide a rich body of evidence about how the colonial invasion, violence, violations, oppression of Indigenous and other sovereign peoples (in the name of the land, labour, resources, trade, and power) resulted in radical challenges to social and ecological systems. Evidence does not, typically, include the "fleshy stories" that elders tell younger family members about how once (before colonisation, hydro-electric dams, introduced species, pollution, and commercial fisheries) there was an abundance of fish that swam in the waterways. Nor do they tell stories of the fish that were caught, cooked, preserved and feed to family, friends, visitors and kin members. As Todd writes, the evidence used to record the Anthropocene precludes "the flash of a school of minnows in the clear prairie lakes I intimately knew as a child … the succulent white fish my stepdad caught from us from the Red Deer River when I was growing up" (Davis and Todd 2017, p. 767). Such lessons (evidence of human-freshwater-biological relationships) have been:

deeply erased from dominant (non-Indigenous) public discourse in Alberta [Canada] and I had not recognised the implicit ways fish were woven into my own life as more than food. This is the thing about colonisation: it tries to erase the relationships and reciprocal duties we share across boundaries, across stories, across species, across space, and it inserts new logics, new principles, and new ideologies in their place. (Todd 2016c)

But these interwoven and ongoing relationships between human and more-than-human beings and the cumulative consequences of radical social and environmental changes are precisely the risks we face in the Anthropocene, particularly in the context of freshwater crises. The Anthropocene exacerbates the existing socio-economic inequities, patterns of political marginalisation and inequitable power relations that sought to divide Indigenous peoples from their lands and waters, fish and other taonga (the word for treasured possessions in Māori), ancestors and wāhi tapu (sacred sites), all of which are interwoven with their identities, values, laws, governance structures, and ways of knowing. The stories, then, that we collectively tell about the beginnings of the Anthropocene epoch connect how we conceptualise the relations that we maintain with our waters (be it rivers, springs, wetlands, springs, lakes, seas and oceans) both now and in the future. Put simply, the naming of the Anthropocene era and its start date has implications for not just how we know the world (or plural worlds), but also this understanding holds real world (material) consequences (consequences that impact human and more-than-human bodies, waters, and lands) (Davis and Todd 2017).

It is critical, then, that we consider the appropriateness of the concept of the Anthropocene and its potential role in challenging or naturalising settler-colonial histories. If the Anthropocene began in 1610 (the Americas), 1800 (Australia), or even 1840 (Aotearoa), European colonialism is inextricably connected to all dates (indeed, it was not universally experienced in place at the same time, in the same ways). The Anthropocene is, however, the histories of ecological imperialism, of violence, of slavery, and coal (Bashford 2013, p. 347). Thus, the commencement of the Anthropocene should not be told merely as the history of the Industrial Revolution and coal, energy transitions, and the emergence of petro-carbon economies, instead it was (and is) the histories of European colonisation, the exploitation of resources were not just confined to the use of fossil fuels but also of forests, lands, waters, biota, and peoples (specifically non-Europeans).

The past—the beginning of the Anthropocene—was and is "not a foreign country at all; it is homeland" that Indigenous, colonial, and (post) colonial states continue to be embodied, to contest, and seek out better ways of living with others (be they human and more-than-human actors) (Bashford 2013, p. 347). Since settler colonisation violently and physically uprooted Indigenous peoples, whānau (the word for extended family in Māori), altered landscapes and waterscapes, and human/more-than-human relations, Indigenous peoples might be considered amongst the victims or survivors of the Anthropocene. Scholars argue that achieving justice (social, environmental, water, climate) for and by Indigenous peoples necessitates concentrating how environmental changes are intertwined with (and is anticipated on) settler-colonial practices (Davis and Todd 2017; Erickson 2020; Todd 2016a, b; Whyte et al. 2019; Zahara 2017). We suggest that decolonisation rests at the fore of rethinking freshwater governance and management, specifically in the context of settler states. Decolonisation requires us not only to rethink the temporalities of the Anthropocene (when the Anthropocene began and for whom?) but also querying and contesting the knowledge(s), values, and practices that underpin dominant ways of governing and managing rivers.

First Nation philosopher Kyle Powys Whyte, of the Citizen Potawatomi Nation, argues that the Anthropocene should be seen as part of the deliberate interventions and endorsements of colonial processes which refuse to recognise specific place-based and reciprocal relations between humans, lands, waters, and more-than-human beings. In settler-colonial societies, the drainage of wetlands, the damming and straightening of rivers, the clearance of forests, and the importation of exotic biota remade the Indigenous worlds of the Americas, Australia, and Aotearoa New Zealand into a vision of a new displaced (supposedly improved on) version of Europe, radically altering the ecosystems (Bacon 2019; Parsons and Nalau 2016). As elsewhere, in Aotearoa settler-colonialism was (and is) characterised by process of terraforming. As Whyte argues, "industrial settler campaigns erase what makes a place ecologically unique in terms of human and nonhuman relations, the ecological history of a place, and the sharing of the environment by different human societies" (Whyte 2016a, p. 8). It involved the dispossession and displacement of Indigenous communities, and peoples being forced to adapt to radically different environments, climatic conditions, ecosystems, biota, and socio-economic conditions. These processes of coupled social and ecological transformations can be understood as a "preview of what it is like to live under the conditions of the Anthropocene" (Davis and Todd 2017, p. 771). Thus, as Whyte, Davis and Todd make clear, the contemporary environmental crises that are defined through the term the Anthropocene should be interpreted as a perpetuation of, as opposed to a definitive division from, previous periods that commenced with colonialism and encompass advanced capitalism.

From this perspective, the Anthropocene, and the uneven and highly inequitable impacts of climate change on the poor and marginalised populations around the globe, can be comprehended not just as a catastrophic and unfortunate accident, but instead as a deliberate consequence of the extension of colonial logics (Whyte 2016a, p. 12). The interventions and violence of colonialism (against humans, more-than-humans, lands, and waters) wretched apart, disrupted and re-modelled the landscapes, waterscapes, and seascapes in the places that we currently dwell in were hit with successive tidal waves or seismic shocks.

Settler-Colonialism

Settler-colonialism consists of structures of domination employed to exploit Indigenous people and other populations. As with different formulations of colonialism, settler-colonialism is interwoven with and frequently overlaps with other types of domination (including capitalist exploitation, chattel slavery, and imperialism) (Coombes 2006; Veracini 2010). Settler-colonialism is, as the work of scholars including Veracini demonstrates (Veracini 2010), distinct from other forms of colonialism in that settler aspirations are to supplant Indigenous peoples as a means to take their land and resources, rather than primarily the control to labour and resources (Coombes 2006; Hiller 2017; Veracini 2010).

Settlers, scholars argue, are considered distinct from colonialists and immigrants (Pulido 2017a; Veracini 2010).

What makes settler-colonial societies distinct from other colonial societies is their aspirations to supplant indigenous peoples as a means to take their land and resources, rather than primarily the control to labour and resources (Arvin et al. 2013; Coombes 2006; Veracini 2013). Settlers are a unique category of migrant made through acts of conquest, not merely the process of migrating somewhere (Mamdani 2001, 2015). Settlers are founders and contributors to new political orders; they carry (and seek to assert) their sovereignty with them, unlike migrants who are in some regards appellants that encounter an already established political order (Veracini 2010). Settler-colonialism, Veracini (2010) observes, are collectives that claim both a "specific sovereign charge and a regenerative capacity". Settlers, unlike other migrants, "remove" themselves to a new location to "establish a better polity, either by setting up an ideal social body or by constituting an exemplary model of social organisation" (Veracini 2010). What makes settler-colonial societies distinct from other colonial societies is their aspirations to supplant indigenous peoples as a means to take their land and resources, rather than primarily the control to labour and resources (Arvin et al. 2013; Coombes 2006; Veracini 2013). Settlers are a unique category of migrant, Mahmood Mamdani argues, made through acts of conquest, not merely the process of migrating somewhere (Mamdani 2001, 2015). Settlers are, Veracini observes, founders and contributors to new political orders; they carry (and seek to assert) their sovereignty with them, unlike migrants who are in some regards appellants that encounter an already established political order (Veracini 2010). Settler-colonialism, Veracini (2010) observes, are collectives that claim both a "specific sovereign charge and a regenerative capacity". Settlers, unlike other migrants, "remove" themselves to a new location to "establish a better polity, either by setting up an ideal social body or by constituting an exemplary model of social organisation" (Veracini 2010).

Settler-colonialism is defined as a "situation" and is not therefore restricted to any specific group, place, or time period. Caroline Elkins and Susan Pedersen's theoretical definition of settler-colonialism emphasises "institutionalised settler privilege (especially as it relates to land allocation practices) and a binary of settler-native structures (especially as it relates to settler capacity to dominant government" (Elkins and Pedersen 2005, p. 18). Settler-colonialism forms are part of global history and are not limited to white or Anglo-European settler societies, or to settler minorities who occupied colonial places; however, for this chapter we refer to European settler-colonialism. Veracani suggests interpreting the settler-colonial phenomenon as a "Lacanian (imaginary-symbolic-real)". Firstly, "there is an imaginary spectacle, an ordered community working hard, and living peacefully". Secondly, "there is the symbolic and ideological backdrop: a moral and regenerative world that supposedly epitomise settler traditions" (Veracini 2010, p. 75) (the 'outback' of Australia, the 'wilderness' or 'frontier' of the US and Canada, and the 'backblocks' of Aotearoa). Lastly, "there is the real: expanding capitalist order associated with the need to resettle a growing number of people", and the creation and incorporation of new products and markets. Such narratives of settlement as improving and producing productive land-scapes functioned to legitimise the arrival (invasion) and continued residence (occupation) by settlers, as well as delegitimise Māori claims (of prior settlement, and continuity of place-attachments, knowledge, land tenure and usage practices, and the broader rights to resources and governance). Indeed, it is through the narrative of 'closer settlement', economic progress, and ecological 'improvement' (of land, rivers, and biota) that the settler-colonial subject always seeks to call itself into being (and justify its presence). Such narratives of legitimisation, found in other settler societies including Canada and Australia, rely on a kind of historylessness or historical forgetting. Accordingly, settler-colonial gaze was one that interpreted the landscapes and waterscapes of the Waipā through the lens of new beginnings and imagined (modernising, advancing, and always improving) communities under construction, but yet to fully come into being (Veracini 2007, 271–272).

Key facets of settler-colonialism are the structures of domination designed "deliberately to exploit one or more groups of people for the sake of one or more groups of people's benefits and aspirations" (Whyte et al. 2019, p. 325). This domination includes legislation and institutions as well as numerous other types of "behavioural complicity in the maintenance of power and privilege" (Whyte et al. 2019, p. 325).

Settler-colonial societies are premised on narrating settlement as inherently peaceful processes (with acts of violence only sporadic and unavoidable) and the erasure of Indigenous spaces, bodies, and ways of being. Central to this erasure of colonial violence and Indigenous dispossession are what Pulido terms "transition narratives", discourses that assist in making the past more palatable. Settler-colonialism, therefore, requires that we acknowledge the "whitewashing associated with hegemonic representations of colonisation" and re-centre attention to Indigenous peoples' experiences (Pulido 2017a, p. 2).

In this book, we focus on one settler-colonial state—Aotearoa New Zealand—and one Indigenous people—Māori; however, the key ideas are relevant to both settler states and other nations (including so-called post-colonial) in which the cultures, values, and wellbeing of the original inhabitants (the first nations) were displaced by those of more recent and numerous colonial settlers. Aotearoa New Zealand is a location in which, despite the language of post-colonialism, the decolonising project is (from the perspective of Indigenous people) just beginning. Situated within settler-colonial structures, laws, river governance and management approaches, social norms and practices are processes and assumptions that allow settlers to retain domination over the Indigenous peoples of those lands and waters.

A wealth of scholars, activists, and writers outline how settler-colonialism around the world is a form of violence that interrupts human connections with their environments. From a Māori worldview, based on one's genealogical relationships (whakapapa) to all living and non-living things within their rohe (traditional lands and waters), violent actions are ones that disrupt or diminish the life force (mauri) of human and more-than-human beings, which includes the whenua (land), awa (rivers), taniwha (supernatural beings that live in water bodies), fish and other biotas. Eve Tuck and K. Wayne Yang similarly observe, writing about the Canadian context, that "the disruption of Indigenous relationships to land represents a profound epistemic, ontological, cosmological violence" (Tuck and Yang 2012, p. 5). Indeed we argue, adding our voices to the countless other Indigenous scholars and non-Indigenous allies, that appraisals of the history of colonial interactions with the whenua, wai, and awa reveal a history of violence wherein land, water, plants, animals, and minerals are accessed, but not learnt from nor treated as part of the whole (Bacon 2019; Tuck and Yang 2012; Watts 2013, p. 26; Whyte 2018). The settler-colonial state-supported projects to drain wetlands and the creation of resource-intensive industries (most notably pastoral agriculture) with devastating impacts that amounted to environmental violence. J.M. Bacon suggests that "colonial ecological violence" was (is) a process of disrupting Indigenous social relationships (Bacon 2019, p. 59). We investigate one component of how settler-colonialism carried out environmental injustices through acts that interrupted and destroyed Indigenous peoples connections with their waterscapes. The dimension focuses on how specific legislation, policies, institutional arrangements and actions (that all formed part of the process of settler-colonialism) worked strategically to undermine Māori resilience. A wealth of scholars, activists, and writers outline how settler-colonialism around the world is a form of violence that interrupts human connections with their environments. From a Māori worldview, based on one's genealogical relationships (whakapapa) to all living and non-living things within their rohe, violent actions are ones that disrupt or diminish the life force (mauri) of human and more-than-human beings, which includes the whenua (land), awa (rivers), taniwha (supernatural beings that live in water bodies), fish and other biotas. Scholars Eve Tuck and K. Wayne Yang similarly observe, in the Canadian context, that "the disruption of Indigenous relationships to land represents a profound epistemic, ontological, cosmological violence" (Tuck and Yang 2012, p. 5). Indeed we argue, adding our voices to the countless other Indigenous scholars and non-Indigenous allies, that appraises of the history of colonial interactions with the whenua (land), wai (water), and awa (river) reveal a history of violence wherein land, water, plants, animals, and minerals are accessed, but not learnt from nor also treated as part of the whole (Bacon 2019; Tuck and Yang 2012; Watts 2013, p. 26; Whyte 2018). The settler-colonial state-supported projects to drain wetlands and the creation of resource-intensive industries (most notably pastoral agriculture) produced devastating impacts that amounted to environmental violence. J.M. Bacon suggests that "colonial ecological violence" was (is) a process of disrupting Indigenous social relationships (Bacon 2019, p. 59). We investigate one component of how settler-colonialism carried out environmental injustices through acts that interrupted and destroyed Indigenous peoples connections with their waterscapes. The dimension focuses on how specific legislation, policies, institutional arrangements and actions (that all formed part of the process of settler-colonialism) worked strategically to undermine Māori resilience.

In this book, we highlight examples of how Māori experiences of settler-colonialism resulted in a plurality of environmental injustices, which were tied to the Anthropocene (Curley 2019; Winter 2019). Rather than an extraordinary one-off environmental disaster—an oil spill, nuclear disaster—we demonstrate that environmental injustices are often creeping and cumulative; a form of slow violence against bodies (be they human or ecological, physical or metaphysical) (Bacon 2019; Davies 2019; Marino 2017). Accordingly, as we later demonstrate in Chaps. 5 and 6, actions to address these environmental injustices are likewise multiple, ongoing, and heterogeneous activities that require (in many instances) a fundamental reconfiguring on the ontological and epistemological privileging of Western knowledge, values, and practices (challenging the supremacy of settler-colonialism and whiteness in socio-natures). We argue that Māori environmental (in)justices (which are not restricted to just whenua/land nor wai/water) cannot be disconnected from the historical (and ongoing) injustices of settler-colonialism. Only as countless other examples from Indigenous peoples around the globe similarly demonstrate, the past continues to shape the present and future (Ahmad 2019; Proulx and Crane 2020; Whyte 2016b, 2017).

Indigenous Environmental Justice

Throughout the book, we seek to explore and extend discussions of environmental justice (and what constitutes environmental injustice) beyond its framings within Western liberal worldviews, philosophies, and legal systems to include indigenous ontologies and epistemologies. Geographers and other scholars concerned with Indigenous struggles about sovereignty over natural resources and decision-making about environmental risks highlight that "justice for one group may mean injustice for another occupying a different political-geographical position" (Ishiyama 2002,

p. 5). Earlier examinations of environmental justice, such as the work of Schlosberg and Ishiyama, seek to clarify processes of defining local struggles for autonomy that are grounded in different Indigenous communities' self-determination in political and economic decision-making in interrelated and paradoxical ways (Ishiyama 2002; Schlosberg and Carruthers 2010; Whyte 2016b). The majority of these studies, however, concentrated on the North American context and did not consider the role of ontological and epistemological differences between Indigenous and Western cultures in accounts of justice/injustice (Álvarez and Coolsaet 2018; Rodríguez and Inturias 2018). We argue that there is an additional layer of complexity to thinking about Indigenous environmental justice (IEJ) due to the constant tensions between Indigenous and Western (settler/European/White) worldviews and the ways in which humans and more-than-human relate to one another. Throughout this book, we seek to explore and extend discussions of environmental justice (and what constitutes environmental injustice) beyond its framings within Western liberal worldviews, philosophies, and legal systems to include indigenous ontologies and epistemologies. Geographers and other scholars concerned with indigenous struggles about sovereignty over natural resources and decision-making about environmental risks highlight that "justice for one group may mean injustice for another occupying a different political-geographical position" (Ishiyama 2002, p. 5). Earlier examinations of environmental justice, such as the work of Schlosberg and Ishiyama, seek to clarify processes of defining local struggles for autonomy that are grounded in different Indigenous communities' self-determination in political and economic decision-making in interrelated and paradoxical ways (Ishiyama 2002; Schlosberg and Carruthers 2010; Whyte 2016b). The majority of these studies, however, concentrated on the North American context and did not consider the role of ontological and epistemological differences between Indigenous and Western cultures in accounts of justice/injustice (Álvarez and Coolsaet 2018; Rodríguez and Inturias 2018). We argue that there is an additional layer of complexity to thinking about IEJ due to the constant tensions between Indigenous and Western (settler/European/White) worldviews and the ways in which humans and more-than-human relate to one another.

The second goal of this book is to bridge the divisions between environmental justice, Indigenous history and geography, and decolonial scholarship (Agyeman et al. 2016; Álvarez and Coolsaet 2018; Daigle 2016; Nirmal 2016; Schlosberg and Carruthers 2010; Whyte 2016b). Environmental justice scholars recommend moving beyond the simplistic mapping of environmental harms (distributive justice) to ask questions about the political, economic, and social dimensions of why and how unjust landscapes manifest and persist across spatial and temporal scales (Barnhill-Dilling et al. 2020; Boone and Buckley 2017; Keeling and Sandlos 2009; Mohai and Saha 2015; Pulido 2017b; Schlosberg 2003). With their focus on the impacts of colonisation of Indigenous peoples, the works of scholars from across the critical social sciences and humanities (drawing of the lenses of settler-colonialism and decolonial theory) provide essential contributions to the study of environmental justice (Adamson 2017; McGregor 2018a; Ulloa 2017; Whyte 2016a, 2017; Winter 2018, 2019). In particular, examining the histories of settler-colonialism and Indigenous agency assists us in seeing more clearly the cumulative impacts of uneven power dynamics and the consequences in terms of how environmental injustices occur in particular locations.

Environmental justice was developed first by scholars interested in the distribution of environmental harms and goods (distributive-based justice) in the context of minority communities in the United States of America (Bullard 2018; Mitchell et al. 1999; Pellow 2004; Pulido and Peña 1998). The language used by environmental justice borrowed from the civil rights movement. Later environmental justice scholarship extends theories to procedural- and recognition-based accounts of environmental justice. In particular, Schlosberg and Carruthers point out that environmental justice (a sub-discipline, a theory, and a social movement) also includes Indigenous peoples' struggles for self-determination, resource sovereignty, and recognition of their cultural identities (Schlosberg and Carruthers 2010). However, increasing numbers of Indigenous scholars critique the mainstream environmental justice literature for rearticulating colonial discourses. Māori philosopher Christine Winter argues that the underlying assumptions of environmental justice remain underpinned by Western liberal justice theories that perpetuate the settler-colonial project, which is designed (amongst other things) "to suppress and destroy Indigenous peoples." She challenges the universality of justice theorising and seeks to demonstrate the need to seek input from Indigenous ontologies to "fashion more robust imaginings" of intergenerational Indigenous environmental justice to respond to the crises facing the globe (Winter 2018, p. 13). Along similar lines, Whyte writes that "for Indigenous peoples, environmental justice is rooted in one society's interference with and erasure of another society's way of experiencing the world as infused with responsibilities" and their collective continuance (Whyte 2016b, p. 159). Whyte continues, "environmental injustice is rooted in how social institutions are structured and operationalised in ways that favour powerful and privileged populations" (Whyte 2016b, p. 159). Along similar lines, Chicano geographer Laura Pulido observes that, often, policies are a vital avenue by which environmental injustices are facilitated by the state (Pulido 1996, 2017a, b).

In the context of indigenous environmental justice, a plethora of scholars identifies the relationships between colonialism and injustices experienced by Indigenous peoples (Coombes 2013; McGregor 2018a; Schlosberg and Carruthers 2010; Watson 2014; Winter 2018). While it is critical to observe the diversity of colonialism, the existence of "colonial situations" persists in supposedly post-colonial countries, and how colonialism continues to be acted on and evolve (particularly in settler-colonial contexts such as Aotearoa, Canada, the US and Australia) (Ahmad 2019; Álvarez and Coolsaet 2018; Balaton-Chrimes and Stead 2017; Bargh 2018; Bell 2018; Maldonado-Torres 2016; McGregor 2018b). In the North American context, Kyle Whyte highlights the close connections between current ecological crises and the socio-cultural, political, and economic interventions (violence, dispossession, genocide) made by colonial societies against Indigenous nations. With the ongoing challenges of settler-colonialism in settler-states, the worsening impacts of environmental degradation and climate change on communities around the globe, and the failure of settler-states to adequately fulfil their responsibilities to address the root causes of environmental injustices facing Indigenous communities, Indigenous voices within environmental justice theorising and activism are being more critical (as demonstrated most recently in the Dakota Pipeline Protest) (Davies 2019; Proulx and Crane 2020; Whyte 2017). Sioux Nation people's protests (supported by other Indigenous peoples as well as non-Indigenous allies) against the construction of the Dakota oil pipeline across their lands and waters (located in the US Midwest) highlight how one event or action of the settler-colonial state against an Indigenous people is not a singular environmental injustice but part of the ongoing process (Gilio-Whitaker 2019; LeQuesne 2019; Proulx and Crane 2020; Whyte 2017). The Sioux Nation protest was not merely a response to building a pipeline across their land, but also a protest about the litany of social and environmental injustices that did and are still occurring as part of settler colonisation. Likewise, as we will highlight in the rest of this book, the histories and present-day lived realities of Māori environmental injustices are interwoven with multiple stories and experiences of losses and damages generated by settler-colonialism. However, we demonstrate how since the commencement of colonisation, Māori (individually and as part of wider collectives iwi/hapū/whanau, and pan-tribal alliances) have consistently protested, resisted, and challenged settler-colonial intrusions into their rohe, and also sought ways to adapt to colonial disruptions, mitigate damage to their livelihoods and environments, and take advantage of new circumstances. Thus, this book is about the loss and destruction of Indigenous landscapes and waterscapes, environmental degradation and racism, but also of Māori agency, resistance, and new more hopeful (ontologically and legally pluralistic) freshwater geographies.

One of the key aims of this book is to fill some of the theoretical gaps that exist within environmental justice and water justice scholarship and explore decolonial spaces for indigenous ontology and epistemology in the context of freshwater governance and management. Environmental justice and water justice, climate justice and social justice are closely connected and overlapping concepts (LeQuesne 2019; Mascarenhas 2007; Mohai 2018; Stensrud 2016). However, most scholars do not consider the intersections between different (in)justices nor how indigenous worldviews, centred on holism, kincentrism and relationality, can challenge and expand how scholars, activists, policy-makers and freshwater management practitioners discuss and attempt to take action to address polluted waterways. In the disciplinary fields of freshwater governance and management, most researchers elect to employ the term water justice (Jackson 2018; McGregor 2015; Robison et al. 2018). However, in this chapter (and the rest of the book) we deliberately chose to frame our discussion as environmental justice not water justice because water (from Māori worldview) is not separate from the environment (taiao). In keeping with our decolonising agenda, we promote mātauranga Māori (Māori knowledges) ways of seeing wai as part of a holistic system in which whenua, awa, repo (wetlands), moana (sea), and tangata whenua (people of the land) are all connected through relationships.

In this book, we also seek to draw on the diverse and frequently disparate works of literature on Indigenous geographies, environmental justice, Indigenous histories, and political ecology to provide a more nuanced and multidimensional account of the histories, politics, and geographies of the freshwater governance and management in Aotearoa. Instead of just describing and situating the distribution of environmental risks, we examine the historical, social, political, and economic processes that gave rise to environmental injustices for the Māori iwi Ngāti Maniapoto, whose rohe included the upper and middle reaches of the Waipā River and its environs. Our research demonstrates that rather than singular environmental injustice, the environmental degradation of the Waipā River was and is evidence of the multiple layers of environmental injustices experienced by not only Māori and now—to some extent—all peoples who live in Waipā catchment. The root cause of these injustices rests with settler colonisation but also in the unchecked capitalistic drive for endless growth (through the accumulation of resources and agricultural expansion) and Western ontologies' unhealthy division of people from nature (what Ghosh calls the "Great Derangement") (Ghosh 2018; Nightingale et al. 2019). These environmental injustices, which lasso the ecological and social (what some scholars dub "socio-natures") together in tightly bound coils, include the historical loss of land and other resources by iwi and by extension the economic basis of their livelihoods; the environmental degradation of their rohe; actions (or inactions) by governments to deny them the capacities to participate in decision-making processes meaningfully; and the settler-colonial state (be it local government or central government) actions to exclude, suppress and fail to acknowledge Māori cultural identities, values, and knowledge.

Whakatauki (Proverb)
Hutia te rito o te harakeke
Kei whea te tauranga o te mako e ko?
He aha te mea nui o tenei ao?
Maku e ki atu
He tangata, He tangata, He tangata.

If you remove the heart of the flax
Where will the bellbird then rest?
If you should ask me
What is the main thing in this world?
I will say to you
It is people; it is people, it is people.

At the heart of this book, then, is the Waipā River and indigenous Māori iwi and hapū who connect to the river and consider it to be an ancestor and members of their whānau (extended family). The iwi and hapū of the Waipā River all affiliate to the Tainui waka (canoe) and include Waikato/Tainui, Ngāti Maniapoto, Ngāti Apakura, Ngāti Hikairo and Ngāti Mahanga (Muru-Lanning 2016; Waitangi Tribunal 2018). The upper and middle reaches of the Waipā River (from the mountains north of Te Kuiti to Puniu River near Te Awamutu) is the domain of Ngāti Maniapoto, whose iwi was one of several that developed from the original Pacific migrants from the Tainui waka. Many of Ngāti Maniapoto's hapū intermingled with the hapū of Waikato-Tainui (also descended from the Tainui waka) in the middle reaches of the Waipā River (where the forest Te Nehenehenui met the southern floodplains of the Waipā and Waikato Rivers), and at the north-west coastline (at the harbours of Whāingaroa and Kāwhia). Hapū associated with Ngāti Maniapoto and Waikato-Tainui also intermingled with those of Ngāti Raukawa (whose rohe lies to the north-east). All three iwi shared common whakapapa (genealogical or ancestral) links to the Tainui waka. Iwi groups who were not descended from Tainui waka also held varying degrees of overlapping influence with Ngāti Maniapoto and Waikato-Tainui within parts of their traditional lands and waters. To the south-east of Te Rohe Pōtae lie the territories occupied by Ngāti Tūwharetoa. Southwards, around the present-day township of Taumarunui, were people who affiliated to hapū of the upper Whanganui River. In the south-west, in the Mōkau River catchment, were hapū associated with northern Taranaki. All these different groups of people, however, shared close links in the various territories, with groups possessing customary interests and rights of usage that overlapped. Accordingly, given our geographical focus, we adopted a case study approach for our research.

Case studies are nothing new to scholars of human geography, environmental justice, or river management; case studies use space to analyse particular and complex phenomena located within a real-world context (Yin 2013). However, we chose adopted a case study approach to examine the intricate and interconnected social, cultural, economic, political and environmental implications of radical environmental changes that occurred within river systems of Aotearoa as a consequence of colonisation. We concentrate, in many instances, primarily on the Waipā River and the rohe (traditional lands and waters) of the Māori iwi Ngāti Maniapoto. However, we do draw linkages with other parts of Aotearoa and another iwi. We consciously chose to employ a case study approach because it provided us with the capacity to undertake an in-depth analysis of the complex historical, geographical, political, legislative, and socio-cultural processes that shaped Māori relationships and experiences of more than two centuries of changes within their rohe.

In keeping with the transdisciplinary focus of the research, we employ research methods from human geography, Indigenous and historical studies. Empirical data collection included archival-based research as well as oral histories, semi-structured interviews and participatory observation. The first half of the book is based on empirical data derived from historical archival-based research that augmented with memoirs and oral histories collected by ourselves and other researchers (held in public collections throughout Aotearoa) with people who lived in the Waipā catchment during the late nineteenth and early twentieth centuries. The use of historical sources, in particular archival collections, is widespread amongst historians, historical geographers and environmental justice scholars (Bolin et al. 2005; Boone and Buckley 2017; Boone and Modarres 1999; Keeling and Sandlos 2009; Pellow 2004; Pulido 1996). As Boone and Buckley (2017, p. 222) observe, historical research approaches, which includes archival-based research as well as oral histories, provide an "important and useful mechanism for understanding the origins, causes, and legacies of present-day environmental injustices, both for the process and the outcome". The remaining second half of the book employs widely used human geography research methods, including primary data (semi-structured interviews) and secondary data (analysis of legislation, policy, and planning documents), to examine the efforts of iwi, the New Zealand Government (the Crown), and local governments (district and regional councils) to address the freshwater degradation in the Waipā River and other rivers within Aotearoa (See Appendix 1, which includes tables detailing the list of interviewees).

We examine the fluid, liminal and challenging to define places that comprise the historical and contemporary waterscapes of the Waipā—watery landscapes or muddy waterscapes—by eschewing the strict methodological moralising of individual disciplines and instead embrace transdisciplinary approaches. Accordingly, throughout this book, we draw on historical geography, environmental history, Indigenous geography, and global environmental change scholarship to weave together a history of Pākehā (used, in this book, as a reference to New Zealand Europeans) and Māori imaginaries and interactions with wetlands of the Waipā River catchment within Aotearoa.[1] We collected and analysed a wide range of data including archival records (newspapers, government documents, maps and photographs), interviews (unstructured life histories and semi-structured interviews), and Māori oral traditions including waiata (songs), whakataukī (proverbs) and pepeha (recitations linking people to place), and pakiwaitara (legend or story). In particular, we concentrated on materials focussing on the floodplains of the Waipā River and its tributaries, which formerly included extensive wetlands and now includes the towns of Otorohanga, Te Kuiti, Te Awamutu, Kihikihi,

---

[1] A note about the terms used throughout the book. Prior to the arrival of Europeans, Māori did not possess a collective identity and instead identified through affiliations to tribal groups. The term "maori" meant ordinary or normal and became used by Māori people (following their encounters with Europeans) to distinguish themselves from Pākehā (the word derives from pakepakeha that translates as fair-skinned beings). Accordingly, Pākehā and Māori were established, and continue to be employed in the twenty-first century, as relational terms. The word Pākehā when used as an adjective is now employed to indicate non-Māori; when employed as a noun, describes people of New Zealand European descent. In the book, we use Māori and Pākehā to refer to the two major ethnic groups in Aotearoa New Zealand (Ballantyne 2012; Fisher 2014; Parsons and Nalau 2016).

Pirongia, and Ngaruawahia. It is the area where we spent a great deal of time as children and adults, and where two of us (Fisher and Crease) trace their whakapapa to. Through these diverse materials, we can see glimpses of the radical changes in the ways that rivers, wetlands, and lands were understood before and after the settler colonisation arrived (invaded) the area.

Each of the authors of this book has a claim to Māori whakapapa along with other cultural identities and ancestry that have shaped their (our) experiences both personally and professionally. Each has experienced (and continue to experience) the effects of colonisation whether it be in the form of overt racism or exclusion based on racialised assumptions about what knowledge and ways of knowing count, or through the historical loss of land suffered by ancestors in the form of raupatu (confiscation) and dispossession through myriad government policies and settler actions. They are also descended from settlers who arrived in Aotearoa from as early as the 1830s and as recently as the 1980s.

Fisher is of Ngāti Maniapoto (Ngāti Paretekawa) and Waikato-Tainui (Ngāti Mahuta) descent (through her mother) and affiliates to the Waipā River through Mangatoatoa marae on the banks of the Puniu River. Mangatoatoa is the marae to which she feels most closely connected, as this is the marae she has visited the most throughout her life. Mangatoatoa is also the marae at which she first learned about the Treaty settlement process for the Waipā River. As she learned more and became more engaged in discussions about the Waipā, she visited another marae along the river to which she also affiliates (Te Keeti, through her mother's childhood visits and holidays; Te Kotahitanga through her great-great-grandfather, Tanirau Patea; Tarewaanga through her great-great-great-great grandmother, Taupoki). Her hybrid identity as coloniser/colonised arises through her Pākehā father (English, Scottish, Swedish, French), who was also raised in the King Country (Te Rohe Potae), and her Māori mother (English, Irish) and especially her Irish ancestor, Thomas Power. She travelled to Te Rohe Potae in the 1840s to oversee the development of flour mills and the introduction of colonial agricultural practices.

Born in the Philippines to a Filipino mother and a New Zealand father of Māori and Pākehā ancestry, Roa Petra Rodriguez Crease is a hybrid who does not neatly fit into one culture or another. She visibly (through skin colour and bodily features) occupies the middle ground and lives (co-exists) in different worlds. Yet at the same time feels excluded from each world since she is not one or the other, neither looking like a 'typical Filipino' (or Māori or Pākehā) and not being able to speak the languages of her mother (Tagalog and Ilocano) or her father's Māori tūpuna (ancestors). Her experience is by no means unique but rather one that many people of mixed heritage speak and write about, which highlights how colonisation is a process premised on the subjugation of non-Western identities and the privileging of (whitewashing) European identities (Alcoff 2018; Bell 2009; Bird 1999; Brablec 2020; Connor 2019; King et al. 2018; Wanhalla 2015). Her family histories attest to the diverse manifestations of colonisation, with the Philippines (like Aotearoa) a product of Euro-Western colonisation (first Spanish and then the US), and where historical injustices associated with colonisation continues to mediate how Filipinos live (including with the consequence of environmental degradation associated with capitalist exploitation) (Crease et al. 2019; McKenna 2017; Moran 2015). Like the majority of Māori (84 per cent) Crease grew up in the urban centre (Auckland City) in an environment far removed from the landscapes and waterscapes of the Waipā. Accordingly, her knowledge of her Māori ancestors derived from occasional mentions and visits to whānau living (and who are buried) near the township of Te Awamutu. It was only through researching this book that she learnt not only of her whakapapa to Ngāti Maniapoto and Te Awa o Waipā but also the histories, mātauranga and tikanga (laws) of her tūpuna. The research included in this book, therefore, is the start of a journey for Crease that involves her seeking to decolonise herself and consider how her multiple identities inform how she relates to and seeks to engage with different environments (both in Aotearoa and in the Philippines) to achieve environmental justice.

Like her fellow co-authors, Parsons' ancestry is a kaleidoscope of different threads from across the globe. Her mother's ancestors came to Aotearoa from Scotland, the Channel Islands, and Denmark in the 1850s and 1860s. Her father's ancestors (Lebanese and Jewish) arrived later, at the end of the nineteenth century, fleeing the violence of colonialism and prejudice, and married into Māori whānau living in Tāmaki Makaurau/ Auckland (but originally from Northland iwi Ngāpuhi) whose experiences of colonisation, dispossession, and the marginalisation echoed those of the incomers. Problematically placed into single category by others (sometimes labelled as Pākehā, Lebanese, Māori, White, non-White, disabled, or able-bodied), Parsons (like Crease) recognises that she is not a singular identity but instead occupies an ambiguous in-between-ness. However, the enduring desire to classify individuals into narrow categories is part of the enduring legacies of Western ontologies and epistemologies that attempts to divide the world (and its peoples) into binaries (self/other, nature/culture, civilised/primitive, West/the Rest, white/black, land/water). Indigenous knowledges, in contrast, are holistic and relational ways of thinking that emphasis the connectivity between individuals and collectives, humans and more-than-humans, and health and wellbeing. Thus, while the book initially started as a project to decolonise freshwater management and governance through incorporating Indigenous knowledge, it became far more this as Parsons and the other authors commenced their watery (physical and discursive) journeys along the Waipā River. It became clear that for Parsons, to consider how to decolonise rivers required scholars (herself included) to challenge their assumptions and preconceived notions of rivers, water, and health.

Recalling her childhood spent walking and running through water-logged fields and swimming in murky waters of its rivers and lakes of the Waikato and Bay Plenty regions, she realised that she was often told by elders (doctors, teachers, professors) that her regular engagement with these muddy blue spaces was not only undesirable but also potentially disease-inducing. Pākehā doctors and nurses informed Parsons (and her family members) that frequent bouts of sickness (bacteria pneumonia, asthma, autoimmune disease) could be traced to the unhealthy environment; not only was her childhood home located too close to a river, it was built on former wetlands, and surrounded by the Waikato air laden with moisture (rain and fog). Even her first-year geography lecturer warned her that bog-dwellers such as herself ended up with autoimmune diseases as a consequence of dampness-inducing negative immunological responses within the bodies. These personal anecdotes can be read on the one hand as evidence of how poorly informed experts who resorted to pseudoscience for explanations rather than acknowledging scientific uncertainty, but on the other hand, highlight how the many water-infused -scapes (aero- water-scapes and land-scapes) of the contemporary Waikato were (and are still) viewed by Pākehā as problematic spaces. Indeed, it seems that the earlier settler-colonial environmental anxieties (outlined in Chap. 3) about the region's wetlands generating ill-health amongst residents did not merely evaporate following the advent of new knowledge (germ theory) and technologies. Instead, ideas merged, fluxed, and flowed into later generations, which directly and indirectly influence the ways in which individuals and communities perceive and interact with particular blue (or formerly blue) spaces. Accordingly, in this book, wetlands are deliberately woven into discussions of rivers, with attention drawn to the need to reassert the holism of freshwater systems, of kinship, and hauora (health).

The Organisation of the Book

The book is organised into three sections (Parts 1–3) that all emphasise the temporal and spatial connectivity between places, peoples, biota, and other more-than-human actors that comprise the freshwater systems. In particular, the relationships between past, present and future accounts of social and environmental changes are circled and interwoven throughout the book, drawing on Māori understandings of time as a temporal loop and of human-environment relationships as one based on whakapapa. Part One recounts the historical waterscapes of the Waipā River that charts stories of change and loss, adaptation and resilience, and the creation of multiple environmental injustices for Māori *iwi* and *hapū*. Part Two charts the emergence of contemporary freshwater co-governance and co-management arrangements in Aotearoa and considers how legal and ontological pluralism can address freshwater degradation and indigenous environmental justice. Lastly, Part Three examines efforts to restore the Waipā River and what freshwater management and restoration mean in the context of changing climate conditions.

In the following chapters, we consider how the shifting social, cultural, political and economic landscapes and waterscapes of the Waipā River were directly linked to the history of local environmental changes in the catchment, and how different generations of people, and different groups of people interacted with and sought to manage the river and its connected ecosystems (Bonnell 2014; Lavau 2011). The consequences of these changes (intended and unintended) and the lessons (in the form of memories and narratives) people took away from their interactions with the river offer essential insights into changing human relationships with environments, and the particular pressures and contingencies at work in Aotearoa. We demonstrate how specific histories of human interactions with freshwater and terrestrial ecosystems in the Waipā River, focusing mainly on the impacts of wetland-drainage, land-use changes, and past management approaches, continue to shape how people perceive and relate to the river in the present-day.

References

Adamson, J. (2017). Roots and Trajectories of the Environmental Humanities: From Environmental Justice to Intergenerational Justice. *English Language Notes, 55*(1–2), 121–134.

Agyeman, J., Schlosberg, D., Craven, L., & Matthews, C. (2016). Trends and Directions in Environmental Justice: From Inequity to Everyday Life, Community, and Just Sustainabilities. *Annual Review of Environment and Resources, 41*(1), 321–340.

Ahmad, N. B. (2019). Mask off – The Coloniality of Environmental Justice. *Widener Law Review, 25*, 195.

Alcoff, L. (2018). Comparative Race, Comparative Racisms. In J. J. E. Gracia (Ed.), *Race or Ethnicity?: On Black and Latino Identity* (pp. 170–188). New Haven: Cornell University Press.

Álvarez, L., & Coolsaet, B. (2018). Decolonising Environmental Justice Studies: A Latin American Perspective. *Capitalism Nature Socialism*, 1–20.

Arvin, M., Tuck, E., & Morrill, A. (2013). Decolonising Feminism: Challenging Connections Between Settler Colonialism and Heteropatriarchy. *Feminist Formations, 25*(1), 8–34.

Bacon, J. M. (2019). Settler Colonialism as Eco-Social Structure and the Production of Colonial Ecological Violence. *Environmental Sociology, 5*(1), 59–69.

Balaton-Chrimes, S., & Stead, V. (2017). Recognition, Power and Coloniality. *Postcolonial Studies, 20*(1), 1–17.

Ballantine, D. J., & Davies-Colley, R. J. (2014). Water Quality Trends in New Zealand Rivers: 1989–2009. *Environmental Monitoring and Assessment, 186*(3), 1939–1950.

Ballantyne, T. (2012). *Webs of Empire: Locating New Zealand's Colonial Past.* Wellington: Bridget Williams Books. Retrieved April 21, 2017, from https://books.google.co.nz/books?hl=en&lr=&id=o7ipY4SmDqoC&oi=fnd&pg=PR2&dq=related:rMTvJd8z9nQJ:scholar.google.com/&ots=9MSbl8yiyI&sig=vt6yqzedN4BM9d9ozDdAX8pKrz8.

Bargh, M. (2018). Māori Political and Economic Recognition in a Diverse Economy. In *The Neoliberal State, Recognition and Indigenous Rights* (p. 293).

Barnhill-Dilling, S. K., Rivers, L., & Delborne, J. A. (2020). Rooted in Recognition: Indigenous Environmental Justice and the Genetically Engineered American Chestnut Tree. *Society & Natural Resources, 33*(1), 83–100.

Bashford, P. A. (2013). The Anthropocene is Modern History: Reflections on Climate and Australian Deep Time. *Australian Historical Studies, 44*(3), 341–349. https://doi.org/10.1080/1031461X.2013.817454.

Bell, A. (2009). Dilemmas of Settler Belonging: Roots, Routes and Redemption in New Zealand National Identity Claims. *The Sociological Review, 57*(1), 145–162.

Bell, A. (2018). A Flawed Treaty Partner: The New Zealand State, Local Government and the Politics of Recognition. In D. Howard-Wagner, M. Bargh, & I. Altamirano-Jimenez (Eds.), *The Neoliberal State, Recognition and Indigenous Rights: New Paternalism to New Imaginings* (pp. 77–92). Canberra: ANU Press.

Biggs, B. J., Duncan, M. J., Jowett, I. G., Quinn, J. M., Hickey, C. W., Davies-Colley, R. J., & Close, M. E. (1990). Ecological Characterisation, Classification, and Modelling of New Zealand Rivers: An Introduction and Synthesis. *New Zealand Journal of Marine and Freshwater Research, 24*(3), 277–304.

Bird, M. Y. (1999). What We Want to Be Called: Indigenous Peoples' Perspectives on Racial and Ethnic Identity Labels. *American Indian Quarterly, 23*(2), 1–21.

Bolin, B., Grineski, S., & Collins, T. (2005). The Geography of Despair: Environmental Racism and the Making of South Phoenix, Arizona, USA. *Human Ecology Review, 12*(2), 156–168.

Bollen, C. (2015). Managing the Adverse Effects of Intensive Farming on Waterways in New Zealand – Regional Approaches to the Management of Non-Point Source Pollution. *New Zealand Journal of Environmental Law, 19*, 207–239.

Bonnell, J. L. (2014). Reclaiming the Don: An Environmental History of Toronto's Don River Valley. University of Toronto Press.

Boone, C. G., & Buckley, G. L. (2017). Historical Approaches to Environmental Justice. In *The Routledge Handbook of Environmental Justice* (pp. 222–230). London; New York: Routledge.

Boone, C. G., & Modarres, A. (1999). Creating a Toxic Neighborhood in Los Angeles County: A Historical Examination of Environmental Inequity. *Urban Affairs Review, 35*(2), 163–187.

Boulton, A. J., Scarsbrook, M. R., Quinn, J. M., & Burrell, G. P. (1997). Land-Use Effects on the Hyporheic Ecology of Five Small Streams Near Hamilton, New Zealand. *New Zealand Journal of Marine and Freshwater Research, 31*(5), 609–622.

Brablec, D. (2020). Who Counts as an Authentic Indigenous? Collective Identity Negotiations in the Chilean Urban Context. *Sociology.* https://doi.org/10.1177/0038038520915435.

Bradford, L. E., Bharadwaj, L. A., Okpalauwaekwe, U., & Waldner, C. L. (2016). Drinking Water Quality in Indigenous Communities in Canada and Health Outcomes: A Scoping Review. *International Journal of Circumpolar Health, 75*(1), 32336.

Bullard, R. D. (2018). *Dumping in Dixie: Race, Class, and Environmental Quality.* Routledge.

Castree, N. (2014). The Anthropocene and Geography III: Future Directions: The Anthropocene and Geography III. *Geography Compass, 8*(7), 464–476.

Connor, H. D. (2019). Whakapapa Back: Mixed Indigenous Māori and Pākehā Genealogy and Heritage in Aotearoa/New Zealand. *Genealogy, 3*(4), 73.

Coombes, A. E. (2006). *Rethinking Settler Colonialism: History and Memory in Australia, Canada, New Zealand and South Africa.* Manchester: Manchester University Press.

Coombes, B. (2013). Māori and Environmental Justice: The Case of 'Lake' Otara. In E. J. Peters & C. Andersen (Eds.), *Indigenous in the City: Contemporary Identities and Cultural Innovation* (pp. 334–354). Vancouver: UBC Press.

Crease, R. P., Parsons, M., & Fisher, K. T. (2019). 'No Climate Justice Without Gender Justice': Explorations of the Intersections Between Gender and Climate Injustices in Climate Adaptation Actions in the Philippines. In T. Jafry (Ed.), *Routledge Handbook of Climate Justice* (pp. 359–377). Oxon: Routledge.

Crutzen, P. J. (2002). The "Anthropocene." *Journal de Physique IV (Proceedings)*, *12*(10), 1–5. https://doi.org/10.1051/jp4:20020447.

Cunningham, M. (2014). *The Environmental Management of the Waipa River and Its Tributaries. Case-Study Commissioned by the Waitangi Tribunal for Te Rohe Potae District Inquiry (Wai 898).* (District Inquiry Research Report No. A150 (Wai 868). Wellington: Waitangi Tribunal.

Curley, A. (2019). "Our Winters' Rights": Challenging Colonial Water Laws. *Global Environmental Politics, 19*(3), 57–76.

Daigle, M. (2016). Awawanenitakik: The Spatial Politics of Recognition and Relational Geographies of Indigenous Self-Determination. *The Canadian Geographer/Le Géographe canadien, 60*(2), 259–269.

Davies, T. (2019). Slow Violence and Toxic Geographies: 'Out of Sight' to Whom? *Environment and Planning C: Politics and Space.* https://doi.org/10.1177/2399654419841063.

Davis, H., & Todd, Z. (2017). On the Importance of a Date, or Decolonising the Anthropocene. *ACME: An International E-Journal for Critical Geographies, 16*, 4.

de Leeuw, S. (2017). Poisoned Perfection: Welling Concerns About Arsenic, Drinking Water, and Public Health in Rural Newfoundland. *Canadian Family Physician, 63*(8), 628–631.

Deitz, S., & Meehan, K. (2019). Plumbing Poverty: Mapping Hot Spots of Racial and Geographic Inequality in US Household Water Insecurity. *Annals of the American Association of Geographers, 109*(4), 1092–1109.

Duncan, R. (2017). Rescaling Knowledge and Governance and Enrolling the Future in New Zealand: A Co-production Analysis of Canterbury's Water Management Reforms to Regulate Diffuse Pollution. *Society & Natural Resources, 30*(4), 436–452.

Elkins, C., & Pedersen, S. (2005). *Settler Colonialism in the Twentieth Century: Projects, Practices, Legacies.* London; New York: Routledge.

Erickson, B. (2020). Anthropocene Futures: Linking Colonialism and Environmentalism in an Age of Crisis. *Environment and Planning D: Society and Space, 38*(1), 111–128.

Fisher, K. T. (2014). Positionality, Subjectivity, and Race in Transnational and Transcultural Geographical Research. *Gender, Place & Culture, 1*–18.

Ghosh, A. (2018). *The Great Derangement: Climate Change and the Unthinkable.* London: Penguin UK.

Gilio-Whitaker, D. (2019). *As Long as Grass Grows: The Indigenous Fight for Environmental Justice from Colonisation to Standing Rock.* Boston: Beacon Press.

Hiller, C. (2017). Tracing the Spirals of Unsettlement: Euro-Canadian Narratives of Coming to Grips with Indigenous Sovereignty, Title, and Rights. *Settler Colonial Studies, 7*(4), 415–440.

Hughes, A. O., & Quinn, J. M. (2014). Before and After Integrated Catchment Management in a Headwater Catchment: Changes in Water Quality. *Environmental Management, 54*(6), 1288–1305.

Hulme, M. (2011). Reducing the Future to Climate: A Story of Climate Determinism and Reductionism. *Osiris, 26*(1), 245–266.

Ishiyama, N. (2002). *Environmental Justice and American-Indian Sovereignty: Political, Economic, and Ethnic Struggles Regarding the Storage of Radioactive Waste.* Thesis, Rutgers, the State University of New Jersey.

Jackson, S. (2018). Indigenous Peoples and Water Justice in a Globalizing World. In K. Conca & E. Weinthal (Eds.), *The Oxford Handbook of Water Politics and Policy.* New York: Oxford University Press.

Julian, J. P., de Beurs, K. M., Owsley, B., Davies-Colley, R. J., & Ausseil, A.-G. E. (2017). River Water Quality Changes in New Zealand over 26 Years: Response to Land Use Intensity. *Hydrology and Earth System Sciences, 21*(2), 1149–1171.

Keeling, A., & Sandlos, J. (2009). Environmental Justice Goes Underground? Historical Notes from Canada's Northern Mining Frontier. *Environmental Justice, 2*(3), 117–125.

Kelly, J. M. (2017). Anthropocenes: A Fractured Picture. In J. M. Kelly, P. Scarpino, H. Berry, J. Syvitski, & M. Meybeck (Eds.), *Rivers of the Anthropocene* (pp. 1–18). University of California Press. https://doi.org/10.1525/luminos.43.a.

King, P., Hodgetts, D., Rua, M., & Morgan, M. (2018). When the Marae Moves into the City: Being Māori in Urban Palmerston North. *City & Community, 17*(4), 1189–1208.

Knight, C. (2016). *New Zealand's Rivers: An Environmental History.* Christchurch: Canterbury University Press.

Knight, C. (2019). The Meaning of Rivers in Aotearoa New Zealand—Past and Future. *River Research and Applications, 35*(10), 1622–1628.

Larned, S., Snelder, T., Unwin, M., & McBride, G. (2016). Water Quality in New Zealand Rivers: Current State and Trends. *New Zealand Journal of Marine and Freshwater Research, 50*(3), 389–417.

Lavau, S. (2011). Curious Indeed, or Curious in Deed? Some Peculiarities of Post-Settlement Relations with an Antipodean River. *Australian Geographer, 42*(3), 241–256. https://doi.org/10.1080/00049182.2011.595671.

LeQuesne, T. (2019). Petro-Hegemony and the Matrix of Resistance: What Can Standing Rock's Water Protectors Teach Us About Organising for Climate Justice in the United States? *Environmental Sociology, 5*(2), 188–206.

Lewis, S. L., & Maslin, M. A. (2015). A Transparent Framework for Defining the Anthropocene Epoch. *The Anthropocene Review, 2*(2), 128–146.

Mahony, M., & Hulme, M. (2018). Epistemic Geographies of Climate Change: Science, Space and Politics. *Progress in Human Geography, 42*(3), 395–424.

Maldonado-Torres, N. (2016). Colonialism, Neocolonial, Internal Colonialism, the Postcolonial, Coloniality, and Decoloniality. In Y. Martínez-San Miguel, B. Sifuentes-Jáuregui, & M. Belausteguigoitia (Eds.), *Critical Terms in Caribbean and Latin American Thought: Historical and Institutional Trajectories* (pp. 67–78). New York: Palgrave Macmillan US.

Mamdani, M. (2001, October). Beyond Settler and Native as Political Identities: Overcoming the Political Legacy of Colonialism. *Comparative Studies in Society and History*. Retrieved March 17, 2018, from https://www.cambridge.org/core/journals/comparative-studies-in-society-and-history/article/beyond-settler-and-native-as-political-identities-overcoming-the-political-legacy-of-colonialism/A1919DC1C4418B5229BBE876C18BFCFB.

Mamdani, M. (2015). Settler Colonialism: Then and Now. *Critical Inquiry., 66*(4), 1039–1055.

Marino, A. (2017). Resisting Slow Violence: Writing, Activism, and Environmentalism. In R. Ciocca & N. Srivastava (Eds.), *Indian Literature and the World: Multilingualism, Translation, and the Public Sphere* (pp. 177–197). London: Palgrave Macmillan UK.

Mascarenhas, M. (2007). Where the Waters Divide: First Nations, Tainted Water and Environmental Justice in Canada. *Local Environment, 12*(6), 565–577.

McDowell, R. W., Dils, R. M., Collins, A. L., Flahive, K. A., Sharpley, A. N., & Quinn, J. (2016). A Review of the Policies and Implementation of Practices to Decrease Water Quality Impairment by Phosphorus in New Zealand, the UK, and the US. *Nutrient Cycling in Agroecosystems, 104*(3), 289–305.

McGregor, D. (2015). Indigenous Women, Water Justice and Zaagidowin (Love). *Canadian Woman Studies, 30*(2–3).

McGregor, D. (2018a). Indigenous Environmental Justice, Knowledge, and Law. *Kalfou, 5*(2), 279.

McGregor, D. (2018b). Mino-Mnaamodzawin: Achieving Indigenous Environmental Justice in Canada. *Environment and Society, 9*(1), 7–24.

McKenna, R. T. (2017). *American Imperial Pastoral: The Architecture of US Colonialism in the Philippines*. Chicago: University of Chicago Press.

Mitchell, J. T., Thomas, D. S., & Cutter, S. L. (1999). Dumping in Dixie Revisited: The Evolution of Environmental Injustices in South Carolina. *Social Science Quarterly*, 229–243.

Moggridge, B. (2018). Where Is the Aboriginal Water Voice Through the Current Murray-Darling Crisis? *Irrigation Australia: The Official Journal of Irrigation Australia, 34*(2), 34.

Mohai, P. (2018). Environmental Justice and the Flint Water Crisis. *Michigan Sociological Review, 32*, 1–41.

Mohai, P., & Saha, R. (2015). Which Came First, People or Pollution? A Review of Theory and Evidence from Longitudinal Environmental Justice Studies. *Environmental Research Letters, 10*(12), 125011.

Moran, K. D. (2015). Beyond the Black Legend: Catholicism and US Empire-Building in the Philippines and Puerto Rico, 1898–1914. *US Catholic Historian, 33*(4), 27–51.

Muru-Lanning, M. (2016). *Tupuna Awa: People and Politics of the Waikato River*. Auckland: Auckland University Press.

National Institute of Water and Atmospheric Research Ltd. (2014). *Maniapoto Priorities for the Restoration of the Waipa River Catchment*. Report Prepared for the Maniapoto Maori Trust Board.

Nightingale, A. J., Eriksen, S., Taylor, M., Forsyth, T., Pelling, M., Newsham, A., et al. (2019). Beyond Technical Fixes: Climate Solutions and the Great Derangement. *Climate and Development, 12*(4), 343–352.

Nirmal, P. (2016). Being and Knowing Differently in Living Worlds: Rooted Networks and Relational Webs in Indigenous Geographies. In W. Harcourt (Ed.), *The Palgrave Handbook of Gender and Development: Critical Engagements in Feminist Theory and Practice* (pp. 232–250). London: Palgrave Macmillan UK.

Paerregaard, K., & Andersen, A. O. (2019). Moving Beyond the Commons/ Commodity Dichotomy: The Socio-Political Complexity of Peru's. *Water Crisis, 12*(2), 12.

Parsons, M., & Nalau, J. (2016). Historical Analogies as Tools in Understanding Transformation. *Global Environmental Change, 38*, 82–96.

Pellow, D. N. (2004). The Politics of Illegal Dumping: An Environmental Justice Framework. *Qualitative Sociology, 27*(4), 511–525.

Proulx, G., & Crane, N. J. (2020). "To See Things in an Objective Light": The Dakota Access Pipeline and the Ongoing Construction of Settler Colonial Landscapes. *Journal of Cultural Geography, 37*(1), 46–66.

Pulido, L. (1996). *Environmentalism and Economic Justice: Two Chicano Struggles in the Southwest*. University of Arizona Press.

Pulido, L. (2017a). Geographies of Race and Ethnicity III: Settler Colonialism and Nonnative People of Color. *Progress in Human Geography, 42*(2), 309–318.

Pulido, L. (2017b). Evolving Racial Formations and the Environmental Justice Movement. In *The Routledge Handbook of Environmental Justice* (p. 2).

Pulido, L., & Peña, D. (1998). Environmentalism and Positionality: The Early Pesticide Campaign of the United Farm Workers' Organising Committee, 1965–71. *Race, Gender & Class*, 33–50.

Robison, J., Cosens, B., Jackson, S., Leonard, K., & McCool, D. (2018). Indigenous Water Justice. *Lewis & Clark Law Review, 22*, 841.

Rodríguez, I., & Inturias, M. L. (2018). Conflict Transformation in Indigenous Peoples' Territories: Doing Environmental Justice with a 'Decolonial Turn'. *Development Studies Research, 5*(1), 90–105.

Salmond, A., Tadaki, M., & Gregory, T. (2014). Enacting New Freshwater Geographies: Te Awaroa and the Transformative Imagination. *New Zealand Geographical Society, 70*(1), 47–55.

Schlosberg, D. (2003). The Justice of Environmental Justice: Reconciling Equity, Recognition, and Participation in a Political Movement. *Moral and Political Reasoning in Environmental Practice, 77*, 106.

Schlosberg, D., & Carruthers, D. (2010). Indigenous Struggles, Environmental Justice, and Community Capabilities. *Global Environmental Politics, 10*(4), 12–35.

Steffen, W., Grinevald, J., Crutzen, P., & McNeill, J. (2011a). The Anthropocene: Conceptual and Historical Perspectives. *Philosophical Transactions of the Royal Society A: Mathematical, Physical and Engineering Sciences, 369*(1938), 842–867.

Steffen, W., Persson, Å., Deutsch, L., Zalasiewicz, J., Williams, M., Richardson, K., et al. (2011b). The Anthropocene: From Global Change to Planetary Stewardship. *AMBIO, 40*(7), 739.

Steffen, W., Broadgate, W., Deutsch, L., Gaffney, O., & Ludwig, C. (2015). The Trajectory of the Anthropocene: The Great Acceleration. *The Anthropocene Review, 2*(1), 81–98.

Stensrud, A. B. (2016). Harvesting Water for the Future: Reciprocity and Environmental Justice in the Politics of Climate Change in Peru. *Latin American Perspectives, 43*(4), 56–72.

Te Aho, L. (2015). *The Waikato River Settlement: Exploring a Model for Co-management and Protection of Natural and Cultural Resources*. Ka Hula Ao Center for Excellence in Native Hawaiian Law, Richardson School of Law.

Retrieved January 6, 2019, from https://researchcommons.waikato.ac.nz/handle/10289/10414.

Todd, Z. (2016a). *'You Never Go Hungry': Fish Pluralities, Human-Fish Relationships, Indigenous Legal Orders and Colonialism in Paulatuuq, Canada.* PhD, University of Aberdeen, Aberdeen. Retrieved July 4, 2019, from http://digitool.abdn.ac.uk:80/webclient/DeliveryManager?pid=231448.

Todd, Z. (2016b). An Indigenous Feminist's Take on the Ontological Turn: 'Ontology' Is Just Another Word for Colonialism: An Indigenous Feminist's Take on the Ontological Turn. *Journal of Historical Sociology, 29*(1), 4–22.

Todd, Z. (2016c). From Fish Lives to Fish Law: Learning to See Indigenous Legal Orders in Canada. *Somatosphere.* Accessed November 23, 2020, http://somatosphere.net/2016/from-fish-lives-to-fish-law-learning-to-see-indigenous-legal-orders-in-canada.html/.

Tuck, E., & Yang, K. W. (2012). Decolonisation Is Not a Metaphor. *Decolonisation: Indigeneity, Education & Society, 1*(1), 1–40.

Ulloa, A. (2017). Perspectives of Environmental Justice from Indigenous Peoples of Latin America: A Relational Indigenous Environmental Justice. *Environmental Justice, 10*(6), 175–180.

Veracini, L. (2007). Historylessness: Australia as a Settler Colonial Collective. *Postcolonial Studies, 10*(3), 271–285. https://doi.org/10.1080/13688790701488155.

Veracini, L. (2010). *Settler Colonialism: A Theoretical Overview.* Basingstoke; New York: Springer.

Veracini, L. (2013). 'Settler Colonialism': Career of a Concept. *The Journal of Imperial and Commonwealth History, 41*(2), 313–333.

Waitangi Tribunal. (2018). *Te Mana Whatu Ahuru: Report on Te Rohe Pōtae Claims Pre-Publication Version Parts I and II.* Wellington: Unpublished.

Wanhalla, A. (2015). *In/Visible Sight: The Mixed-Descent Families of Southern New Zealand.* Wellington: Bridget Williams Books.

Watson, I. (2014). *Aboriginal Peoples, Colonialism and International Law: Raw Law.* London: Routledge.

Watts, V. (2013). Indigenous Place-Thought and Agency Amongst Humans and Non Humans (First Woman and Sky Woman Go On a European World Tour!). *Decolonization: Indigeneity, Education & Society, 2*(1) Retrieved May 16, 2020, from https://jps.library.utoronto.ca/index.php/des/article/view/19145.

Whyte, K. P. (2016a). *Our Ancestors' Dystopia Now: Indigenous Conservation and the Anthropocene.* (SSRN Scholarly Paper No. ID 2770047). Rochester, NY: Social Science Research Network. Retrieved June 12, 2020, from https://papers.ssrn.com/abstract=2770047.

Whyte, K. P. (2016b). Indigenous Experience, Environmental Justice and Settler Colonialism. In B. Bannon (Ed.), *Nature and Experience: Phenomenology and the Environment* (pp. 157–174). Lanham: Rowman & Littlefield. Retrieved January 30, 2020, from http://www.ssrn.com/abstract=2770058.

Whyte, K. P. (2016c). Is It Colonial Déjà Vu? Indigenous Peoples and Climate Injustice. In *Humanities for the Environment* (pp. 102–119). London: Routledge.

Whyte, K. P. (2016d). Indigenous Environmental Movements and the Function of Governance Institutions. In T. Gabrielson, C. Hall, J. Meyer, & D. Schlosberg (Eds.), *Oxford Handbook of Environmental Political Theory* (pp. 563–580). Oxford: Oxford University Press.

Whyte, K. P. (2017). The Dakota Access Pipeline, Environmental Injustice, and U.S. Colonialism. *Red Ink: An International Journal of Indigenous Literature, Arts, & Humanities, 19*(1) Retrieved May 29, 2020, from https://ssrn.com/abstract=2925513.

Whyte, K. P. (2018). Settler Colonialism, Ecology, and Environmental Injustice. *Environment and Society, 9*(1), 125–144. https://doi.org/10.3167/ares.2018.090109.

Whyte, K. P., Talley, J. L., & Gibson, J. D. (2019). Indigenous Mobility Traditions, Colonialism, and the Anthropocene. *Mobilities, 14*(3), 319–335.

Wilcock, R. J., Nagels, J. W., Rodda, H. J., O'Connor, M. B., Thorrold, B. S., & Barnett, J. W. (1999). Water Quality of a Lowland Stream in a New Zealand Dairy Farming Catchment. *New Zealand Journal of Marine and Freshwater Research, 33*(4), 683–696.

Wilson, N. J. (2014). Indigenous Water Governance: Insights from the Hydrosocial Relations of the Koyukon Athabascan Village of Ruby, Alaska. *Geoforum, 57*, 1–11. https://doi.org/10.1016/j.geoforum.2014.08.005.

Winter, C. J. (2018). *The Paralysis of Intergenerational Justice: Decolonising Entangled Futures.* Retrieved January 11, 2020, from https://ses.library.usyd.edu.au/handle/2123/18009.

Winter, C. J. (2019). Does Time Colonise Intergenerational Environmental Justice Theory? *Environmental Politics*, 1–19.

Wutich, A. (2009). Water Scarcity and the Sustainability of a Common Pool Resource Institution in the Urban Andes. *Human Ecology, 37*(2), 179–192. https://doi.org/10.1007/s10745-009-9227-4.

Yin, R. K. (2013). *Case Study Research: Design and Methods.* Sage Publications. Retrieved July 2, 2017, from https://books.google.co.nz/books?hl=en&lr=& id=OgyqBAAAQBAJ&oi=fnd&pg=PT243&dq=case+study+2010+social+s cience+research+methodology&ots=FaN7g9j72e&sig=h2XLUHg8qXQ1i4 TpvasC_BRxnig.

Zahara, A. (2017, March 14). Difference in the Anthropocene: Indigenous Environmentalism in the Face of Settler Colonialism. *Discard Studies.* Retrieved June 10, 2020, from https://discardstudies.com/2017/03/14/ difference-in-the-anthropocene-indigenous-environmentalism-in-the-face-of-settler-colonialism/.

2

Environmental Justice and Indigenous Environmental Justice

In this chapter, we outline the four essential ideas or proposals that provide the theoretical framework of this book. Firstly, the dominant framings and articulations of environmental justice (EJ) do not account for the complexities of Indigenous intergenerational environmental justice. Secondly, scholars and decision-makers need to consider what EJ is and how it can be taken into account in the context of environmental governance and management that goes beyond a narrow framing of justice as distributive equity, procedural inclusion, or recognition of Indigenous rights and consider the intersecting and interacting processes that underpin environmental (in)justices faced by Indigenous peoples. Thirdly, the theoretical discussion of EJ needs to recognise Indigenous sovereignties, cultures, and identities through Indigenous ontologies and epistemologies rather than through Western liberal thought and governance approaches. And lastly, the theoretical underpinnings of the study of Indigenous environmental justice (IEJ) need to incorporate intergenerational considerations.

These four ideas or arguments allow us to consider and explore the theoretical and empirical gaps within the literature on EJ. Besides, it provides us space to explore how a diversity of different scholars (Indigenous

© The Author(s) 2021
M. Parsons et al., *Decolonising Blue Spaces in the Anthropocene*, Palgrave Studies in Natural Resource Management, https://doi.org/10.1007/978-3-030-61071-5_2

and Indigenous allies) from a wide array of fields (human geography, political science, sociology, anthropology, history, Indigenous studies, environmental management, economics, philosophy, climate change adaptation) are calling for pluralistic accounts of justice that take into account local contexts, legal orders, and ontologies. Indeed, EJ always existed in the complex, overarching framework that was interwoven with the goals of social justice. As Taylor (2000) argues, in the USA context, the social injustices experienced by African-Americans and Indigenous Nations (slavery, discrimination, genocide, land confiscation) were also types of environmental injustice; policies and practices that resulted in social injustices also influence how communities were able to engage with environments and access resources (Taylor 2000). Accordingly, it is vital to highlight the ways EJ as a field of academic study and movement lassos the environmental and social together, particularly in the context of Indigenous EJ (as we will demonstrate later in this book through our case study of the Waipā River).

EJ: Distributive Justice

Early EJ research employed a distributive justice lens to examine the inequitable distribution of environmental risks and the physical proximity of specific communities to the environmental risk (Walker 2009). EJ (EJ) scholar trace origins of EJ (as a movement and a field of study) to Warren County (North Carolina USA) where a hazardous waste storage site (Polychlorinated Biphenyl PCB) was established near low-income Black communities despite widespread community protests. A wealth of subsequent different studies, beginning with Warren County, elsewhere in the USA and around the globe, investigated the differential exposure of communities to hazardous and toxic facilities (Bevc et al. 2007; Bullard 1993; Burwell and Cole 2007; Greife et al. 2017; Pastor et al. 2001; Wilson et al. 2012). These studies widely found, in a diversity of local and national contexts, that marginalised populations (ethnic minority groups, low-income, lower-caste, undocumented migrants, Indigenous peoples) were significantly more likely to live near environmental risks than the privileged populations (an ethnic majority, high-income, higher-caste, homeowners, citizens, settlers) (Arcury and Quandt 2009; Gordon et al. 2010; Salazar 2009; Salazar and Alper 2011; Vickery and Hunter 2016). The focus in these studies was on the distribution of environmental risks (or the environmental "bads" or negative impacts of environmental hazards) across populations and geographical areas using statistical and spatial analyses (Bell and Ebisu 2012; Fisher et al. 2006; Kingham et al. 2007; Pearce et al. 2006). Later research expanded beyond just the placements of environmental risks (such as polluted waters and contaminated soils) and examined where environmental "goods" or positives (such as clean water and land) was located (Caney 2008; Holifield et al. 2017).

In the case of Warren Country scholars declare it an example of environmental racism, referring to intentional, overt, and malicious acts of EJ against specific ethnic groups (specifically non-White) (Figueroa 2001; Pulido 2017). Scholar Bullard argues that such environmental racism was (and still is) widespread in the context of the USA (Bullard 2002). Indeed, as the work of other scholars attests to, racism remains a persistent feature of environmental governance, management, planning, and decision-making processes in many different colonial contexts, including the settler nation of Aotearoa.

EJ scholars argue that racism plays a critical factor in environmental planning and decision-making processes in the US and other settler nations. In Aotearoa few academic studies explicitly examine the distribution of environmental injustices across populations and areas (Coombes 2013; Pearce and Kingham 2008; Pearce et al. 2011; Rixecker and Tipene-Matua 2003); one study found that 40 per cent of low-income neighbourhoods in Wellington were exposed to environmental harms compared to 10 per cent of high-income areas (Salmond 1999). However, Māori activists and community leaders frequently speak out about issues pertinent to discussions of environmental racism; environmental harms (pollution generating factories, hazardous waste disposal sites, contaminated lands and waters) are frequently being located in poorer non-Pākehā (chiefly Māori and Pasifika) neighbourhoods. Greensill (2010) cites the example of the town of Kawerau situated in the Bay of Plenty near where the lead author (Parsons) grew (Greensill 2010). The population of Kawerau (according to the 2018 census) was 7146 people of whom 62 per cent identified as Māori; Kawerau is one of the only three areas in contemporary Aotearoa with a Māori-majority populace (the others being Ōpōtiki and Wairoa). The town is the site of Aotearoa's largest paper mill, established in 1953, which generates substantive pollutants (released into the air and water and deposited onto land). In particular, the mill discharges a toxic mixture of wastewater and solid materials directly into the Tarawera River. Dubbed the "Black Drain", the local iwi (Ngāti Rangitihi) reported that they could no longer collect customary food sources (fish, watercress, and birds) due to biodiversity loss as well as health risks associated with eating contaminated foods from the river, similarly, they no longer swam in the river due to the danger it posed to their health (Davison 2009; Dodd 2010). Also, certain types of millworkers (particularly those involved in processing tasks) are more likely to be exposed to toxic chemicals during daily; processing jobs (lower-paid and supposedly lower-skilled) are overwhelming held by Māori, whereas managerial roles (higher-paid and supposedly higher-skilled) are held by Pākehā. Accordingly, the Kawerau example could be read as an example of environmental racism and the inequitable distribution of environmental harms (however, further in-depth studies are needed).

A wealth of scholars now critiqued early EJ research for framing of EJ solely in terms of distributive equity (Mills 2015; Schlosberg 2003, 2004). Distributive justice is based on the assumption that if everyone is given equal access to environmental goods and balanced exposure to environmental harms then no environmental injustice occurs (Schlosberg 2004) So, for example, if a toxic waste dump is located an equal distance from Indigenous and White communities then there would be no environmental injustice accordingly to this framing of EJ as distributive justice (Sze and London 2008). However, such a framing of EJ ignores the social, cultural, and institutional contexts in which environmental injustices take place and the historical and contemporary systematic acts of discrimination against marginalised populations (including Indigenous peoples and other non-White non-Indigenous communities in settler-colonial societies, members of lower-incomes and lower-castes in India, and formerly colonised peoples throughout the Global South) which all play substantial roles in creating environmental injustices. A distributive-framing of EJ (along with environmental racism) therefore misses crucial opportunities to critique the parts of colonialism and capitalism in its relation to different subjectivities and how it creates place-based and culturally-situated environmental injustices (Hendlin 2019; Jackson 2018; Swyngedouw and Heynen 2003; Whyte 2014, 2016a). More recent EJ scholarship demonstrates that the narrow focus on equitable distribution largely ignores the broader social, cultural, and institutional contexts in which environmental injustices take place and the role that capitalism, colonialism, and patriarchy plays in legitimising, driving and deepening environmental inequities (Álvarez and Coolsaet 2018; McGregor 2015; Swyngedouw and Heynen 2003; Sze and London 2008; Walker 2009). Later EJ research draws attention to the need to consider how procedures (policies, decision-making processes, and participation) and recognition (of cultural differences) play in EJ.

Procedural Justice

Scholars draw attention to the need to consider procedural justice to combat the issues associated with distributive justice. Procedural-based EJ focuses on decisions and the decision-makers involved with environmental management decisions. In early EJ research, there was an unspoken assumption that the decision-makers where institutions of power (for example government agencies and energy companies) with communities (mostly poor and non-white communities) as the helpless victims of these decisions (Antadze 2018; Pitea 2009). Walker notes in several works that agencies such as the Environmental Protection Agency (US) developed policies and procedures to facilitate community input in decision making and hold guilty agencies responsible. However, as the works of Banisar et al. (2011) and others argued that these spaces of public participation, in the form of submissions, and public ways, did not yield the outcomes that communities hoped for (Banisar et al. 2011; Paloniemi et al. 2015). Often these spaces were controlled by either government agencies or the companies themselves, who were committed to focusing on their agendas rather than on community needs. This reflects a broader scholarship on public participation in environmental management, informed by the work of Arnstein, Tritter and McCallum, who argue that the majority of government attempts to include the public (or specific social groups) are superficial, and there remain considerable constraints on communities capacities to participate in the decision-making process (Arnstein 1969; Tritter and McCallum 2006).

A wealth of scholarship exploring public participation in environmental management and EJ builds on the seminal work of Arnstein, specifically the article "A Ladder of Citizen Participation" (Arnstein 1969; Boone and Buckley 2017; Carpentier 2016; Connor 1988; Hurlbert and Gupta 2015; Ross et al. 2002). Arnstein (1969) argues that participation is the "cornerstone" of democracy; however, marginalised communities demand a form of involvement that goes beyond simply just being consulted about decisions and be involved in and shape the decisions. Such participation, as Arnstein notes, calls for a redistribution of power (from the powerful to the marginalised groups within society) to enable those who are marginalised to join the conversation to determine how information is shared and ultimately encourage social reform that allows previously marginalised communities to benefit (Arnstein 1969). Arnstein breaks down participation into a ladder which is broken up into eight different steps. The steps are then grouped into three categories (Non-participation, Degrees of Tokenism and Degrees of Citizen Power); which range from the no or little public participation in decision-making (Non-participation) to some public participation (Degrees of Tokenism), and finally, a significant amount of participation and the capacities to shape government decisions (Degrees of Citizen Power).

Arnstein's participation ladder is not without criticism amongst scholars (Carpentier 2016; Hart 2008). Indeed, those in positions of power in a society are often highly resistant to giving up any power and, as the work of feminist and anti-racist scholars demonstrates, the continuation of patriarchal structures as well as racism and other discriminatory beliefs effectively set up roadblocks to specific groups' achieving higher levels of participation (what Arnstein terms "Degrees of Citizen Power") in decision-making (Azmanova 2012; Crease et al. 2019; Pulido and De Lara 2018; Schlosberg 2003; Sen 1995; Tschakert and Machado 2012). The roadblocks for marginalised social groups being able to participate in environmental governance and management, as our later analysis of co-governance arrangements for the Waipā River (Chap. 7), include lack of access to appropriate financial resources, technologies, and training, as well as public participation forums being designed to fit the intellectual, cultural, and political traditions of the dominant social group (and in doing so re-articulating the state's exclusion of Indigenous knowledge, values, and practices). Thus, it is not just limited resources and capacities that create barriers to marginalised groups participating in environmental management decision-making processes; it is also the failure of the state to recognise different cultures' values, knowledges, and ways of life. Indeed, as Blue et al. (2019) recently highlights, participatory practices and justice are closely related (Blue et al. 2019) and, as the work of Nancy Fraser also demonstrates (Fraser 1990, 1995, 2007, 2009), people's abilities to participate in decision-making processes are influenced by a range of economic, political and socio-economic factors that extend beyond distributive and procedural and also include recognition of cultural differences.

Recognition Justice

Other scholars advocate for thinking about EJ as recognition and respect of individual and communal cultural differences (Barnhill-Dilling et al. 2020; Fraser 1995). Particularly in the context of water security, ecosystem restoration, and biodiversity conservation, recent scholarship examines the discursive and practical constraints of the dominant Western liberal framings of distributive and procedural EJ (He and Sikor 2015; Martin et al. 2016; Sikor et al. 2014; Sze 2018). Instead, Martin (2016), Sze (2018) and other scholars (Barnhill-Dilling et al. 2020) argue that recognition is a critical part of justice and a "necessary precondition for participating in environmental decisions" (Barnhill-Dilling et al. 2020, p. 84).

A lack of recognition, Schlosberg (2004) and Adger et al. (2011), of the impacts of environmental degradation and risks faced by specific communities, can detrimentally affect both the material and cultural wellbeing of individuals and communities' (Adger et al. 2011; Schlosberg 2012; Schlosberg and Carruthers 2010). If, for instance, national or local governments do not acknowledge that existence of specific environmental harms, hazards, or risks (be it the pollution of waterways or the impacts of climate change) that are occurring within their jurisdictions, they are likely to be apathetic to the environmental risks and take limited actions to mitigate those risks. Similarly, if governments, interest groups, leaders, and the citizenry as a whole do not recognise that marginalised populations, including Indigenous peoples, (within their nation-states and around the world) are the most at risk (most vulnerable) to the negative impacts of environmental hazards (including water pollution, a tropical cyclone or the effects of climate change), resources are unlikely to be directed at assisting those groups (Rydin 2006; Schlosberg and Collins 2014). Hurricane Katrina, which devastated the US city of New Orleans in 2005, is a glaring example of this and is widely analysed by justice scholars. In New Orleans, a natural hazard was transformed into a disaster when distributive injustices (environmental racism against the Black population) coincided with procedural and recognitional injustices (inequitable institutional arrangements, planning regimes, legal systems and economic structures) to marginalise the lives, bodies, and ways of life of individual people (Black/African-American residents) over others based on race (White residents). The hurricane became a large-scale disaster and was a consequence of flood levees failing and flooding predominately Black neighbourhoods, resulting in the deaths of more than 1800 people (the majority of whom were Black). Yet, scholars concur that these deaths were mostly avoidable and a consequence of actions to address the multiple social and environmental injustices faced by Black communities in New Orleans (Bullard and Wright 2008; Miller and Rivera 2009; Rohland 2018).

Scholar highlight how the settler state's failure (or misrecognition) of Indigenous communities (by marginalising their knowledge, values, ways of life and excluding it from decisions) contributes to environmental injustices (Barnhill-Dilling et al. 2020; Holifield 2012; Holifield et al. 2017). Examples of misrecognition extend beyond the misrecognition of the culture and includes the misrecognition of land and water (and Indigenous people's relationships with their properties, waters, and biota). Such misrecognition of lands consists of the common practice whereby settler nation-states (and settlers) devalued indigenous lands, labelling it 'wastelands', 'unusable', or 'undesirable' (until the land was no longer held by Indigenous peoples). These labels make it easier for settler state, settlers, and companies to justify the placement of environmental harms or risks in undesirable or marginal lands (Barnhill-Dilling et al. 2020; Holifield 2012; Walker 2009). Misrecognition, however, is only one part of the framing of EJ as recognition. The other part is recognitional justice, which is critical for Indigenous people, are the capacities of people to determine their interpretation of what environmental (in)justice is (Jackson 2008; Lowitt et al. 2019; Whyte 2011). Indeed, for Indigenous peoples who already possess or want treaties and laws that acknowledge and enforce their self-determination rights and tribal sovereignty. Even when settler states recognise indigenous peoples' rights of self-determination, their capacities to make decisions and enact their sovereignty are often under-minded by the settler state and other outside organisations (Holifield 2012; Ranco 2008). While most scholars agree that both procedural justice and justice as recognition are essential to EJ, many scholars also declared that procedural justice and recognition alone do not provide enough to guarantee EJ.

Recognition can consist of an affirmation of a group's cultural difference and identity and/or strategies that are directed at overcoming institutional harms that prevent meaningful engagement with political and social institutions. Recognition-informed actions include those that aim to address or mitigate injustices against Indigenous peoples through strategies termed affirmation actions (such as educational scholarships and provision of welfare). Through projects that aim to transform Indigenous-non-Indigenous relationships (such redistribution of the benefits and altering modes of production), recognition approaches are primarily directed at social and cultural changes including the "deconstruction" of principal arrangements of socio-cultural representation in ways that recognise and "change everyone's social identities" (Fraser and Honneth 2003, pp. 12–13).

Critique of Recognition

Dene political theorist Coulthard (2007, 2014), writing in the context of settler-colonial Canada and Indigenous nations, critiques the idea that the relationships between the settler-nation and Indigenous peoples are transformed through the "politics of recognition" (2007, p. 438). Recognition, Coulthard defines in terms of to the affirmative acknowledgement "of societal, cultural differences" and "freedom and wellbeing of marginalised individuals and groups living in ethnically diverse states" (Coulthard 2007, p. 438). Coulthard maintains that recognition-based conceptualisations of justice, emerging from Western liberal pluralism, aim to reconcile Indigenous sovereignty claims (which range from complete nation-state sovereignty to limited self-determination) with the sovereignty of the nation-state through a compromise of sorts. The state recognises Indigenous cultural identities and engages in projects aimed at improving and reconfiguring the relationships of Indigenous peoples with the nation-state. Coulthard (2007) notes that the "politics of recognition" in its present form simply reproduces "the very configurations of colonial power that Indigenous peoples' demands for recognition have historically sought to transcend" (Coulthard 2007, p. 439). Indeed, Coulthard (2014) observes that despite different Indigenous peoples in Canada achieving recognition through legislation, Treaties, and other formal agreements with the federal and provincial governments, the Canadian courts continue to declare that the settler-state possess the right to make decisions about environmental management and developments within Indigenous landscapes and waterscapes. The vast majority of government-sponsored projects, including the construction of infrastructure and settlements as well as hydroelectric, forestry, agricultural and mining ventures, is justified and rationalised so long as each project is "'consistent with the special fiduciary relationship' between the Canadian government and the indigenous peoples" (Coulthard 2007, p. 451).

Other academics, following on from the work of Coulthard, similarly demonstrate how existing neoliberalism (in Aotearoa, Australia, Canada and beyond) has influenced and constrained the forms of recognition proposed by the state as a method to address social injustices experienced by Indigenous peoples as a consequence of settler colonialism (Azmanova 2012; Bargh 2018; Bell 2018; McCormack 2018). Avril Bell, for instance, highlights how:

> At its Hegelian roots, recognition theory is about the struggle to achieve a relationship of equals between two subjects. To recognise the subjectivity of another is to recognise their equal and autonomous status as self-determining people worthy of respect. (Bell 2018)

What prevails is (in the words of Jakeet Singh) "recognition from above" in which the state "is the arbiter of just and unjust claims for recognition from subordinate groups" (Singh 2014a, p. 47). Aside from deciding what types of recognition are on offer, the state also spells out the provisions of recognition. For instance, while the state may legally acknowledge Indigenous rights and identities, as a range of critical humanities and social science scholars demonstrate, those rights and identity are frequently essentialised in ways that enable the state's economic interests in the era of neoliberalism (Bargh 2018; Coombes et al. 2012; Coulthard 2014; Singh 2014b, 2019).

Avril Bell's examination of how local governments in Aotearoa recognise Māori provides a sharp critique of how neoliberal politics influenced and constrained the form of recognition on offer by the state (Bell 2018). She highlights how the central government of Aotearoa (the Crown) now officially recognises that Māori and the Crown are Treaty partners (as encapsulated in Aotearoa's founding document Te Tiriti o Waitangi/the Treaty of Waitangi), the legislation governing local government explicitly states that local authorities are not the Crown and are not Treaty partners with Māori. Since environmental governance is highly devoted in Aotearoa, the failure to legally include local government as Treaty partners means that local government authorities routinely misrecognise Māori interests, only allow for Māori participation in planning that is tokenism, and make no attempts to achieve distributive equity (Bell 2018; Ryks et al. 2010). Accordingly, local government is, in Bell's view, emblematic of the failure of the New Zealand Crown to adequately recognise Māori as full Treaty partners (which we discuss further in Chap. 2) (Bell 2018). We will pick up on Bell's analysis further in our review of the management of freshwater within the Waipā River catchment (Chaps. 4, 5, and 6) and highlight how the current politics of recognition, within the context of freshwater management, does not challenge the settler state to reform itself. Indeed, we echo Bell's argument that when the two arms of government (central and local) are assessed together, the settler state is "not a fit subject for recognition politics" (that is to impose "recognition from above") (Bell 2018, p. 78). At the local level, the state suffers from ongoing historical amnesia (continuing to frame local histories as one of peaceful settlement and continuous progress) and, more generally, makes endless statements that emphasise the rights of Māori iwi (as Treaty partners and as tribal authority-holders referred to as mana whenua); the importance of incorporating mātauranga (Māori knowledge) and tikanga (laws and principles) into freshwater management, yet at the same time taking actions that are opposed to their statements; put simply, local governments' frequently say one thing while doing another.

Furthermore, although Indigenous identity is recognised, the articulation of Indigenous peoples' inclusion within neoliberal economies endeavours to foreclose other alternative economic arrangements. While we do not, in this book, focus on economics, it is nevertheless important to acknowledge this significant critique of recognition-based justice. Scholars highlight the fundamental need for local arrangements that allow for Indigenous peoples to be agents of recognition thereby gaining control over the redistributive of revenues and expenditures directed at addressing Indigenous peoples' socio-economic disadvantage and marginalisation, and in doing so promote Indigenous peoples' inclusion and address injustices; this is how "recognition from below" takes place, "when people in dominated social positions turn away from institutionalised power hierarchies, shaping their own social orders without approval or permission of any authority beyond themselves" (Williams 2014, p. 10). As Williams observes, these "processes of the state self-constituting power", realised through formal political movements or acts of resistance, also involve struggles for recognition, but the "agents of recognition" are Indigenous peoples rather than the state. Evidence of what Coulthard terms "recognition from below" which he defines as the: strategies of 'self-recognition' through which colonised or dominated subjects "critically revalu[e], reconstruct ... and redeploy ... culture and tradition" and, through such a process, transform their own subjectivities and consciousness as political agents (Coulthard 2007, p. 456). Significantly, many scholars examine the dynamic and complex trajectories of neoliberalism within settler-nations and highlight how neoliberal governance frequently involves a shift in state recognition of Indigenous interests and demonstrates what is needed to create situations where recognition from below is possible. For instance, Will Sanders argues (in the context of Australia but equally applicable to other settler-nations) that what is needed in contemporary Indigenous policymaking is some re-recognition of decolonisation as a means to address continuing Indigenous socio-economic disadvantage (Sanders 2018). He goes onto suggest that labelling and framing are significant, and it is critical to continue to insist articulating and acting on the process of decolonisation (even if we live in the age of neoliberalism) because it keeps alive the central ideas about the critical need to recognise Indigenous interests and demands for justice.

Such ideas can also be extended to thinking about IEJ as there are concerns that the state continues to be the arbiter decider of what and how Indigenous rights and interests in water (land, seas, and so forth) are recognised (as we demonstrate in Chap. 4). What this means, as we explore in-depth in Chap. 9 (which explores river restoration), is the nuances and complexities of Indigenous interests in their local environments, which includes their use of natural resources and environmental stewardship across successive generations as well as deliberative forms of place-based and kinship-centred governance, are frequently overlooked in favour of recognition formats that fit the needs (worldviews and governance structures) of the settler state rather than Indigenous peoples' themselves. In doing so, the plethora of intergenerational environmental injustices experienced by Indigenous peoples is frequently overlooked by the narrow "recognition from above" models employed by the states. However, we demonstrate the potential to disrupt the narrow conceptualisations of recognition and extend it to include multiple ontologies and legal orders. We suggest that there is an emerging middle ground between a settler state and Indigenous political agendas in Aotearoa, which imperfect, in the context of the emergence of co-governance and co-management arrangements over rivers (and mountains) (outlined in Chaps. 7 and 8) does present the potentialities of addressing environmental injustice through governance structures and management approaches under-pinned situated within Māori ways of knowing and beyond.

Beyond Recognition: Indigenous Ontologies and Epistemologies

There is a fundamental need, Māori philosopher Christine Winter argues, for accounts of environmental justice to move beyond Western liberal thought to meaningfully include Indigenous ontologies and epistemologies (Winter 2018, 2019a, b). One way of doing this would be to expand the dimensions of recognitional justice to embrace ontological and epistemological pluralism. Winter identifies some of the differences between Indigenous and Western intellectual traditions (see Figs. 2.1 and 2.2) that

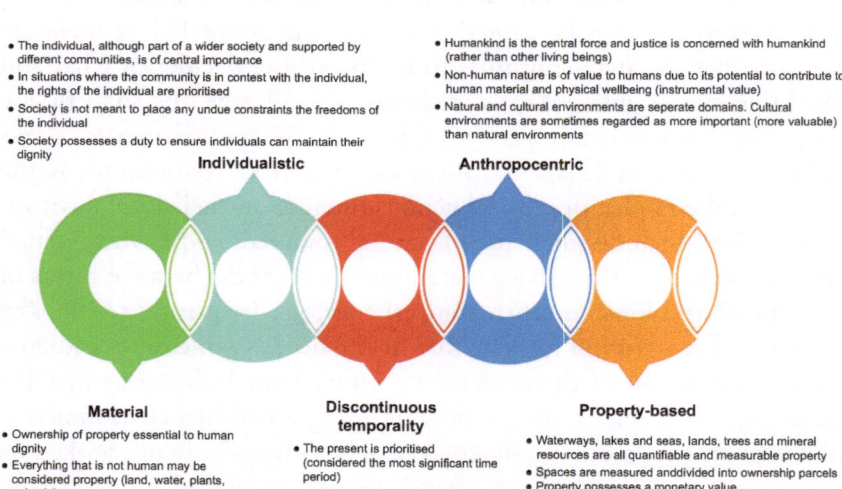

- The individual, although part of a wider society and supported by different communities, is of central importance
- In situations where the community is in contest with the individual, the rights of the individual are prioritised
- Society is not meant to place any undue constraints the freedoms of the individual
- Society possesses a duty to ensure individuals can maintain their dignity

Individualistic

- Humankind is the central force and justice is concerned with humankind (rather than other living beings)
- Non-human nature is of value to humans due to its potential to contribute to human material and physical wellbeing (instrumental value)
- Natural and cultural environments are seperate domains. Cultural environments are sometimes regarded as more important (more valuable) than natural environments

Anthropocentric

Material
- Ownership of property essential to human dignity
- Everything that is not human may be considered property (land, water, plants, animals)
- Property rights give owners the right to extract financial value from their property with few limiting prohibitions
- Property rights can be transferred through various mechanisms (selling, leasing, and gifting) with property laws about interpersonal obligations not about property itself

Discontinuous temporality
- The present is prioritised (considered the most significant time period)
- The past is less important than the present-day
- The future is potential time that current inhabitants may inhabit. However, the majority of decision-making ficuses on the present and near-present rather than medium-or longer-term futures

Property-based
- Waterways, lakes and seas, lands, trees and mineral resources are all quantifiable and measurable property
- Spaces are measured anddivided into ownership parcels
- Property possesses a monetary value
- Property can be transferred between parties for money or other resources
- Spaces are lived in by people as property and considered "culture" spaces, whereas uninhabited spaces are "natural" spaces
- Property contains resources that can be accessed by people for the accumulation of material wealth

Western Worldviews: Te Ao Pākehtā

Fig. 2.1 Key features of Western worldviews that pertain to discussions of EJ

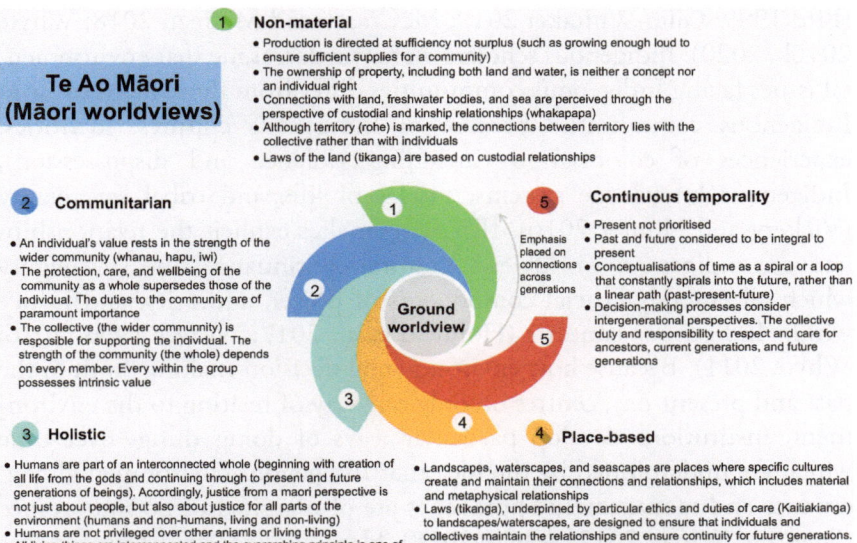

Te Ao Māori (Māori worldviews)

1 Non-material
- Production is directed at sufficiency not surplus (such as growing enough loud to ensure sufficient supplies for a community)
- The ownership of property, including both land and water, is neither a concept nor an individual right
- Connections with land, freshwater bodies, and sea are perceived through the perspective of custodial and kinship relationships (whakapapa)
- Although territory (rohe) is marked, the connections between territory lies with the collective rather than with individuals
- Laws of the land (tikanga) are based on custodial relationships

2 Communitarian
- An individual's value rests in the strength of the wider community (whanau, hapu, iwi)
- The protection, care, and wellbeing of the community as a whole supersedes those of the individual. The duties to the community are of paramount importance
- The collective (the wider community) is responsible for supporting the individual. The strength of the community (the whole) depends on every member. Every within the group possesses intrinsic value

5 Continuous temporality
- Present not prioritised
- Past and future considered to be integral to present
- Conceptualisations of time as a spiral or a loop that constantly spirals into the future, rather than a linear path (past-present-future)
- Decision-making processes consider intergenerational perspectives. The collective duty and responsibility to respect and care for ancestors, current generations, and future generations

Emphasis placed on connections across generations

Ground worldview

3 Holistic
- Humans are part of an interconnected whole (beginning with creation of all life from the gods and continuing through to present and future generations of beings). Accordingly, justice from a maori perspective is not just about people, but also about justice for all parts of the environment (humans and non-humans, living and non-living)
- Humans are not privileged over other aniamls or living things
- All living things are interconnected and the overarching principle is one of ensuring there is balance between living things
- Cultural identities are intertwined with non-human dimensions of the environment
- Value is situated as integrated with the whole (symbiotic relationships)
- Nature is everything and everywhere (rather than something seperate from people and out there)

4 Place-based
- Landscapes, waterscapes, and seascapes are places where specific cultures create and maintain their connections and relationships, which includes material and metaphysical relationships
- Laws (tikanga), underpinned by particular ethics and duties of care (Kaitiakianga) to landscapes/waterscapes, are designed to ensure that individuals and collectives maintain the relationships and ensure continuity for future generations
- Laws are both prescriptions about behaviour but also expressions of particular identities, and ways of being and thinking about the world
- Places, including one's land (whenua), awa (river), and mauga (mountain), are at the heart of tribal identity; with each person, whanau, hapu, and iwi connected to specific spaces through whakapapa (genealogy)
- Actions tha diminish the life force (mauri) of culturally important landscapes and waterscapes also damage the wellbeing of iwi and hapu who whakapapa to the

Fig. 2.2 Key features of Indigenous worldviews

inform our later discussions of EJ. Recent research by Indigenous scholars, including Winter, McGregor and Whyte, documents instances of environmental injustices suffered by Indigenous communities, which are tied to the continued dominance of Western worldviews (including framings of what constitutes justice as summarised in Fig. 2.1) that are premised on nature/culture binaries (already critiqued by a plethora of scholars) (McGregor 2018a; Todd 2016; Whyte 2018; Winter 2019a, b).

Despite how well-intended the EJ scholarship is, the dominant EJ framework being used by scholars (and applied to Indigenous communities around the globe) continues to neglect the unique experiences of Indigenous communities and their collective trauma under colonialism (Whyte 2016, 2017). A wealth of indigenous and non-indigenous academics call for the colonial structures that underpin EJ (as a movement and a field of study) to be overthrown to allow space to both acknowledge and enact the knowledge, rights, and sovereignty of indigenous peoples

(Bird 1999; Gilio-Whitaker 2019; McCreary and Milligan 2018; Whyte 2016b, 2020). Indigenous scholars, in particular, argue that environmental issues facing Indigenous communities differ from those faced by non-Indigenous communities because of Indigenous cultures, identities, experiences of colonisation (including violence and dispossession), Indigenous knowledge systems, modes of life, and tribal sovereignty (Vickery and Hunter 2016). IEJs (IEJ) makes explicit the relationships between indigenous worldviews, cultural continuance, and sovereignty which all embody crucial components of power, authority, and justice within Indigenous contexts (Holifield et al. 2017; Weaver 1996, 2016; Whyte 2011). Because how environmental decision-making, both in the past and present-day, centres on only one way of relating to the environment, institutions develop particular ways of doing things over time which are underpinned by the idea that Indigenous environmental governance and management approaches are of marginal or no importance (Steel and Whyte 2012; Whyte 2018). Here, the lens of IEJ provides us with the opportunity to acknowledge both Indigenous sovereignty and indigenous worldviews as rooted in justice-oriented freshwater governance management and decision-making.

Māori worldviews, which exist on a continuum that is increasingly incorporating Western liberal individualism as well as Māori collectivism, continue to resonate in and shape Māori people's lives and their engagement with their awa as we outline in later chapters. Like other Indigenous people who live within the borders of settler-colonial states, Māori iwi (tribes), hapū (sub-tribes), and whānau (extended family) endure despite the social, cultural, economic, political, and ecological marginalisation they experienced as a consequence of settler colonialism. Indeed, the histories of Māori and other Indigenous cultures over the last two hundred plus years of colonisation offers us all (Indigenous and non-Indigenous alike) essential lessons about what constitutes a life well-lived and how to maintain cultures, identities, a sense of belongingness and connectivity, and pursue a good life (one that holds value to you) in the face of radical (seemingly Earth-shattering) social, economic, political, cultural, and environmental changes. Indeed, the populations of Indigenous peoples around the globe experienced a massive loss of life as a consequence of infectious disease outbreaks linked to the arrival of colonisers bringing with them new diseases; smallpox, for instance, killed between 60–80 per cent of the Indigenous peoples of the Americas and Australia (far more than colonial violence ever did). While Indigenous peoples, like all peoples around the globe, are facing the COVID-19 pandemic, it is worth remembering the long history of past experiences of destruction and loss, and how different ways of thinking about the world can guide daily and future practices for more sustainable and hopeful futures. Indeed, Māori ontology can offer valuable learnings into two theoretical domains—dignity and time—that offer the potential to address both Western and Indigenous demands for EJ and intergenerational justice.

For more than a century, Indigenous worldviews and philosophies were frequently excluded or disparaged, deemed either primitive or a-theoretical by scholars, scientists, and decision-makers alike (Mills 2015; Buckinx et al. 2015; Tully 2000). Despite persistent attempts to erase, replace, and eradicate Indigenous beliefs and worldviews (be it through academia, the legal system, policymaking, media and the education curricula), such values and understandings remain relevant to the lives of many Indigenous peoples. Increasingly, as the emergent co-governance and co-management approaches attest to, settler states are enacting policies (after centuries and decades of protests and campaigns by Māori groups) that recognise Indigenous authority, knowledges and principles (which challenge the supposed universal applicability and superiority of Western liberal thought). These portrayals shape current lives and will affect future generations of Indigenous peoples (as individuals, communities, and societies). A new concept of intergenerational EJ could, however, include and encourage Indigenous and non-Indigenous alike.

Some Indigenous ontologies are characterised as holistic and kin-centric, such as found amongst Māori of Aotearoa, Aboriginal peoples of Australia, and Indigenous peoples throughout North and South America. People are active and co-producing (participatory) players within ecosystems (see Burarrwanga et al. 2013; etc.). Without the wrenching division between humans and nonhumans which characterises Western thought (post the European Enlightenment) (Ghosh 2018), Indigenous peoples exist in a complex and highly dynamic continuum of relationships with natures (physical, ecological and metaphysical). So interwoven are these connections that some scholars include places as co-authors (Country et al. 2016; Suchet-Pearson et al. 2013). As Australian Indigenous scholar Laklak Burarrwanga (an elder from Datiwuy located in North East Arnhem Land) and collaboration with Indigenous and non-Indigenous scholars writes:

> our homeland of Bawaka as co-author. That's because the land, the water, the animals, the plants, the rocks, the thought and songs that makeup Bawaka contribute to what we are saying here in important ways. They speak to us, inform what we do and have guided our thinking and talking). (Burarrwanga et al. 2013, p. Loc 324 of 3120)

Indeed, many Indigenous cultures are therefore located on such intellectual groundings, ontological underpinnings firmly rooted in the lack of distinction between human and nature. It is, therefore, a crucial counterpoint to the Western liberal dichotomy of human-nature, civilised-savage, tamed-wild, productive/wasteful, modern/primitive, from a holistic and connective perspective that situates people as part of nature: "Humans can no more go out of nature than they can go out of their bodies" (Green 2011, p. 132).

Accordingly, this raises several critical questions about freshwater management in the Anthropocene, both in terms of theorising about EJ and actions to address the drivers and implications of freshwater degradation. Western liberal theorises of EJ (which remain dominant within both the international scholarship and policymaking domains) continue to claim neutrally, impartially, and universally. Yet, Indigenous scholars, including Watene and Winter, are challenging the field of EJ to reconsider and extend what constitutes life and dignity supporting environments for all peoples around the globe (including those from non-Western cultures) in the context of changing climate conditions and its intergenerational justice ramifications (Budowle et al. 2019; Spiegel et al. 2020; Watene 2016; Winter 2019b). The critical question is, what does EJ look like if we are to take into account the ontologies of Indigenous peoples in the context of freshwater governance and management? Is it possible to formulate, within the Western liberal theories of justice, an account of EJ (incorporating social, environmental and intergenerational justice) that provides for Indigenous peoples within settler societies? Indeed, are Western liberal theories capable of doing this or are the ontological differences so significant that the conceptualisation of justice is different? We attempt to address some of these questions in the following chapters in this book.

Winter identifies the ways in which the dominant cultures of settler states remain epistemologically ignorant of Indigenous perspectives. It is not possible to describe in-depth all Indigenous worldviews, but we do identify some standard features that differ from those of Western worldviews: non-materiality; a sense of place; communitarianism; holism; and non-linear temporality (summarised in Fig. 2.1). Likewise, other scholars challenge Western articulations of EJ and advocate for Indigenous-informed EJ approaches. There is no agreeable definition of what exactly counts as IEJ; however, McGregor summarises the approach that advocates for "relationships based on environmental justice [that] are not limited to relations between people but consist of those among all beings of Creation" (McGregor 2010, p. 27). Indeed, a common feature of the various IEJ scholarship is a framing of EJ that goes beyond humans (the anthropocentric lens) to include animals, plants, weather, geology, spirits and supernatural beings, and IEJ thus deserves an Indigenous-informed framework (distinct from EJ frameworks employed in Canada, United States, Australia and elsewhere). IEJ as a framework, McGregor et al. (2020) argues, provides a set of logics that moves beyond the myopic anthropogenic lens of Western liberal theorising to recognise and include more-than-human actors as well as the Earth itself (McGregor 2018b; McGregor et al. 2020). For example, in the context of freshwater management and water justice, scholarship exploring Indigenous knowledges and experiences of water injustices highlight how, for many different Indigenous peoples, water is conceptualised as a living, more-than-human entity with responsibilities and duties to maintain the life and wellbeing of itself and other beings, which contrasts markedly from Western understandings of water as a resource and commodity (Jackson 2018; McGregor 2015; Perreault et al. 2012; Stensrud 2016). According to Indigenous ontologies, as we explore further in Chap. 6, issues of water justice and security are not merely about Indigenous peoples (and other social groups) being able to access water equitably (as encapsulated in the United Nations right to-water discourse) but also about justice for water as a more-than-human entity who possesses its own rights and responsibilities, which need to be recognised and provided for (Jackson 2018; McGregor et al. 2020).

In line with other Indigenous-informed approaches to EJ and maintaining the significance of EJ being spatial and temporally located (considering local histories, cultures, and geographies), in the rest of the book we explore Māori (specifically Ngāti Maniapoto) conceptualisations of and responses to environmental injustices. However, we do draw links to other Indigenous peoples' ontologies and framings of justice (with particular emphasis on reciprocal relations, intergenerational responsibilities and more-than-human entities) to highlight the ways in which a growing chorus of EJ scholars and activists are drawing attention to other forms of knowing and being and the limitations of the hegemonic (Eurocentric) EJ paradigm. For instance, Māori emphasise the need to manage environmental resources sustainably (guided by the principle of kaitiakitanga meaning environmental guardianship) to ensure that future generations can use those resources (which we explore in future depth in Chap. 2). A commonly used whakataukī (proverb used within Māori societies to share cultural norms and values) that encapsulated the intergenerational dimension of Māori environmental management:

> Hutia te rito o te harakeke. Kei hea te korimako, e ko? Ki mai ki ahau, he aha te mea nui o te ao? Maku e ki atu He tangata, he tangata, he tangata.
> Pluck the heart from the flax bush - where will the bellbird be? Ask me, what is the most critical thing in the world? I will reply, it is people, it is people, it is people. (Cherrington 2019, p. 53)

While the meaning of this whakataukī is multi-layered, its central message is one of sustainability. It underpins the idea that balance is needed between all elements of the world (humans and more-than-humans) to maintain the health and wellbeing of all (Durie 2006; Rixecker and Tipene-Matua 2003; Walker et al. 2019; Wehi and Lord 2017). Harakeke (the flax bush *Phormium Tenax*) is a prodigious plant that grows throughout Aotearoa and is (and historically was) used for a variety of purposes by Māori (specifically for the weaving of clothing, art, baskets and ropes). Accordingly, efforts are taken to use it sustainably. For instance, the side leaves of a flax plant can be removed, but if the plant's central core is damaged, the plant will die. Likewise, the korimako (bellbird *Anthornis Melanura*) collects nectar from the flowers of flax bush (and is also praised for its beautiful song). So, the death of flax negatively impacts the health of bellbirds. The answer to the question stresses that people must practice reciprocal relationships with the more-than-human beings that share the world(s) with them and emphasises the sustainable use of resources to ensure the wellbeing of current and future generations.

Far across the Pacific Ocean, in the Canadian context, Anishinaabe scholar Deborah McGregor articulates similar ideas in her research into Anishinaabe EJ. She demonstrates how, under Anishinaabe traditions, justice extends to include both current generations as well as the "ancestors of current beings and those yet to come (at least as far ahead as seven generations from now)" (McGregor 2010, p. 30). For the Anishinaabek people, environmental management decision-making is required to consider at least seven generations of beings (human and more-than-human). Such conceptualisations of looking seven generations into the future are likewise articulated in various Canadian and US Indigenous peoples' declarations about their rights and responsibilities for their waters, including the Water Declaration of the Anishinabek, Mushegowuk, and Onkwehonwe (2008) and the Tribal and First Nations Great Lakes Water Accord (2004).

In Australian Aboriginal societies, and even longer intergenerational lens is applied to conceptualisations of EJ that reflect different conceptualisations of time (which challenges assumptions of linearity and forward-thinking). In Australia, Australian Aboriginal peoples' occupation traces back more than 50,000 years and Aboriginal clans have been living in their 'Country' (traditional lands and waters) for 2000 generations (something that Western scientists only recently "discovered" but Aboriginal peoples already knew and recounted in their oral histories and traditions). Each Australian Aboriginal people and their specific Country, therefore, are co-constituted. In the words of Winter: "Together they have weathered ice ages, sea-level rise and fall, drought, and storms, extinctions and the flourishings: these changes are recorded in their stories" (2018, p. 127). Within Australian Aboriginal cultures, the land is the source of identity, and everything is interwoven back to and within reciprocal relationships with the land. The Aboriginal people come from the *country*, and they return to it where they reside as ancestors (underpinned by cosmological thinking of Dreamtime and Dreaming). There is more this understanding of reciprocal and intergenerational relationships. The ancestral beings, (more-than-human beings who lived on the Australian continent before humans occupied the landmass), provided the form to the original human beings. That is, as Moreton-Robinson highlights, such ontological relationships centre on the connectivity of ancestral beings with the land and humans as co-constituted and interwoven embodied entities, wherein injustice against one is an injustice against all (Moreton-Robinson 2015, p. 12, 2017). As Aboriginal legal scholar Irene Watson writes:

> The Nunga 'I am' is not like the other, dominant Western subject of being, which is represented by a straight line of thought—beginning, middle and ending. Instead, a Nunga process encircles; within there is a process that allows a person to become one and to begin again. This process is non-hierarchical and non-linear; rather it takes the form of a cycle, of the continuity of being, becoming another cycle, nurntikki [to go on forever]. (Watson 2014, p. 16)

As an Australian Aboriginal person comes from the land and ancestral beings come from the ground before returning to the land and living within the land, from where they (people/ancestors) may arise again in some other form. Accordingly, "when listening to *country* Aboriginal people listen to ancestors, bringing them into the present, including them within an intergenerational, inter-species, inter-form community" (Winter 2018, p. 129). Such listening is an active process wherein Aboriginal people narrated how their whole body is involved in listening. It requires them to interpret the results (what they hear) in light of their specific responsibilities to care for country and past/present/future generations of humans and ancestors (Maclean and The Bana Yarralji Bubu Inc. 2015; Moreton-Robinson 2015; Woodward and Marrfurra McTaggart 2019; Zurba and Berkes 2014). More in-depth understandings of Indigenous philosophies and justice theorising are provided by Deborah McGregor, Kyle Powys Winter and Christine Jill Winter. We offer here just a brief introduction to some of these and other scholars' works to make it clear (in contrast to the Western framing of justice that emphasis universality) that Indigenous peoples can experience injustice differently (to non-Indigenous peoples and other Indigenous peoples). Furthermore, the types of actions that are (or should be) taken to address environmental injustices, therefore, need to take into account these differences (historical, biophysical, socio-cultural, economic, political, and ontological).

Conclusion

In the following chapters of this book, we advocate for thinking about IEJ in intergenerational, pluralistic, and relational terms, which extends to include the material and metaphysical and does not institute strict divisions between humans and more-than-human actors, between land and water, or between past, present, and future generations. We argue and demonstrate how Ngāti Maniapoto environmental injustices were and are not extraordinary one-off events (a flood) or singular causes (a polluting factory) rather injustices build up over time. In this book, therefore, we explore how Māori challenges to settler-colonial governance and management of the Waipā River, along with other river systems in Aotearoa, are examples of Māori iwi and hapū rangatiratanga (chiefly authority) and their cultural continuance, despite their ongoing experiences of settler colonialism (invasion, dispossession, socio-economic and political marginalisation, attempts at cultural assimilation). The existing scholarship on IEJ indicates that the sophisticated practices of historical colonialism and political economy are evidence of indigenous communities' around the globe's ongoing struggles to maintain and re-assert their rights of self-determination. In this book, we argue, that it is not just a struggle over self-determination and the political economy but also a conflict between contrasting worldviews (or ways of thinking about the world—ontologies) and practices (ways of acting in the world—epistemologies) between the Western liberal worldview (Pākehā/White New Zealand) and Māori worldview, which were reflected in how each group conceptualised the nature of the problem, potential solutions, and on-the-ground actions. Furthermore, we demonstrate how, even when government policies were designed (on paper) to protect the environment and allow for Indigenous communities to participate in environmental decision-making, the settler-colonial governments often applied their policies in a way that encouraged environmental degradation and limited community participation and, in doing so, exacerbated Indigenous environmental injustices. The EJ framework, at present, does not sufficiently take into account the influence of settler colonialism on Indigenous peoples and recognise that settler-colonial rule exacerbates and/or causes environmental injustices for Indigenous peoples. Accordingly, we draw on decolonial theory to consider how theorising about IEJ can move beyond the western liberal EJ dogma to Indigenous ontologies and epistemologies (Álvarez and Coolsaet 2018; Barker and Pickerill 2019; Blaney and Tickner 2017; Davis and Todd 2017; Pulido and De Lara 2018; Rose 2004; Smith 1999). Whereas, the dominant framing of EJ (as a movement and body of scholarship) focuses on the human-to-human interactions with the environment as the background, IEJ, as we articulate throughout the rest of this book (from the perspective of three feminist Māori/Pākehā/Other hybrids from Aotearoa), includes the interactions between humans and more-than-humans (nonhumans) on a spiritual, cultural, and temporal level.

References

Adger, W. N., Barnett, J., Chapin, F. S., & Ellemor, H. (2011). This Must Be the Place: Underrepresentation of Identity and Meaning in Climate Change Decision-Making. *Global Environmental Politics, 11*(2), 1–25. https://doi.org/10.1162/GLEP_a_00051.

Álvarez, L., & Coolsaet, B. (2018). Decolonizing Environmental Justice Studies: A Latin American Perspective. *Capitalism Nature Socialism, 31*(2), 50–69.

Antadze, N. (2018). Polyphonic Environmental Planning Processes: Establishing Conceptual Connections Between Procedural and Recognition Justice. *Local Environment, 23*(2), 239–255.

Arcury, T. A., & Quandt, S. A. (2009). *Latinx Farmworkers in the Eastern United States: Health, Safety and Justice.* Cham: Springer Science & Business Media.

Arnstein, S. R. (1969). A Ladder of Citizen Participation. *Journal of the American Institute of Planners, 35*(4), 216–224.

Azmanova, A. (2012). De-gendering Social Justice in the 21st Century: An Immanent Critique of Neoliberal Capitalism. *European Journal of Social Theory, 15*(2), 143–156.

Banisar, D., Parmar, S., de Silva, L., & Excell, C. (2011). Moving from Principles to Rights: Rio 2012 and Access to Information, Public Participation, and Justice. *Sustainable Development Law & Policy, 12*, 8.

Bargh, M. (2018). Māori Political and Economic Recognition in a Diverse Economy. In D. Howard-Wagner, M. Bargh, & I. Altamirano-Jiménez (Eds.), *The Neoliberal State, Recognition and Indigenous Rights* (1st ed., pp. 293–308). Canberra: ANU Press. https://doi.org/10.22459/CAEPR40.07.2018.

Barker, A. J., & Pickerill, J. (2019). Doings with the Land and Sea: Decolonising Geographies, Indigeneity, and Enacting Place-Agency. *Progress in Human Geography, 44*(4), 640–662.

Barnhill-Dilling, S. K., Rivers, L., & Delborne, J. A. (2020). Rooted in Recognition: Indigenous Environmental Justice and the Genetically Engineered American Chestnut Tree. *Society & Natural Resources, 33*(1), 83–100.

Bell, A. (2018). A Flawed Treaty Partner: The New Zealand State, Local Government and the Politics of Recognition. In D. Howard-Wagner, M. Bargh, & I. Altamirano-Jimenez (Eds.), *The Neoliberal State, Recognition and Indigenous Rights: New Paternalism to New Imaginings* (pp. 77–92). Canberra: ANU Press.

Bell, M. L., & Ebisu, K. (2012). Environmental Inequality in Exposures to Airborne Particulate Matter Components in the United States. *Environmental Health Perspectives, 120*(12), 1699–1704.

Bevc, C. A., Marshall, B. K., & Picou, J. S. (2007). Environmental Justice and Toxic Exposure: Toward a Spatial Model of Physical Health and Psychological Well-Being. *Social Science Research, 36*(1), 48–67.

Bird, M. Y. (1999). What We Want to Be Called: Indigenous Peoples' Perspectives on Racial and Ethnic Identity Labels. *American Indian Quarterly, 23*(2), 1–21.

Blaney, D. L., & Tickner, A. B. (2017). Worlding, Ontological Politics and the Possibility of a Decolonial IR. *Millennium, 45*(3), 293–311.

Blue, G., Rosol, M., & Fast, V. (2019). Justice as Parity of Participation. *Journal of the American Planning Association, 85*(3), 363–376.

Boone, C. G., & Buckley, G. L. (2017). Historical Approaches to Environmental Justice. In R. Holifield, J. Chakraborty, & G. Walker (Eds.), *The Routledge Handbook of Environmental Justice* (pp. 222–230). New York; London: Taylor & Francis.

Buckinx, B., Trejo-Mathys, J., & Waligore, T. (2015). Domination Across the Borders: An Introduction. In *Domination and Global Political Justice: Conceptual, Historical and Institutional Perspectives* (pp. 1–36). London; New York: Routledge.

Budowle, R., Arthur, M., & Porter, C. (2019). Growing Intergenerational Resilience for Indigenous Food Sovereignty Through Home Gardening. *Journal of Agriculture, Food Systems, and Community Development, 9*(B), 1–21.

Bullard, R. D. (1993). Race and Environmental Justice in the United States. *Yale Journal of International Law, 18*, 319.

Bullard, R. D. (2002). Confronting Environmental Racism in the Twenty-First Century. *Global Dialogue; Nicosia, 4*(1), 34–48.

Bullard, R. D., & Wright, B. (2008). Disastrous Response to Natural and Man-Made Disasters: An Environmental Justice Analysis Twenty-Five Years after Warren County. *UCLA Journal of Environmental Law and Policy, 26*, 217.

Burarrwanga, L., Ritjilili, G., Ganambarr-Stubbs, M., Maymuru, D., Wright, S., & Suchet-Pearson, S. (2013). They Are Not Voiceless. In J. M. Coetzee, O. Sherman, W. Were, & S. Wyndham (Eds.), *The Voiceless Anthology* (Electronic ed.). Sydney: Allen & Unwin.

Burwell, D., & Cole, L. W. (2007). Environmental Justice Comes Full Circle: Warren County Before and After. *Golden Gate University Environmental Law Journal, 1*, 9.

Caney, S. (2008). Global Distributive Justice and the State. *Political Studies, 56*(3), 487–518.

Carpentier, N. (2016). Beyond the Ladder of Participation: An Analytical Toolkit for the Critical Analysis of Participatory Media Processes. *Javnost-The Public, 23*(1), 70–88.

Cherrington, M. (2019). Environmental Social and Governance Sustainability – Ka Mua, Ka Muri. *Scope: Learning and Teaching, 8*, 51–56.

Connor, D. M. (1988). A New Ladder of Citizen Participation. *National Civic Review, 77*(3), 249–257.

Coombes, B. (2013). Māori and Environmental Justice: The Case of 'Lake' Otara. In E. J. Peters & C. Andersen (Eds.), *Indigenous in the City: Contemporary Identities and Cultural Innovation* (pp. 334–354). Vancouver: UBC Press.

Coombes, B., Johnson, J. T., & Howitt, R. (2012). Indigenous Geographies I: Mere Resource Conflicts? The Complexities in Indigenous Land and Environmental Claims. *Progress in Human Geography, 36*(6), 810–821.

Coulthard, G. S. (2007). Subjects of Empire: Indigenous Peoples and the 'Politics of Recognition' in Canada. *Contemporary Political Theory, 6*(4), 437–460.

Coulthard, G. S. (2014). *Red Skin, White Masks: Rejecting the Colonial Politics of Recognition.* Minneapolis: University of Minnesota Press.

Country, B., Wright, S., Suchet-Pearson, S., Lloyd, K., Burarrwanga, L., Ganambarr, R., et al. (2016). Co-becoming Bawaka: Towards a Relational Understanding of Place/Space. *Progress in Human Geography, 40*(4), 455–475.

Crease, R. P., Parsons, M., & Fisher, K. T. (2019). 'No Climate Justice Without Gender Justice': Explorations of the Intersections Between Gender and Climate Injustices in Climate Adaptation Actions in the Philippines. In T. Jafry (Ed.), *Routledge Handbook of Climate Justice* (pp. 359–377). Oxon; New York: Routledge.

Davis, H., & Todd, Z. (2017). On the Importance of a Date, or Decolonizing the Anthropocene. *ACME: An International E-Journal for Critical Geographies, 16*, 4.

Davison, I. (2009). 'Black Drain' Threat to Food and Exports. *NZ Herald,* August 11. Retrieved May 31, 2020, from https://www.nzherald.co.nz/nz/news/article.cfm?c_id=1&objectid=10590195.

Dodd, M. (2010). Effects of Industry on Maori Cultural Values: The Case of the Tarawera River. *Indigenous Voices, Indigenous Research,* 53–63.

Durie, M. (2006). Measuring Māori Wellbeing. *New Zealand Treasury Guest Lecture Series, 1.*

Figueroa, R. M. (2001). Other Faces: Latinos and Environmental Justice. In L. Westra & B. E. Lawson (Eds.), *Faces of Environmental Racism: Confronting Issues of Global Justice* (pp. 167–184). Lanham: Rowman & Littlefield Publishers.

Fisher, J. B., Kelly, M., & Romm, J. (2006). Scales of Environmental Justice: Combining GIS and Spatial Analysis for Air Toxics in West Oakland, California. *Health & Place, 12*(4), 701–714. https://doi.org/10.1016/j.healthplace.2005.09.005.

Fraser, N. (1990). Rethinking the Public Sphere: A Contribution to the Critique of Actually Existing Democracy. *Social Text, 25/26,* 56–80.

Fraser, N. (1995). Recognition or Redistribution? A Critical Reading of Iris Young's Justice and the Politics of Difference*. *Journal of Political Philosophy, 3*(2), 166–180.

Fraser, N. (2007). Feminist Politics in the Age of Recognition: A Two-Dimensional Approach to Gender Justice. *Studies in Social Justice, 1*(1), 23–35.

Fraser, N. (2009). *Scales of Justice: Reimagining Political Space in a Globalizing World*. New York: Columbia University Press.

Fraser, N., & Honneth, A. (2003). *Redistribution or Recognition?: A Political-Philosophical Exchange*. London; New York: Verso.

Ghosh, A. (2018). *The Great Derangement: Climate Change and the Unthinkable*. London: Penguin UK.

Gilio-Whitaker, D. (2019). *As Long as Grass Grows: The Indigenous Fight for Environmental Justice from Colonization to Standing Rock*. Boston: Beacon Press.

Gordon, L., Payne-Sturges, D., & Gee, G. (2010). Environmental Health Disparities: Select Case Studies Related to Asian and Pacific Islander Americans. *Environmental Justice, 3*(1), 21–26.

Green, H. F. (2011). Cosmology and Earth Jurisprudence. In P. Burdon (Ed.), *Exploring Wild Law: The Philosophy of Earth Jurisprudence* (pp. 126–136). Kent Town: Wakefield Press.

Greensill, A. N. (2010). *Inside the Resource Management Act: A Tainui Case Study*. Thesis, The University of Waikato. Retrieved May 31, 2020, from https://researchcommons.waikato.ac.nz/handle/10289/4922.

Greife, M., Stretesky, P. B., Shelley, T. O., & Pogrebin, M. (2017). Corporate Environmental Crime and Environmental Justice. *Criminal Justice Policy Review, 28*(4), 327–346.

Hart, R. A. (2008). Stepping Back from 'The Ladder': Reflections on a Model of Participatory Work with Children. In A. Reid, B. B. Jensen, J. Nikel, & V. Simovska (Eds.), *Participation and Learning* (pp. 19–31). Dordrecht: Springer Netherlands.

He, J., & Sikor, T. (2015). Notions of Justice in Payments for Ecosystem Services: Insights from China's Sloping Land Conversion Program in Yunnan Province. *Land Use Policy, 43*, 207–216.

Hendlin, Y. H. (2019). Environmental Justice as a (Potentially) Hegemonic Concept: A Historical Look at Competing Interests Between the MST and Indigenous People in Brazil. *Local Environment, 24*(2), 113–128.

Holifield, R. (2012). Environmental Justice as Recognition and Participation in Risk Assessment: Negotiating and Translating Health Risk at a Superfund Site in Indian Country. *Annals of the Association of American Geographers, 102*(3), 591–613.

Holifield, R., Chakraborty, J., & Walker, G. (Eds.). (2017). *The Routledge Handbook of Environmental Justice* (1st ed.). London; New York: Routledge.

Hurlbert, M., & Gupta, J. (2015). The Split Ladder of Participation: A Diagnostic, Strategic, and Evaluation Tool to Assess When Participation Is Necessary. *Environmental Science & Policy, 50,* 100–113.

Jackson, S. (2008). Recognition of Indigenous Interests in Australian Water Resource Management, with Particular Reference to Environmental Flow Assessment. *Geography Compass, 2*(3), 874–898.

Jackson, S. (2018). Indigenous Peoples and Water Justice in a Globalizing World. In K. Conca & E. Weinthal (Eds.), *The Oxford Handbook of Water Politics and Policy.* New York: Oxford University Press.

Kingham, S., Pearce, J., & Zawar-Reza, P. (2007). Driven to Injustice? Environmental Justice and Vehicle Pollution in Christchurch, New Zealand. *Transportation Research Part D: Transport and Environment, 12*(4), 254–263.

Lowitt, K., Levkoe, C. Z., Lauzon, R., Ryan, K., & Sayers, C. D. (2019). 7 Indigenous Self-Determination and Food Sovereignty Through Fisheries Governance in the Great Lakes Region. *Civil Society and Social Movements in Food System Governance,* 145.

Maclean, K., & The Bana Yarralji Bubu Inc. (2015). Crossing Cultural Boundaries: Integrating Indigenous Water Knowledge into Water Governance Through Co-research in the Queensland Wet Tropics, Australia. *Geoforum, 59,* 142–152.

Martin, A., Coolsaet, B., Corbera, E., Dawson, N. M., Fraser, J. A., Lehmann, I., & Rodriguez, I. (2016). Justice and Conservation: The Need to Incorporate Recognition. *Biological Conservation, 197,* 254–261.

McCormack, F. (2018). Indigenous Settlements and Market Environmentalism: An Untimely Coincidence? In D. Howard-Wagner, M. Bargh, & I. Altamirano-Jiménez (Eds.), *The Neoliberal State, Recognition and Indigenous Rights* (Vol. 40, pp. 273–292). Canberra: ANU Press.

McCreary, T., & Milligan, R. (2018). The Limits of Liberal Recognition: Racial Capitalism, Settler Colonialism, and Environmental Governance in Vancouver and Atlanta. *Antipode.* https://onlinelibrary.wiley.com/doi/abs/10.1111/anti.12465.

McGregor, D. (2010). Honouring Our Relations: An Anishnaabe Perspective on Environmental Justice. In J. Agyeman, P. Cole, R. Haluza-DeLay, & P. O'Riley (Eds.), *Speaking for Ourselves: Environmental Justice in Canada* (pp. 27–41). Vancouver: UBC Press.

McGregor, D. (2015). Indigenous Women, Water Justice and Zaagidowin (Love). *Canadian Woman Studies, 30*(2–3), 71–78.

McGregor, D. (2018a). Indigenous Environmental Justice, Knowledge, and Law. *Kalfou, 5*(2), 279.

McGregor, D. (2018b). Mino-Mnaamodzawin: Achieving Indigenous Environmental Justice in Canada. *Environment and Society, 9*(1), 7–24.

McGregor, D., Whitaker, S., & Sritharan, M. (2020). Indigenous Environmental Justice and Sustainability. *Current Opinion in Environmental Sustainability, 43*, 35–40.

Miller, D. S., & Rivera, J. D. (2009). *Hurricane Katrina and the Redefinition of Landscape*. Lanham: Lexington Books.

Mills, C. W. (2015). Race and Global Justice. In B. Buckinx, J. Trejo-Mathys, & T. Waligore (Eds.), *Domination and Global Political Justice: Conceptual, Historical and Institutional Perspectives* (pp. 193–217). London; New York: Routledge.

Moreton-Robinson, A. (2015). *The White Possessive: Property, Power, and Indigenous Sovereignty*. Minneapolis: University of Minnesota Press.

Moreton-Robinson, A. (2017, February 21). Senses of Belonging: How Indigenous Sovereignty Unsettles White Australia. *ABC Religion and Ethics*.

Paloniemi, R., Apostolopoulou, E., Cent, J., Bormpoudakis, D., Scott, A., Grodzińska-Jurczak, M., et al. (2015). Public Participation and Environmental Justice in Biodiversity Governance in Finland, Greece, Poland and the UK. *Environmental Policy and Governance, 25*(5), 330–342.

Pastor, M., Sadd, J., & Hipp, J. (2001). Which Came First? Toxic Facilities, Minority Move-In, and Environmental Justice. *Journal of Urban Affairs, 23*(1), 1–21.

Pearce, J. R., & Kingham, S. (2008). Environmental Inequalities in New Zealand: A National Study of Air Pollution and Environmental Justice. *Geoforum, 39*(2), 980–993.

Pearce, J., Kingham, S., & Zawar-Reza, P. (2006). Every Breath you Take? Environmental Justice and Air Pollution in Christchurch, New Zealand. *Environment and Planning A, 38*(5), 919–938.

Pearce, J. R., Richardson, E. A., Mitchell, R. J., & Shortt, N. K. (2011). Environmental Justice and Health: A Study of Multiple Environmental Deprivation and Geographical Inequalities in Health in New Zealand. *Social Science & Medicine, 73*(3), 410–420.

Perreault, T., Wraight, S., & Perreault, M. (2012). Environmental Injustice in the Onondaga Lake Waterscape, New York State, USA. *Water Alternatives, 5*(2), 485–506.

Pitea, C. (2009). Procedures and Mechanisms for Review of Compliance Under the 1998 Aarhus Convention on Access to Information, Public Participation and Access to Justice in Environmental Matters. In *Non-Compliance Procedures and Mechanisms and the Effectiveness of International Environmental Agreements* (pp. 221–249). The Hague: TMC Asser Press.

Pulido, L. (2017). Geographies of Race and Ethnicity II: Environmental Racism, Racial Capitalism and State-Sanctioned Violence. *Progress in Human Geography, 41*(4), 524–533. https://doi.org/10.1177/0309132516646495.

Pulido, L., & De Lara, J. (2018). Reimagining 'Justice' in Environmental Justice: Radical Ecologies, Decolonial Thought, and the Black Radical Tradition. *Environment and Planning E: Nature and Space, 1*(1–2), 76–98.

Ranco, D. J. (2008). The Trust Responsibility and Limited Sovereignty: What Can Environmental Justice Groups Learn from Indian Nations? *Society and Natural Resources, 21*(4), 354–362.

Rixecker, S. S., & Tipene-Matua, B. (2003). Maori Kaupapa and the Inseparability of Social and Environmental Justice: An Analysis of Bioprospecting and a People's Resistance to (Bio) Cultural Assimilation. In R. D. Bullard, J. Agyeman, & B. Evans (Eds.), *Just Sustainabilities: Development in an Unequal World* (pp. 252–268). London: Earthscan.

Rohland, E. (2018). Adapting to Hurricanes. A Historical Perspective on New Orleans from Its Foundation to Hurricane Katrina, 1718–2005. *Wiley Interdisciplinary Reviews: Climate Change, 9*(1), e488.

Rose, D. B. (2004). *Reports from a Wild Country: Ethics for Decolonisation.* UNSW Press.

Ross, H., Buchy, M., & Proctor, W. (2002). Laying Down the Ladder: A Typology of Public Participation in Australian Natural Resource Management. *Australian Journal of Environmental Management, 9*(4), 205–217.

Rydin, Y. (2006). *Justice and the Geography of Hurricane Katrina.* Pergamon.

Ryks, J., Wythe, J., Baldwin, S., & Kennedy, N. (2010). The Teeth of the Taniwha: Māori Representation and Participation in Local Government. *Planning Quarterly, 177,* 39–42.

Salazar, D. J. (2009). Saving Nature and Seeking Justice: Environmental Activists in the Pacific Northwest. *Organization & Environment, 22*(2), 230–254.

Salazar, D. J., & Alper, D. K. (2011). Justice and Environmentalisms in the British Columbia and U.S. Pacific Northwest Environmental Movements. *Society & Natural Resources, 24*(8), 767–784.

Salmond, K. (1999). Setting Our Sights on Justice: Contaminated Sites and Socio-Economic Deprivation in New Zealand. *International Journal of Environmental Health Research, 9*(1), 19–29.

Sanders, W. (2018). *Reconciling Public Accountability and Aboriginal Self-Determination/Self-Management: Is ATSIC Succeeding?* Canberra, ACT: Centre for Aboriginal Economic Policy Research (CAEPR), The Australian National University. Accessed June 29, 2020, from https://openresearch-repository.anu.edu.au/handle/1885/145482.

Schlosberg, D. (2003). The Justice of Environmental Justice: Reconciling Equity, Recognition, and Participation in a Political Movement. In A. Light & A. de Shalit (Eds.), *Moral and Political Reasoning in Environmental Practice* (pp. 77–106). Cambridge: MIT Press.

Schlosberg, D. (2004). Reconceiving Environmental Justice: Global Movements and Political Theories. *Environmental Politics, 13*(3), 517–540.

Schlosberg, D. (2012). Justice, Ecological Integrity, and Climate Change. In A. Thompson & J. Bendik-Keymer (Eds.), *Ethical Adaptation to Climate Change: Human Virtues of the Future* (pp. 165–183). Cambridge: MIT Press.

Schlosberg, D., & Carruthers, D. (2010). Indigenous Struggles, Environmental Justice, and Community Capabilities. *Global Environmental Politics, 10*(4), 12–35.

Schlosberg, D., & Collins, L. B. (2014). From Environmental to Climate Justice: Climate Change and the Discourse of Environmental Justice. *Wiley Interdisciplinary Reviews: Climate Change, 5*(3), 359–374.

Sen, A. (1995). Gender Inequality and Theories of Justice. In M. Nussbaum & J. Glover (Eds.), *Women, Culture and Development: A Study of Human Capabilities* (pp. 259–273). Oxford: Clarendon Press.

Sikor, T., Martin, A., Fisher, J., & He, J. (2014). Toward an Empirical Analysis of Justice in Ecosystem Governance. *Conservation Letters, 7*(6), 524–532.

Singh, J. (2014a). Recognition and Self-Determination: Approaches from Above and Below. In A. Eisenberg, J. H. A. Webber, A. Boisselle, & G. Coulthard (Eds.), *Recognition Versus Self-Determination: Dilemmas of Emancipatory Politics* (pp. 47–74). Toronto: UBC Press.

Singh, J. (2014b). *Beyond Free and Equal: Subalternity and the Limits of Liberal-Democracy.* Thesis, University of Toronto. Retrieved May 29, 2020, from https://tspace.library.utoronto.ca/handle/1807/65490.

Singh, J. (2019). Decolonizing Radical Democracy. *Contemporary Political Theory, 18*(3), 331–356.

Smith, L. T. (1999). *Decolonising Methodologies.* New York: Zed Books.

Spiegel, S. J., Thomas, S., O'Neill, K., Brondgeest, C., Thomas, J., Beltran, J., et al. (2020). Visual Storytelling, Intergenerational Environmental Justice and Indigenous Sovereignty: Exploring Images and Stories amid a Contested Oil Pipeline Project. *International Journal of Environmental Research and Public Health, 17*(7), 2362.

Steel, D., & Whyte, K. P. (2012). Environmental Justice, Values, and Scientific Expertise. *Kennedy Institute of Ethics Journal, 22*(2), 163–182.

Stensrud, A. B. (2016). Harvesting Water for the Future: Reciprocity and Environmental Justice in the Politics of Climate Change in Peru. *Latin American Perspectives, 43*(4), 56–72.

Suchet-Pearson, S., Wright, S., Lloyd, K., Burarrwanga, L., & Country, B. (2013). Caring as Country: Towards an Ontology of Co-Becoming in Natural Resource Management. *Asia Pacific Viewpoint, 54*(2), 185–197.

Swyngedouw, E., & Heynen, N. C. (2003). Urban Political Ecology, Justice and the Politics of Scale. *Antipode, 35*(5), 898–918.

Sze, J. (2018). *Sustainability: Approaches to Environmental Justice and Social Power.* New York: NYU Press.

Sze, J., & London, J. K. (2008). Environmental Justice at the Crossroads. *Sociology Compass, 2*(4), 1331–1354.

Taylor, D. E. (2000). The Rise of the Environmental Justice Paradigm: Injustice Framing and the Social Construction of Environmental Discourses. *American Behavioral Scientist, 43*(4), 508–580.

Todd, Z. (2016). An Indigenous Feminist's Take on the Ontological Turn: 'Ontology' Is Just Another Word for Colonialism: an Indigenous Feminist's Take on the Ontological Turn. *Journal of Historical Sociology, 29*(1), 4–22.

Tritter, J. Q., & McCallum, A. (2006). The Snakes and Ladders of User Involvement: Moving Beyond Arnstein. *Health Policy, 76*(2), 156–168.

Tschakert, P., & Machado, M. (2012). Gender Justice and Rights in Climate Change Adaptation: Opportunities and Pitfalls. *Ethics and Social Welfare, 6*(3), 275–289.

Tully, J. (2000). Struggles over Recognition and Distribution. *Constellations, 7*(4), 469–482. https://doi.org/10.1111/1467-8675.00203.

Vickery, J., & Hunter, L. M. (2016). Native Americans: Where in Environmental Justice Research? *Society & Natural Resources, 29*(1), 36–52.

Walker, G. (2009). Beyond Distribution and Proximity: Exploring the Multiple Spatialities of Environmental Justice. *Antipode, 41*(4), 614–636.

Walker, E. T., Wehi, P. M., Nelson, N. J., Beggs, J. R., & Whaanga, H. (2019). Kaitiakitanga, Place and the Urban Restoration Agenda. *New Zealand Journal of Ecology, 43*(3), 1–8.

Watene, K. (2016). Valuing Nature: Māori Philosophy and the Capability Approach. *Oxford Development Studies, 44*(3), 287–296.

Watson, I. (2014). *Aboriginal Peoples, Colonialism and International Law: Raw Law.* London: Routledge.

Weaver, J. (1996). *Defending Mother Earth Native American Perspectives on Environmental Justice*. Maryknoll, NY: Orbis Books.

Weaver, J. (2016). Ko Te Whenua Te Utu: Land is the Price: Essays on Maori History, Land, and Politics. *The Journal of New Zealand Studies, 22*, 124–126.

Wehi, P. M., & Lord, J. M. (2017). Importance of Including Cultural Practices in Ecological Restoration. *Conservation Biology, 31*(5), 1109–1118.

Whyte, K. P. (2011). The Recognition Dimensions of Environmental Justice in Indian Country. *Environmental Justice, 4*(4), 199–205. https://doi.org/10.1089/env.2011.0036.

Whyte, K. P. (2014). Indigenous Women, Climate Change Impacts, and Collective Action. *Hypatia, 29*(3), 599–616.

Whyte, K. P. (2016). Indigenous Experience, Environmental Justice and Settler Colonialism. In B. Bannon (Ed.), *Nature and Experience: Phenomenology and the Environment* (pp. 157–174). Lanham: Rowman & Littlefield. Retrieved January 30, 2020, from http://www.ssrn.com/abstract=2770058.

Whyte, K. P. (2016a). Is It Colonial Déjà Vu? Indigenous Peoples and Climate Injustice. In *Humanities for the Environment* (pp. 102–119). London: Routledge.

Whyte, K. P. (2016b). Indigenous Environmental Movements and the Function of Governance Institutions. In T. Gabrielson, C. Hall, J. Meyer, & D. Schlosberg (Eds.), *Oxford Handbook of Environmental Political Theory* (pp. 563–580). Oxford: Oxford University Press.

Whyte, K. P. (2017). The Dakota Access Pipeline, Environmental Injustice, and U.S. Colonialism. *Red Ink: An International Journal of Indigenous Literature, Arts, & Humanities, 19*(1) Retrieved May 29, 2020, from https://ssrn.com/abstract=2925513.

Whyte, K. P. (2018). Settler Colonialism, Ecology, and Environmental Injustice. *Environment and Society, 9*(1), 125–144.

Whyte, K. P. (2020). Too Late for Indigenous Climate Justice: Ecological and Relational Tipping Points. *WIREs Climate Change, 11*(1), e603.

Williams, M. S. (2014). Introduction: On the Use and Abuse of Recognition in Politics. In A. Eisenberg, J. H. A. Webber, A. Boisselle, & G. Coulthard (Eds.), *Recognition Versus Self-Determination: Dilemmas of Emancipatory Politics* (pp. 3–20). Vancouver: UBC Press.

Wilson, S. M., Fraser-Rahim, H., Williams, E., Zhang, H., Rice, L., Svendsen, E., & Abara, W. (2012). Assessment of the Distribution of Toxic Release Inventory Facilities in Metropolitan Charleston: An Environmental Justice Case Study. *American Journal of Public Health, 102*(10), 1974–1980.

Winter, C. J. (2018). *The Paralysis of Intergenerational Justice: Decolonising Entangled Futures* (PhD Thesis). University of Sydney. Retrieved from https://ses.library.usyd.edu.au/handle/2123/18009.

Winter, C. J. (2019a). Decolonising Dignity for Inclusive Democracy. *Environmental Values, 28*(1), 9–30.

Winter, C. J. (2019b). Does Time Colonise Intergenerational Environmental Justice Theory? *Environmental Politics*, 1–19.

Woodward, E., & Marrfurra McTaggart, P. (2019). Co-Developing Indigenous Seasonal Calendars to Support 'Healthy Country, Healthy People' Outcomes. *Global Health Promotion, 26*(3_suppl), 26–34.

Zurba, M., & Berkes, F. (2014). Caring for Country Through Participatory Art: Creating a Boundary Object for Communicating Indigenous Knowledge and Values. *Local Environment, 19*(8), 821–836.

3

'The past is always in front of us': Locating Historical Māori Waterscapes at the Centre of Discussions of Current and Future Freshwater Management

Kia whakatōmuri te haere whakamua: I walk backwards into the future with my eyes fixed on the past.
—*A common Māori proverb*

Variations of the common whakataukī '*the past are always in front of you*' encompass Māori understandings of time and the importance of knowing one's whakapapa (genealogy) and histories. Such understandings extend to current discussions of the freshwater crisis, climate change and the Anthropocene. Just as rivers cannot be separated into components (river-stream-wetland) and instead must be viewed as ki uta ki tai (from the mountains to the sea), from a mātauranga (Māori knowledge) perspective, the impacts of colonisation cannot be de-coupled from local and global environmental changes (which contribute to past, present, and future social-environmental injustices). A Ngāti Maniapoto waiata catalogued in 1930 references Rukutia twisting flax fibre to create a thread: "*Tenei to tohu; Ka mau ki au; Miria mai, e; Te miri o Rukutia*", which was translated by Apirana Ngata as "Henceforward your landmarks; Are firmly imprinted within me; Come with your caress; The caress of Rukutia" (Ngata 2004, p. 293). The act of weaving also mimics the layering of whakapapa, one generation laid onto another, and highlights the interrelationship between landscapes and peoples. We, thus, begin this story by examining historical waterscapes of the Waipā River catchment and its peoples.

In what follows, we outline the historical and socio-economic landscapes of the tangata whenua of Te Waipā Awa (the Waipā River) before and immediately after the coming of Pākehā (New Zealand European ethnic group). It is, as with all historical works, just one history that can be told, interpreted, and retold in numerous other ways, and which is always partial and subjective. One part of this history resides in Te Ao Māori (the world of Māori), and the other part lies in Te Ao Pākehā (the world of Pākehā aka the settler world) and associated cultural identities (Parsons et al. 2019; Salmond 2017). Although Māori and Pākehā who lived (and continue to live) in the Waipā catchment share commonalities in their respective engagements with their environments, their understandings of and relationships with place are based on different understandings of the environment (ontologies) as well as different modes of living and environmental management practices (epistemologies). The relationships between iwi (tribes) and the Waipā River were/are a complex network of metaphysical and physical connections (see Fig. 3.1). Te Ao Pākehā, on the other hand, situated the Waipā River as a resource or commodity that could be controlled and used to further human prosperity and development activities.

Te Ao Māori (The Māori World)

The catchment of the Waipā River was home to Māori people for many hundreds of years before when Pākehā first began their visits to Aotearoa (Barrett 2012; Collins et al. 2012; Ellison et al. 2012; Tauriki et al. 2012). Whakapapa ties successive generations of people in the Waipā catchment to Tainui waka (canoe), which was one of the first waka to bring Māori to Aotearoa from their homeland of Hawaiki, in the South Pacific. Early Pacific migrants explored, named and storied, cultivated, settled and defended the lands and waters from Tāmaki Makaurau (modern-day

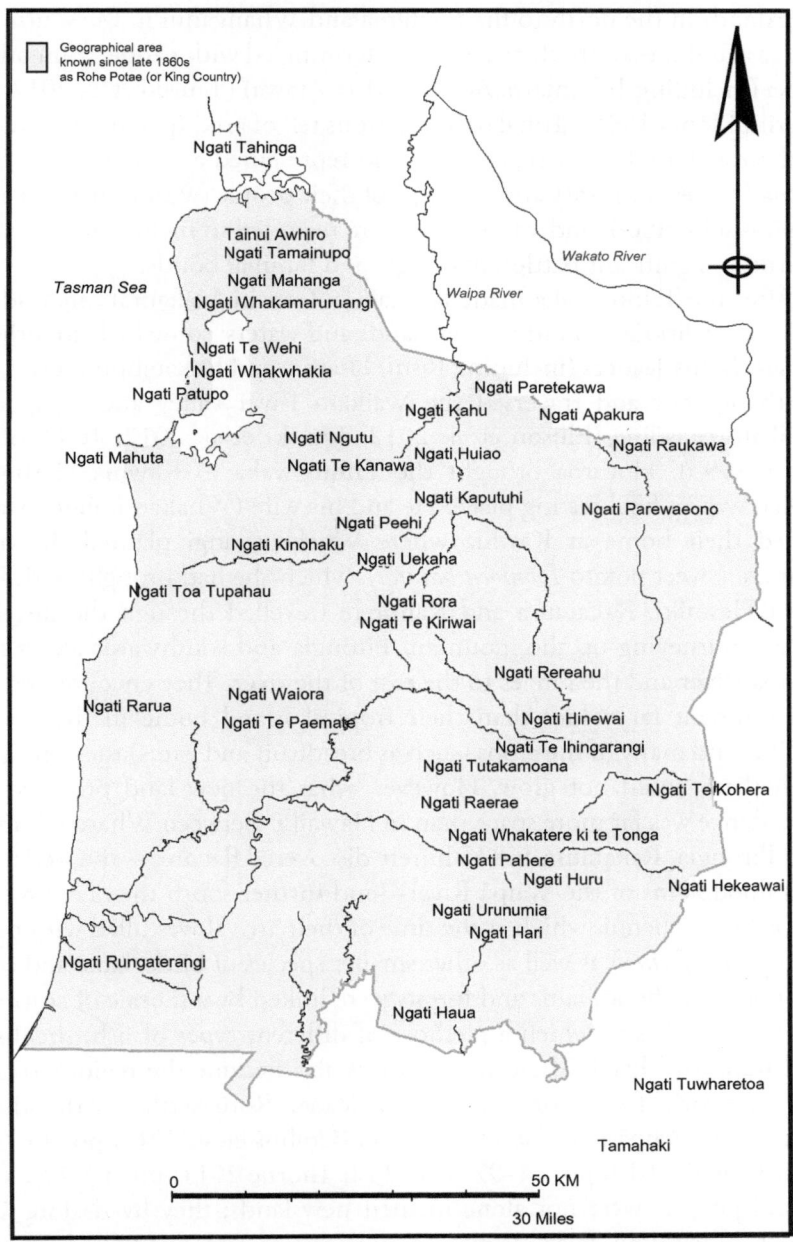

Fig. 3.1 Map showing different iwi and hapū within the Waipā River catchment

Auckland) in the north to the Poutama and Whanganui in the south. In the east and the south, they met and intermingled with people from other waka (including Tokomaru, Aotea, and Te Arawa) (Tauriki et al. 2012; Te Hurinui Jones 1995). Tribal oral traditions tell of ariki (paramount chief) and rangatira (chief or high rank) who represented and maintained the mana (power, authority and prestige) of their people (within their iwi and hapū—sub-tribes), and whose lives were recorded to include remarkable journeys, significant battles, marriages and familial bonds.

After the Tainui waka made landfall at Tāmaki Makaurau, its leaders made expeditions around the new lands and waters; some by land, others by sea. Many leaders (including Rōtū, Hiaroa, and Rakataura) journeyed south on foot and traversed the Waikato River valley and along the Waikato coastline (Ellison et al. 2012; Tauriki et al. 2012; Te Hurinui Jones 1995). Hoturoa brought the Tainui waka to Kāwhia Harbour, which was its final resting place. He and his wife (Whakaotirangi) established their home at Kāwhia where Whakaotirangi planted the first kūmara (sweet potato *Ipomoea batatas*), which she had brought with her from Hawaiki. Rakataura and Kahurere travelled through the interior lands, journeying up the mountain Pirongia and southwards along the Waipā River and the ranges to the east of the river. They encountered an environment far colder than their tropical island home in the South Pacific, and many of the crops (such as breadfruit and yams) they brought with them could not grow. However, what the new land provided in abundance was far more space than in Hawaiki. Between Wharepūhunga and Pirongia, Rakataura and Kahurere discovered flat area—the wetlands and floodplains of the Waipā River—and further south the dense forest of Te Nehenehenui, which at the time of their arrival was filled with moa (*Dinornithiformes*) as well as other smaller species of birds, bats, and edible flora. The floodplains and forests were linked by a mosaic of streams, rivers, and lakes in which a plethora of different types of fish, shellfish, and waterfowl lived. These watercourses also became the region's transport network, for Māori. The Tainui leader Rōtū settled with others besides the Waipā River at Whatawhata (Collins et al. 2012, pp. 49, 55; Tauriki et al. 2012, pp. 94–97, 110–112; Thorne 2011, pp. 47–49). The Tainui peoples were not alone in their new lands; they lived alongside others—including fairy people (patupaiarehe) and fair-haired people (urukehu)—who lived on the slopes of the mountains of Pirongia and Kakepuku, and in the forests of Tūhua and Pureora (Barber 1978; Cowan 1901; Ellison et al. 2012).

Descendants of Tainui waka initially lived in small family groupings and moved from location to location to allow them to access resources at different times of year (Anderson et al. 2012; Anderson 2002). For example, for hunting or fishing activities, harvesting berries, as well as the cultivation of kūmara and taro (root crops) (Bassett et al. 2004; Furey n.d.; Horrocks et al. 2008; Leach 2005). Initially, Māori hunted moa and other abundant fauna; however, unsustainable harvesting practices, rising human populations, loss of flora, and the introduction of rats resulted in rapid decline in moa numbers (Anderson 2003; Gumbley et al. 2004; Hogg et al. 2017; Worthy and Swabey 2002). Accordingly, the focus of the Māori economy adapted and diversified, and included the year-round harvesting of flora (berries, fern roots) and fauna (fish, eels, shellfish, birds), as well as the cultivation of crops. Horticultural techniques used in Hawaiki were adapted to suit the colder and wetter conditions of the Waikato and Te Rohe Pōtae (The King Country), where it was far more challenging to grow kūmara and taro than in warmer tropical climates. Over centuries, the social structure changed as individual families began to work collaboratively, probably as a way to share the division of labour (collecting foodstuffs, defending territories). Thus, hapū were formed.

Early Māori settlement in the Waipā catchment following the landing of the waka Tainui established the tūpuna (ancestors) of the various hapū and iwi within today's landscape. Important events led to the appearance of one significant tūpuna: Maniapoto, after whom the iwi Ngāti Maniapoto is named. The development of Ngāti Maniapoto as a separate iwi coincided with other groups (who traced their lineage to the Tainui waka) branching out to establish their iwi distinct from Tainui. These groups all continued to engage with one another but maintained separate rohe (tribal lands and waters) (Ellison et al. 2012; Tauriki et al. 2012).

The principal socio-cultural and political grouping from the 1500s through until the 1840s was the hapū (Anderson et al. 2012; Anderson 2002). Hapū were made up of several different family (whanau) groups who lived nearby each other, were united by common ancestry, and who worked together under leader(s) (rangatira) for the benefit of all hapū members. Natural resources were managed communally, which included the production and sustainable harvesting of resources from lands and waters. Likewise, waka, pā whawhai (often shortened to pā, which are fortified settlements used during conflicts), and other things were communal property held by collectives rather than individuals. Nonetheless, rights overlapped between hapū, with people living in neighbouring areas often related and able to assert their ancestral relationships within another hapū territory. The nature of these territorial rights depended on the specific connections, but usually included rights to seasonal visits or to occupy an area; seasonal or permanent rights to use particular resources (foods, water, cultivations); and rights to travel through space safely. Records highlight how the various hapū of Ngāti Maniapoto moved around seasonally to access the diversity of food sources within tribal boundaries (moving north and south, from inland to the coastal areas). Oral histories recount how people came from far and wide for the annual tuna heke (the migration of freshwater eels) as well as seasonal fishing trips. The rivers and streams were crucial transport routes that allowed people to move around the area and transport themselves and their foodstuffs around the region (see Fig. 3.2) (Tauriki et al. 2012).

Knowledge, Values and Guiding Principles

Pre-colonial Māori in the Waipā catchment, as elsewhere, held their distinct understanding about the origins of the world, how the universe operated, and the nature of all beings (both living and non-living) (Anderson et al. 2012; Collins et al. 2012; Salmond 2017; Tauriki et al. 2012; Waitangi Tribunal 2018). This understanding extended to their own identities, cultural traditions, and histories; their ways of governing, establishing and maintaining laws, assessing whether the behaviour was correct or incorrect, and systems of social organisation; their values, beliefs, and social norms; their approach to determining rights to resources (including land, water and biota); their approaches to managing their interactions with the environment; and their ways to deal with relationships between people (interpersonal as well as inter-whanau, hapū, and iwi). How Māori understood themselves and their position within the world centred on recognising that a host of different interacting human, biophysical, and

Fig. 3.2 Photograph showing Māori canoe and bridge in a stream (somewhere between Te Kuiti and Te Awamutu) in one of the Waipā River's numerous wetlands circa 1890. The vegetation consists primarily of manuka that was cultivated and harvested by Māori for building materials, to make medicines, and to produce perfumes. (Source: Ref/1/2-045762-F. Alexander Turnbull Library, Wellington, New Zealand)

spiritual forces shaped the world, their environments, and their day-to-day lives (Mead 2016). At the heart of Māori ways of knowing and thinking (their ontology) is the connectivity of all things (human and more-than-humans). The principle of whanaungatanga (which translates as kinship, relationship, and sense of familial connection) highlights this interconnection and holds that all things (animate and inanimate, living and dead, past and present) were (and are always) linked together (Haar 2009; Harmsworth et al. 2016). All the entitlements, obligations, beliefs, and values and the power to lead and make decisions, originated from the domain of ancestor-gods and fundamentally from Te Korekore (see Fig. 3.3: Whakapapa spiral). All relationships (social and environmental) were interposed through the spiritual realm.

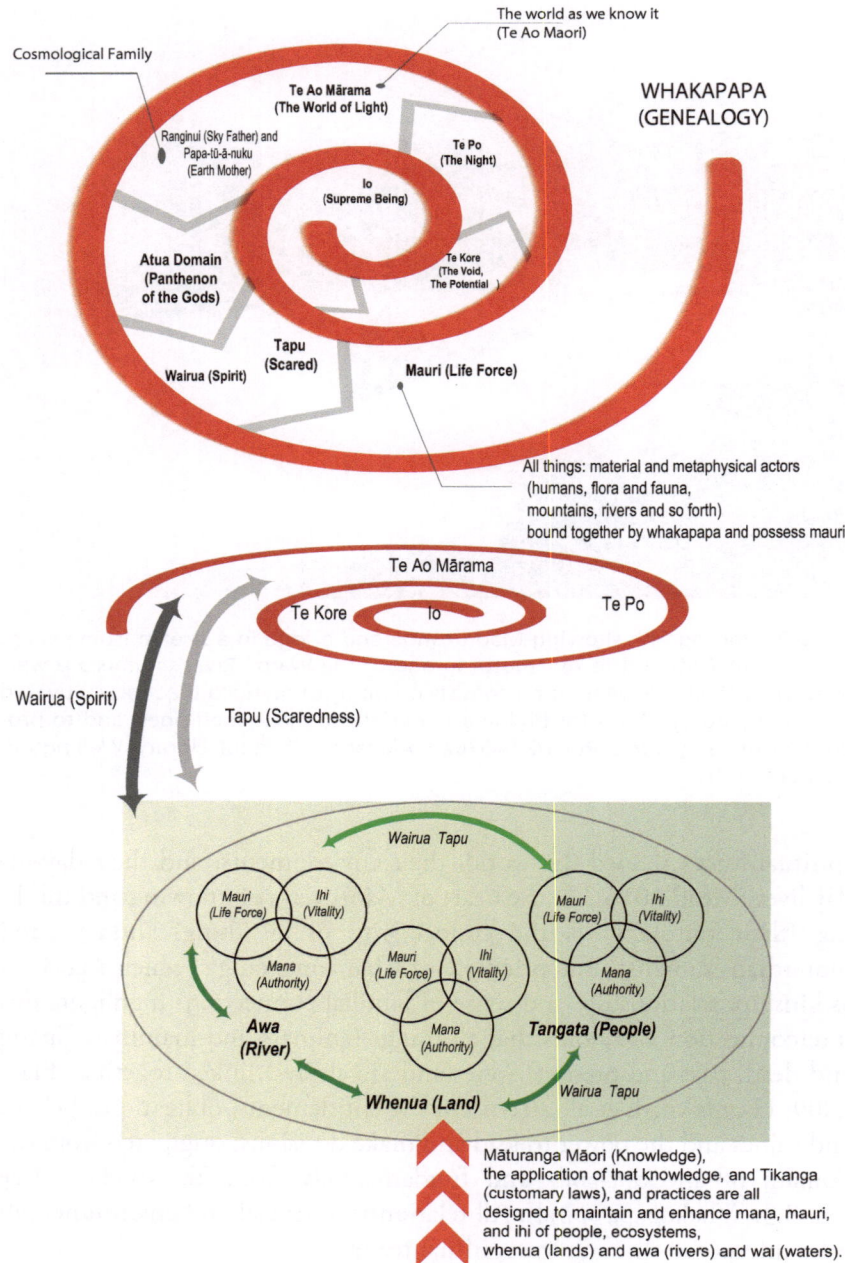

Fig. 3.3 Whakapapa spiral

Tainui oral traditions describe Te Korekore (also shortened to Te Kore), as the Void or original nothingness that existed before everything (including time) (Ellison et al. 2012; Te Hurinui Jones 1995). All of creation (including natural and supernatural beings) came out of Te Kore, which was powered by Io-matua-te-kore (the parentless one or supreme being) the ultimate energy and consciousness believed to be the constant, endless, and original parent of all things and knowledge.[1] From Io and Te Kore came the two gods Papa-tū-ā-nuku (Earth Mother) and Ranginui (Sky Father), who were linked together in such a tight embrace that no light could get through (Salmond 2017). It was into this dark and cramped space that their children lived and some of them eventually conspired to prize their parents apart so that they could live in the light). The children were ultimately successful and became the Atua (gods) of Māori and the progenitors for every part of the world. Tāne-mahuta (God of the Forest) became the atua of all aspects of trees insects, birds, and rocks, and all other aspects of forests. Haumiatiketike became the atua of the plants that could be collected as food for people, while Rongomātāne (the first-born child) became the god of peace as well as cultivated foods. Tangaroa was responsible for governing the oceans as well as all aquatic life, Tāwhirimātea the winds, Rūaumoko the seasons as well as earthquakes and volcanoes. Human beings were the descendants of the god of war (Tūmatauenga). In Tainui oral traditions, however, the spark of life came from Hani (the male essence) and Puna (the female essence) who fashioned   Tiki-i-āhua-mai-i-Hawaiki   (Tiki-who-was-fashioned-in-Hawaiki) and Tiki-apoa from the earth (limestone clay) (Ellison et al. 2012; Te Hurinui Jones 1995).

---

[1] There is disagreement amongst scholars and Māori leaders alike as to whether Io existed in Māori societies prior to contact with Europeans. Some scholars point to the existence of Io in the oral traditions collected by Te Whatahoro Jury from two Wairarapa priests (tohunga) Nēpia Pōhūhū and Te Mātorohanga. Others, including Māori scholar Te Rangi Hīroa (Peter Buck), argue that Io emerged following contact with Europeans and exposure to Christian teachings. By establishing Io as the supreme god of Māori, Hīroa and others argue, tohunga (and other members of Māori society) sought to reconcile their own religious beliefs (which included a pantheon of gods) with that of Christianity's single god (Buck 1950, p. 526).

Tainui traditions tell the whakapapa of how the two ancestors (Tiki-i-āhua-mai-i-Hawaiki and Tiki-apoa Whiro) begot Whiro, who produced the explorer Toi, and from Toi came Whatonga. At least 20 generations took place before the Tainui waka arrived in Aotearoa, and another 50 generations were said to have occurred since then. The critical point is that Māori (both pre-colonial and post-colonisation) saw themselves and all parts of the world as being direct descendants of the gods, who in turn sprung from Te Kore (the Void) and the Io (the parentless one, supreme being and original energy) (see Fig. 3.3: Whakapapa spiral). Every part of the universe was, thus, interrelated, and all relationships were organised through whakapapa (lines or layers of genealogical progression). Every person was an envoy and assistant to all those who came before them—both their human tūpuna (ancestors) and their ancestor-gods—back to their common source (Te Kore and Io); similarly, they were/are representatives and ancestors for future generations (Parsons and Fisher 2020).

The holistic system of thinking and knowledge, commonly referred to as Mātauranga Māori (Māori knowledge) and tikanga Māori (laws and principles), established and maintained a matrix of kin-centric relationships between all living and non-living actors (Ataria et al. 2018; Hikuroa 2016; Mead 2016). Included in this web of connections and overlapping rights and responsibilities were both humans and biota, biophysical and metaphysical entities, gods and ancestors. Māori groups were directed by their system of law (tikanga), decision-making authority, and a suite of fundamental concepts (Mead 2016). Key Māori principles, including tapu, mana, whanaungatanga, manaakitanga, tikanga, utu, tuku, and kaitiakitanga, are outlined in Fig. 3.4.

Within their complex, dynamic, and principled society, which was governed by tikanga, Māori developed and sustained ongoing and substantive ties with te taiao (the natural environment) of their rohe, which included their whenua (lands), awa (rivers), maunga (mountains), and repo (wetlands). Māori worldviews (which were multiple rather than singular due to the tribal nature of society but shared similar principles) were holistic and kin-centric wherein all things were interconnected together through whakapapa. They were linked to every part of their local taiao (environment) through their whakapapa (Collins et al. 2012, p. 323). Each maunga, awa, roto (lake), repo, and other features of the

Kaitiakitanga

Guardianship or custodianship of the environment which is connected through whakapapa to mana whenua (people with spiritual authority over lands/waters). Māori commonly refer themselves as kaitiaki (guardians) of their rohe (tribal lans/waters). Ritenga (customs or protocols or laws of an area) is closely connected with kaitiakatanga. Concepts such as tapu (scared), rahui (restriction of access), noa (normal or unrestricted access) are employed by kaitiaki to manage people's use of resources. In addition, Ki uta ki tai (mountains to sea) approach is a whole-of-landscape approach used by iwi (such as Ngāi Tahu) in the present-day to manage interconnected ecosystems and resources

Manaakitanga

Expectations and duties of hau (reciporcity) and care that based on recongising mana (power and authority) of others, reciprocal responsilitoes and duties to other whanau/hapu/iwi as well as other people not connected by whakapapa.

Mātauranga

Spiritual embodiment (includes spiritual beliefs, capacity to exercise those beliefs, and maintain spiritual wellbeing). Wairua refers to spiritual essence that is carried within a person and other living beings. Wairua also refers to the unseen domain that connections people to the future, the present, and the past, and denotes a state of connectedness with the world. Wairua is governed by tapu (scared or set apart) with the mechanisms of tapu applied to places and resources (such as protection over a resource to avoid over-exploitations).

Whakapapa

Genealogy, ancestral lineage, layers, or connections that connects humans with more-than-humans beings (including plants, animals, rivers, lands, mountains, and gods). Whakapapa is the basis on which Māori society is organised, which people's geneological connections with members of their whanau (extended family), hapu (sub-tribe) and iwi (tribe) based lines of descent to particular ancestors (tūpuna) as well as to specific tribal areas (rohe).

Te Ao Māori

Rangatiratanga

The authority of a chief, political authority and control, self-determination, sovereignty. Rangatiratanga is inter-related with mana (power, prestige, sovereignty) and ability to exercise decision-authoritty authority over one's rohe, act as kaitiaki (guardians) and sustain the mauri (life force) and wairua (spirit) of the people (whānau, hapū, iwi and others) and other living things

Whānaungatanga

Kinship connections and the importance of relationships and networks, which are underpinned by notion of connectivity between all peoples and things. The principle governs relationships between māori (as individuals and members of iwi, hapū and whānau) and other people (including non-Māori).

Kotahitanga

Collective unity of māuri as whanau (family), hapū (sub-tribe), and iwi (tribe), and supporting whanaungatenge, leadership and resilience.

Wairuatanga

Spiritual embodiment (includes spiritual beliefs, capacity to enact tose beliefs, and maintain wellbeing). Wairua refers to spirit that is carried within a person.Sometimes more-than-human beings are said to possess wairua (such as rivers or supernatural beings). Wairua also can refer to the unseen domain which connects people to the future, the present, and the past, and denotes a state of interconnection and relationships with multiple worlds. Wairua is governed by tapu (scared, not ordinary, restricted) with the mechanisms of tapu applied to places and things (such as protection over a section of land or forest or waterways to ensure that it is not over-exploited by people).

Fig. 3.4 Key principles and concepts of Te Ao Māori

landscape was a component of the web of whakapapa relationships that began with Io and Te Kore, carried on to the creation of the gods and all living beings, continued through the migrations of Māori ancestors from Hawaiki to Aotearoa, and maintained in every generation since then. Within the Waipā River catchment, as elsewhere in Aotearoa, the landscapes and waterscapes were also the source of people's cultural identities, and their collective health and wellbeing.

Just as Māori was linked to one another through their whakapapa, so too were they connected to whenua, bodies of wai (water), biota, and other parts of the taiao. Whenua was the source of identity and life, with whakapapa creating a sense of place wherein each person lived and felt a sense of belongingness to their rohe. Bodies of wai were similarly crucial to Māori lives, livelihoods, and identities, with ancestral connections to specific rivers, streams, lakes, springs, wetlands, lagoons, estuaries, and seas forming essential parts of people's embodied waterscapes which sustained their lives (providing them with drinking water, foodstuffs, medicines, and other vital resources) but also their identities. When the people left the Tainui waka and travelled the region giving names to the physical features of lands and waterways, they were making claims to those areas and establishing their authority over it (in this way, naming was the equivalent of the European custom of raising a flag to take possession of a foreign land). Each subsequent generation of Māori re-articulated their ancestors' relationships and established their ones, which were transmitted down through names, songs (waiata), sayings (pepeha and whakataukī) and other kōrero that linked their tribal territory to those of important tūpuna (Belgrave et al. 2011, p. 59; Collins et al. 2012, pp. 109, 114; Durie 1994, pp. 61–63; Tauriki et al. 2012, pp. 120–121). Through such methods, individuals were able to understand and describe in-depth their ancestral linkages with their rivers, mountains, flora and fauna, and other parts of the environment. Moreover, they were able to understand the complex network of occupation and usage rights they (and their whanau and hapū) possessed, and how the different parts of their landscapes and waterscapes needed to be managed (according to tikanga and the various principles, most notably kaitiakitanga).

Waterscapes of the Waipā

Te Ao Māori and ways of life for different Māori communities, before British colonisation, centred on the use of resources within their fisheries (freshwater, estuaries and marine) and forests, as well as horticultural cultivations (climate-dependent) (Anderson 2002; Anderson et al. 2012; Stokes 1988). The population, living as whanau and hapū, were concentrated along rivers and river mouths, beside lakes and estuaries—in environments that provided a myriad of flora and fauna to harvest.

Water plays a central role in Māori cosmological accounts. In one waiata (song), for instance, water materialises when Papa-tū-ā-nuku and

Ranginui are coupled together in an embrace. When their son, Tane Mahuta disconnected Papa-tū-ā-nuku and Ranginui, they cried for one another and expressed their loss and sorrow at their separation. The tears of Ranginui fell onto the land and created the lakes and rivers, which included the Waipā River (Salmond 2017). The tears of Papa-tū-ā-nuku rose as mists. From this cosmological account, freshwater was not something singular but rather dual. Water as groundwater, aquifers, and springs (which emerges from the ground) was classified as wai mā (clear and pure waters) and was used in practices to cleanse or purify (tapu-raising practices/procedures). Water as rainfall was identified as something distinct as wai mangu (dark waters). These two waters together were known as wai rua (meaning two spirits or waters) that provide the life force for all living beings. Oral traditions tell how female and male ancestors' relationships were characterised by periods of intimacy and connection, and conflict and division. Their descendants could similarly take care of or undermine the hau ora (the energy or 'breath of life') of living beings (both physical and metaphysical), which includes both lands and waters (Salmond 2017, p. 300).

A wealth of historical, Māori oral traditions and archaeological studies demonstrate how the wetlands of the Waipā River catchment (and other freshwater systems around Aotearoa) were key locations where Māori whanau, hapū and iwi lived, gardened, fished, and sought refuge in times of conflict (Cromarty and Scott 1995; W. Shawcross 1968; Tauriki et al. 2012). Oral histories from Māori who affiliate to the iwi of Ngāti Apakura and Ngāti Maniapoto, whose rohe includes the area surrounding Te Awamutu, recount how their ancestors lived, managed and cultivated specific wetland-centred waterscapes. In addition to extensive kumara and taro cultivations, local hapū lived in kainga (villages) and pā located within the wetlands. Ngāti Apakura elders describe how their ancestors created Lake Ngāroto in the mid-eighteenth century through damming streams to create the vast expanse of water (Belgrave et al. 2011; Borell and Joseph 2012; Tauriki et al. 2012). Their ancestors built the island in the middle of the lake using large logs as the foundations (secured to the lakebed), over which they laid bundles of tree (mānuka) branches and reeds (rāupo), and soil. Over generations, the soil was slowly built up over the one-acre, until it formed a large mound that became the home of 200 people (Borell and Joseph 2012; Tauriki et al. 2012). Archaeological studies support these oral traditions and identified at least five pā located around the lake edge as well as those constructed on an island (Cassels 1972a; Internal Affairs Undated; Pick 1967; W. Shawcross 1968). Significantly, the island pā site was, archaeologists, suggest, continuously occupied by Māori from its creation until residents were forced to flee during the 1863 invasion (Belgrave et al. 2011, pp. 36–41; Tauriki et al. 2012). The continuous occupation was notably different from typical Māori practices of using pā only intermittently (seasonally or during times of warfare) and highlights the capacities of the lake and surrounding wetlands to provide consistent supplies of resources (which did not require them to relocate to other places to access food resources as was the standard practice amongst Māori communities). Archaeological evidence from elsewhere in the Waikato region records the existence of numerous wetland pā, artefacts including waka, buildings, and horticultural tools, which demonstrates the continued occupation and resource usage within the area (Boswijk and Johns 2018; Hogg et al. 2017; Lorrey et al. 2018; Pick 1968).

Wetlands like Ngāroto were significant resource extraction spaces, not limited to mahinga kai (food gathering sites) (Anderson 2016; Cassels 1972b; Kennedy 2017; Robb 2014; Tāne 2017, p. 47; Tauriki et al. 2012). A diversity of flora was harvested: harakeke (*Phormium Tenax*— New Zealand Flax) was used to make rope, baskets, mats, and clothing; kuta (*Elochiris sphacelata*—bamboo spike sedge) used for insulating and weaving; raupō (*Tayphia Orientalis*) for thatching, construction, and food (the pollen could be used to make a form of bread); besides, other sources of food included the berries of Kahikatea and Matai trees, fungi, and fern-roots (Forster 2012, pp. 39, 121–145; Parsons and Nalau 2016; Pond 1997). Numerous fauna were also harvested from wetlands, notably prized tuna (*Anguilla spp.*—short and long-tailed freshwater eels), inanga (*Galazias spp.*—whitebait), and koura (freshwater crayfish—*Paranephrops planifrons*), as well as numerous fish, shellfish, and aquatic bird species (Barrett 2012; Collins et al. 2012; Downes 1918; Tauriki et al. 2012). Crops were grown on higher areas of land; traditionally taro (*Colocasia esculenta*) and kūmara, and later with the arrival of Pākehā, wheat, potatoes, and other crops.

Ngāti Apakura also describes how Lake Ngāroto was the home of the atua Uenuku (god of rainbows), who was brought on the Tainui waka on its journey from their ancestral home of Hawaiki to Aotearoa in the form of stone (Borell and Joseph 2012). In Ngāti Apakura, Ngāti Maniapoto and Waikato/Tainui oral traditions, Uenuku was also considered the god of food gathering (Borell and Joseph 2012; Davidson et al. 2011; Ellison et al. 2012; Marsden 2003; Smith 2011; Taumoefolau 1996; Tauriki et al. 2012). Ngāti Apakura elders recalled how historically (prior to 1863) before anyone undertook fishing activities they would seek to satisfy Uenuku. The first eel or fish caught would be given to Uenuku as a way to ensure the health and wellbeing of themselves (and their whanau and hapū) as well as their environments (to maintain the mauri or life force of all beings that made up their rohe) (Borell and Joseph 2012).

Māori from the Waipā catchment describes the wealth of food sources that their tūpuna were able to access from the lands and waters (Waitangi Tribunal 2018, pp. 43–44). The network of land, water, and wetlands of Waipā Valley (where the Waipā River join with the Waikato River) was known as the "great food bowl". Similarly, Te Kawa wetlands were referred to as the "pātaka kai" (food store) of tangata whenua (Waitangi Tribunal 2018, p. 44). Pureora Forest (east of Te Kuiti at the headwaters of the Waipā River) and other forested areas were essential sites for harvesting berries and birds, and all the rivers, streams, and lakes of the region provided plentiful supplies of tuna (Belgrave et al. 2011; Cunningham 2014; Tauriki et al. 2012).

The capacities of people to take care of freshwater and land, sea and wetlands, people and other living beings depended on their whakapapa relationships to specific landscapes and networks of exchange between whānau, hapū and iwi. During different times of the year, groups travelled across the land, along the rivers, and sea, and followed the pathways of their ancestors (Anderson et al. 2012). Their knowledge of their ancestors allowed them to harvest specific biota in specific places at particular times of the year; indicators of seasonal change such as cloud formations and phenological signs (flowering and fruiting of plants, the behaviour of animals) were used to inform people's livelihoods (digging fern root, planting and harvesting taro and kūmara crops).

Waste and Water: The Two Should Never Mix

Historically, Ngāti Maniapoto communities (like those of other iwi) within the Waipā River catchment relied on their ecosystems for their physical and spiritual wellbeing (as we outlined in Chap. 2). They harvested food to eat and collected water to drink from their local environments, and therefore those resources needed to be both healthy and plentiful. As the size of communities grew (during the Classic Māori period the 1600s-1800s) and fortified settlements (pa) became more commonplace, Māori developed a range of comprehensive governance and management strategies, which were designed to regulate and mitigate the negative impacts of human activities on the environment that included material and metaphysical (more-than-human) actors (Anderson et al. 2012; Ballara 1998). Strategies included: rāhui, a temporary restriction placed by a chief on people accessing and using a particular area, which included a prohibition on harvesting activities (could be placed following an accidental death or because a need to conserve resources); as well as practices surrounding tapu (sacred, prohibited, to be set apart, not ordinary) and noa (usual, ordinary, safe, not subject to restrictions) (Bambridge 2013; Best 1904; Mead 2016). An array of regulations were rigidly enforced about how each different types of waste products (food, human, animal, and other material waste) were dealt with, which depended on the source of the waste (shellfish middens, wood shavings from carvings, human effluent, hair and nail clippings, disposal of the dead and so forth). Early Pākehā (explorers, missionaries, and researchers) noted that Māori did not use any type of manure on their cultivations and missionary attempts to use "such substances on their kitchen gardens [Māori] bought against them … a charge of high opprobrium" (Best 1931, p. 131). It was subsequently reported (by Firth) that Captain James Cook praised the way Poverty Bay Māori managed waste products and contrasted it to European practices at the time (Firth 2012, pp. 94, 312–314). Traditionally, then, Māori kainga, marae, and pa complexes ensured that there were strict zones of activities designed to uphold Māori values (tapu and noa) and protect people and their wider environment (including their more-than-human kin) from harm caused by contamination of waste products. Māori waste management practices centred on the physical separation and designation of certain areas for specific waste products (human waste like faeces, urine, and menstrual waste, the washing of clothes and bodies, food waste) (Mead 2016; Pauling and Ataria 2010; Waitangi Tribunal 1993).

Areas for waste disposal were kept spatially segregated from important spaces related to sustaining the community (such as food cultivation and harvesting, food preparation, and the collection of drinking water supplies) as well as significant activities and people (Mead 2016; Pauling and Ataria 2010). Traditionally, this was done by ensuring that the toilet, kitchen and living areas within a Māori settlement were physically detached from one another, as well as the deliberate placement of a toilet on sloping land or beside a cliff, and by demarcating different areas of waterways for bathing, harvesting kai, and collecting drinking water (Ataria et al. 2018; Marsden 2003; Mead 2016; Pauling and Ataria 2010).

The vital principle of tapu underpinned waste management strategies. Tapu translates into English as something sacred, prohibited, special, not ordinary, and something that needs to be set apart; the antithesis of tapu is noa that refers to normal, ordinary and unrestricted. Tapu was (and still is) of a temporary (such as a rāhui or temporary ban on harvesting) or intrinsic nature (such as a burial ground and human waste products). However, all things in the world are considered to possess tapu, inherited from atua and holding considerable (spiritual) power (Mead 2016). Tapu provides the spirituality and metaphysical connections between people, plants, animals, whenua, awa, and taiao. When different types of tapu interact, the results can be destructive (such as illness or death). So, systems of control are designed to manage these interactions to achieve positive outcomes (maintaining good health of a community). In the case of human waste and body parts (which are considered very tapu) waste management practices are designed to protect the tapu (sacredness) and mana (power) of human beings, awa, and whenua. However, following colonisation, the established system whereby Māori managed waste products were no longer disrupted and displaced, as communities were dispossessed, migrated (voluntary or forcibly or a combination of both) to new areas, and new modes of living adopted. Yet, Māori understandings of waste and the need to ensure the principle of tapu were maintained persisted into the twentieth and twenty-first century. In the second half of the twentieth century, as we later examine in Chap. 5, Ngāti Maniapoto iwi members campaigned against local government authorities practices of discharging treated human waste (from town sewage treatment plants) into the Waipā River and its tributaries on the basis that human waste (even if handled by scientific methods) was still tapu. Accordingly, all water became polluted once it was in contact with human waste, and iwi fought for alternative modes of waste disposal to be implemented (which were more in keeping with their tikanga).

Te Ao Māori at the Time of European Contact

Broadly, Aotearoa New Zealand history, over the last two hundred plus years, is reflective of broader processes of European imperialism and colonialism around the globe, which involved phases of contact, colonisation (using military, socio-cultural, legal, and ecological mechanisms), and development. The first recorded contact between Māori and Pākehā occurred in 1642 when the Danish explorer, Abel Tasman, gave the islands the name Nova Zeelandia (New Zealand). In 1769, nearly 130 years later, an English explorer (Captain James Cook) and his crew's supposed 'discovery' of Aotearoa heralded the first wave of European imperial activities from the 1790s onwards; sealers, whalers, and traders arrived (namely from the Australian colony New South Wales) seeking new resources to exploit (see Fig. 3.5).

Pākehā encountered a socio-cultural landscape made up of heavily mountainous terrain with a temperate climate, extensive and dense forests teeming with birdlife, and a vast indigenous Māori population (estimated to be number 80 000 in 1840) (Stokes 2013). The vast majority of Māori lived in the northern half of Te Ika-a-Māui (North Island), which possessed a climate that was warmer and better suited for Māori horticultural techniques and plants. In addition to the cultivation of kūmara (Bassett et al. 2004; Unknown Author 1902), Māori living in warmer regions with access to sufficient freshwater supplies were able to grow taro (Ban 1998; Irwin 2013; Matthews 1985), which included Māori hapū

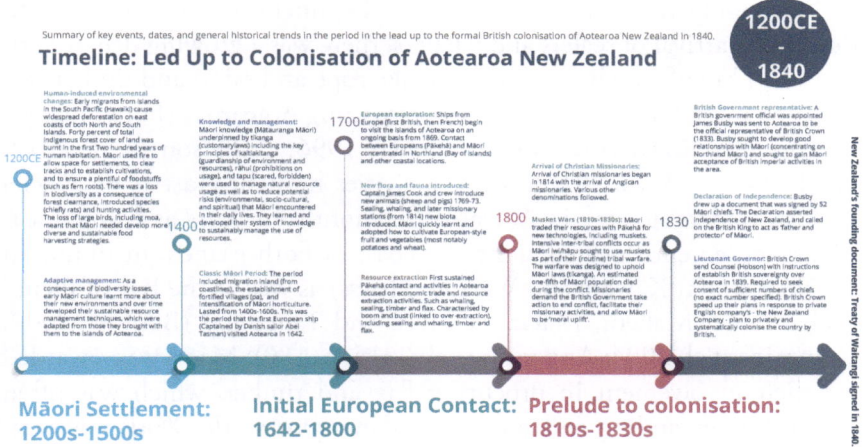

Fig. 3.5 Timeline—Lead up to colonisation

living in floodplains of the Bay of Plenty and the Waikato (Best 1930, 1931; Hargreaves 1959; K. Shawcross 1967).

When Pākehā first began to visit, trade, establish relationships with and make homes with and alongside local hapū in the Waipā (like elsewhere in Aotearoa), they did so under the watchful eye and with the explicit permission of local rangatira. Waipā iwi initially engaged with new arrivals (most notably new biota, knowledge, and technologies) with enthusiasm. It sought to incorporate new plants and animals, agricultural production technologies (flour mills and ploughs), clothing and trading relationships. Māori subsistence-based horticultural operations were extended and modified to engage (to some extent) with the capitalist marketplaces of Pākehā, all. At the same time, Māori sought to maintain their values, knowledge and reciprocal relationships (based on whakapapa and occupancy) with their whenua, awa, maunga and other taonga. Hapū along the Waikato and Waipā Rivers became the major suppliers of agricultural product to the fledgling Pākehā settlement, Auckland; Māori was well on their way to being agricultural entrepreneurs as they shipped their crops and other goods by sea to settlers living in towns in Aotearoa, the Australian colonies, and California.

Throughout the first three decades of the nineteenth century, in addition to the arrival of sealers and whalers, there was a growing demand for flax (the fibre of which was used to make rope and sales) and timber (for shipping masts) from Pākehā who were visiting Aotearoa's shores, including Britain's Royal Navy vessels (Belich 1996; Anderson et al. 2014, pp. 179–181; Petrie 2013). These activities brought clusters of Pākehā (sailors, traders) to stay in Aotearoa for months or years at a time, near Māori settlements located along the coast of both islands, including in the Waikato (at Kāwhia and Waikato Heads) as well as the Bay of Plenty (Whakatāne, Maketū, Matatā, Ōhiwa, and Tauranga) (Boulton 2007; Cummins et al. 2004; Anderson et al. 2014, p. 180). Māori was rewarded for their involvement in procuring flax and timber, which was often physically demanding and sometimes dangerous work. Work was typically organised along hapū lines, as was the majority of Māori economic activity in this time period. The rangatira acted in the role of the subcontractor, who agreed to the terms of work relationships with Pākehā, supervised the work, and received payment, which was distributed amongst the hapū members (Adams 2013; Petrie 2013). While Pākehā presence was restricted to the coastal areas of the Waikato and the King Country in the 1830s, hapū from Waikato and Waipā catchments were harvesting flax that they transported to the coastal kainga of Kāwhia, and Waikato Heads to trade with Pākehā. The flax trade continued to be a feature of the Waikato economy until the start of the twentieth century (Francis 2011).

Divergent Understandings of Land: Rights Versus Ownership

There was no concept of owning land, water, or any other resources in Māori society, in marked contrast with European cultures. When Māori engaged in land transactions with Pākehā, they did so through their cultural lens. Rights to land, water, forests, and other resources were, in Māori culture, collective and highly fluid. Individuals could inherit usage rights to certain areas of land or a river through their parents (both paternal and maternal lines). Rights were shared between different whanau, hapū, and iwi; usage rights often overlapped, with groups able to visit and harvest resources but only from a particular area or at a specific time of year. Often there were disputes between hapū, which created conflict. The fluidity in defining and policing boundaries was further complicated with the arrival of Pākehā, who first sought Māori timber and flax, and later land. Pākehā brought with them their ways of thinking about and managing land, which was premised on the idea that all land boundaries were fixed and final, and that land could be owned (and by extension bought and sold).

In the late 1820s and throughout the 1830s, Pākehā began to seek to purchase land from Māori including in the western harbours of the Waikato region and the river mouths and concentrated on identifying one person who was the "owner" of the land and who could sell them the land (Cummins et al. 2004; Cunningham 2014; Hammer 1991; Luiten 2011; Schnackenberg 1935). In doing so, Pākehā overlooked (frequently deliberately) the complex and dynamic layers of Māori rights to land and resources and imposed their own cultural and legal traditions. Māori similarly engaged with Pākehā for their ends using their worldview. An organising principle of Māori society was the pursuit of mana, whereby people (specifically rangatira) sought to gain and maintain mana through actions. When a Pākehā found to form an agreement (for trade, land or other resources) with one chief (and associated hapū) it served to enhance the mana of that rangatira (and hapū) and diminished rival rangatira and hapū who were not a party to that agreement. Accordingly, many rangatira signed written documents (deed of sale), which transferred ownership (a concept that had no equivalent in Te Reo Māori) of areas of land and received small payments (often in the form of goods). For hapū, the provision of land to Pākehā was based on the understanding that missionaries and merchants would form long-standing and mutually beneficial relationships with Māori. Broader historical scholarship on Māori-Pākehā highlights how, in the period prior to the Te Tiriti o Waitangi/Treaty of Waitangi (1840), Māori communities throughout Aotearoa made land available for Pākehā to live on and use in the context of continuing reciprocal relationships (Brookes 2016; Wanhalla 2015, 2017).

Rangatira expected that Pākehā traders, for instance, would give regular gifts to them (such as weatherboard houses, small boats, tools, and clothing) in return for rangatira giving them the right to stay in the chief's area of authority. Different interpretations of these arrangements existed between Māori and Pākehā. For instance, whaling stations were on land that Pākehā perceived they had "purchased" from Māori. In contrast, rangatira generally interpreted that Pākehā had given them a tuku (gift) in exchange for the rangatira allowing whalers the right to occupy their tribal territory (Anderson et al. 2014, p. 181). The gifts were similar to a form of rent paid to rangatira in return for rangatira giving whalers the right to occupy land as well as supplying labour for whaling operations. Similar arrangements existed for embryonic timber, flax, and horticultural activities (first in the Bay of Islands, and later elsewhere in Aotearoa including Waikato). Vast supplies of food (specifically potatoes) began to be grown by Māori for barter and trade with Pākehā, which generated food surpluses (which were previously not common) and different rangatira organised large hui to display and demonstrate their mana (Anderson et al. 2014, p. 181).

Māori communities, particularly on the western harbours of the Waikato and King Country, were able to expand the quantity and diversity of goods (flax and food) which they could produce and ship to expanding towns (such as Auckland and New Plymouth, and Sydney and Melbourne) and in doing so generate wealth and enhance their mana. When intimate relationships between European men and Māori women resulted in the growth of "half-caste" children, many hapū simply viewed the land on which each family lived remained entrusted to the Māori community (Grimshaw 2002; Paterson 2010; Stevens and Wanhalla 2017). The rights to occupy the land would transfer from the Māori woman to her children. Adequately, the community considered that the Pākehā man held only a right to use the land so long as he stayed in the community (Boulton 2007). Such rights could not be transferred to someone else outside the community; Māori was determined to retain their mana and rangatiratanga. Despite their engagements with Te Ao Pākehā, they remained thoroughly connected and living within Te Ao Māori (Petrie 2013).

Customary Māori governance arrangements were collective and involved different people within hapū acting together through tribal structures, which included rūnganga (tribal councils) and hui (Ballara 1998; Mead 2016). Members of hapū were influenced by the views and decisions of their rangatira; however, people did not restrict their views to what their rangatira thought. Indeed, while rangatira possessed the ability to shape any agreements reached within and between hapū, they did not have cohesive power over their people (that is "the ability to make and ensure compliance with unilateral decisions") (Anderson et al. 2014, p. 182). Rangatira was required to gain the support of their hapū members, and decision-making for Māori was a collective rather than individual process (Ballantyne 2012; Belgrave 2017; Belich 1996; O'Malley 2017). Implicitly, there was an expectation that all significant decisions and agreements would be made following in-depth discussions within the home (amongst whānau members) and on the marae (between members of the hapū) (Salmond 2018). Once a decision was made, it was rangatira who were held responsible for the choice and its consequences (be it good or bad). Accordingly, when taking part in negotiations with Pākehā, rangatira often added a requirement that they needed to take the proposal or decisions back to their hapū for their agreement (Anderson et al. 2014, pp. 180–182).

Christian missionaries began to establish mission stations around the western harbours of the Waikato region (Whaingaroa/Raglan, and Kawhia) and along the Waikato, Mokau and Waipā Rivers in the mid-1820s. In 1834, missionary John Morgan arrived at Mangapouri, located where the Puniu River joined the Waipā River and took up residence. In January 1841, John Morgan took over the mission station at Ōtāwhao (present-day Te Awamutu), which he headed from 1841 to 1863. Morgan saw the rise of Māori economic development by local hapū (O'Malley 2013, 2016). In 1863 (in the lead up to the Waikato Wars), just before the British invasion of the Waikato, Morgan fled after it became known by Māori that he was spying on them on behalf of the British.

The spread of Christianity complicated political mechanisms and tikanga because becoming a Christian (a mihinare) meant Māori were required to observe new rituals and worship a new atua. It also required taking up new codes of personal and inter-personal behaviour that differed from traditional forms of conflict management (such as the practice of taking utu—revenge—which was discouraged under Christian teachings) (Anderson et al. 2012; Waitangi Tribunal 2018). The advice of missionaries was sought for how to resolve conflict and regulate conduct within and between individuals, whanau and hapū (such as who should pay for damage to crops and what the punishments should be imposed if someone committed adultery). As more and more Māori became Christianised in the 1820s and 1830s, missionaries sought to ensure that their congregations followed their codes of conduct (law—ture) and rangatira tried to ensure that new laws were balanced with old (tikanga) (Ballara 1998). For rangatira for the presence of Pākehā visits and economic activities brought numerous gains to them and their hapū, but also brought with them new infectious diseases. These diseases included measles, influenza, syphilis and tuberculosis, which contributed to a sharp demographic decline amongst the Māori population in the mid-to-late nineteenth century (Pool 2015).

Lead up to Colonisation: 1830s

In the late eighteenth and early nineteenth centuries, the British government in London showed little interest in Aotearoa as a colony after the costly experience of the American colonies revolt (Anderson et al. 2014; O'Malley 2013; Orange 2015). The empire was already substantial, and the British government was determined to maintain and guard its colonial rule of the Indian sub-continent, parts of the Caribbean as well as the newly created Australian colonies. Thus, British wealth could be better ensured by "free trade" rather than by acquiring more colonies in the Pacific. Indeed, five decades earlier, Britain rejected Aotearoa as the site of a convict settlement due to the perception that Māori culture was too warlike and likely to violently oppose any British intrusions (leading officials to establish the convict settlement in Botany Bay, Sydney, New South Wales). However, as British subjects were continuing to visit and move to Aotearoa, the British government increasingly recognised in the 1820s and 1830s that it could not ignore Aotearoa entirely.

After years of individual Māori and missionary protests to British colonial officials about the bad behaviour of Pākehā in Aotearoa, the Governor of New South Wales decided that British authority in Aotearoa needed to be strengthened by sending a colonial official. The "British Resident", James Busby, arrived in the Bay of Islands in 1833 with the instructions to cultivate influence amongst rangatira and promote a "settled form of government" and a system of law that would allow Māori courts to manage conflicts (Anderson et al. 2014, p. 184; Stirling 2016). The shift in British Crown policy from 'free trade' to the formal colonisation of Aotearoa can be traced back to (in part at least) the work of Busby and the creation of the Declaration of Independence (He Whakaputanga o te Rangatiratanga o Nu Irene).

Declaration of Independence

James Busby sought to foster cordial relationships between the British Crown and Māori (specifically those in the Bay of Islands region) and prompted the idea of a centralised Māori government (more alike to those of Britain). He drafted the Declaration of Independence of the United Tribes of New Zealand (He Whakaputanga o te Rangatiratanga o Nu Irene), which was translated into te reo Māori by missionary Henry Williams, in the hope it would form the building blocks for a formal Māori government (beginning with rangatira in the Bay of Islands). On 28 October 1835, thirty-four rangatira signed the declaration and signatures continued to be collected until 1839; most signatories came from chiefs in the Bay of Islands; however, Waikato rangatira, including Te Wherewhero (from Ngāti Mahuta hapū part of iwi Waikato-Tainui) also signed (Harris et al. 2014; Orange 2015; Waitangi Tribunal 2014). Busby prompted the idea that the British Crown could protect Māori interests and take care of external problems (such as British or French traders who came to Aotearoa and breached Māori rules). The declaration stated that sovereign power rested with rangatira in collective abilities and that rangatira looked to the English King (then William IV) to protect them against any attempts (such as French imperialism) to challenge their independence. The document also (in the te reo Māori version) requested that Britain act as a parent ("matua") to their infant Māori political state ("to matou Tamarikitanga") (Harris et al. 2014; O'Malley 2017; Waitangi Tribunal 2014). The declaration probably contributed to the emergence of Māori participants' (who took part in discussing and signing the document) sense of being a collective cultural and political grouping (rather than individual hapū and iwi) (O'Malley 2017); however, for most Māori living within the Waipā catchment and elsewhere around Aotearoa (in 1835 and for the proceeding next decade or so after), their daily lives and the very nature of political decision-making within communities continued to be centred on long-standing tribal and hapū identities, values, and practices. The 1835 declaration did not provide any legal basis for British rule or facilitate British immigration or settlement in the country (O'Malley 2017; Stirling 2016), yet this was precisely what many Pākehā living in New South Wales, Britain, and Aotearoa wanted.

By the late 1830s, there were rising tensions between different groups of Māori and Pākehā within Aotearoa (specifically in the Bay of Islands) over land and authority. Land dealings of individual Pākehā and individual rangatira caused conflict within and between hapū (particularly in Northland) as different rangatira sought to maintain and extend their mana through signing land deeds with Pākehā (Anderson et al. 2014, pp. 188–189; O'Malley 2017). By the end of the 1830s, many rangatira (particularly in North Island) began to express concerns about Pākehā land dealings with Māori. In addition, wealthy entrepreneurs in Europe were attempting to purchase Māori land and establish private colonisation schemes (most notably the New Zealand Company led by Edward Gibbon Wakefield).

Busby and missionaries petitioned the British Colonial Office to incorporate Aotearoa into the British Empire, and in August 1839 the British Secretary of State for the Colonies (Lord Normandy) sent a government official to do just that. Lord Normanby issued detailed instructions to the official—Captain William Hobson—as to how he was to proceed, which included Maori approval for the cession of their sovereignty, the broad outline of the need for a treaty to be signed with Maori and the way in which existing purchases by Europeans were to be dealt with (Lord Normanby 1908) The instructions included a requirement that Hobson convinces some (but not all or even a majority) of Māori chiefs to sign a treaty, and that efforts were taken to protect Māori rights. Normanby's instructions drew on the language of the mid-nineteenth century British humanitarian movement, which sought to regulate the behaviour of settlers, protect Indigenous peoples from the worst excesses of colonisation (which was seen to include violence, alcoholism, gambling) and assimilate Indigenous peoples into British culture. Britain already employed such treaties in its dealings with some North America Native American and First Nations peoples (Attwood 2001, 2014; Belich 2009; Jones 2016; Parsonson 2017).

Te Tiriti o Waitangi (The Treaty of Waitangi)

Te Tiriti o Waitangi (The Treaty of Waitangi) was signed by the representative of the British Government (Captain Hobson), various English residents, and more than 500 Māori rangatira in 1840 (Orange 2015). Signatories included rangatira from Waikato-Tainui and Ngāti Maniapoto iwi. (Anderson et al. 2012; Orange 2015). Not all rangatira representing the hundreds of different hapū and iwi in Aotearoa signed the Treaty. The majority of rangatira who did sign the Treaty placed their signatures on the Māori language version (Te Tiriti o Waitangi) rather than the English version (the Treaty of Waitangi). The Treaty recognised both the British Crown and Māori rights in the new colonial order, but there was a lack of clarity about what precisely both sides were signing.

There were significant divergences between the two versions of the Treaty (Fig. 3.6) (Jackson 1993; Orange 2015; Waitangi Tribunal 2014). Both versions of the Treaty contained four parts: a preamble, three articles, and a postscript (see Fig. 3.6). The objectives of the Treaty were outlined in its preamble: the British Crown was to establish a government to protect Māori and Pākehā interests in Aotearoa and centred on the overarching principle (in both Māori and English language versions) on a partnership between the Crown and Māori iwi (and by extension two cultures: Pākehā and Māori) (Tawhai and Gray-Sharp 2011). In Article Two of the Māori version of the document, rangatiratanga (chiefly authority) was acknowledged, and the Crown gave an assurance that it would be secured under the new partnership arrangement. From a Māori perspective, the Treaty kept chiefly authority well-maintained, with the inclusion of British laws and institutions added onto Māori tikanga (rather than the other way around) (Healy et al. 2012; Jackson 1993; Orange 2015; Waitangi Tribunal 2014).

Different articles, different interpretations

FIRST ARTICLE

1

Article 1 outlined the nature of British Crown's authority in Aotearoa. This article that was (and continues to be) the source of ongoing disagreements between the Crown and iwi Māori. In the English version Article 1 explicitly stated that Māori agreed to cede their sovereignty to Britain, whereas in the Māori language version they gave Britain the right of kāwanatanga (governorship): no mention was given of Māori relinquishing their sovereignty rights (mana and rangatiratanga). Questions arose whether the British Crown ever actually acquired sovereignty from Māori. Te Tiriti outlined a form of shared authority between Māori and the Crown that was always qualified by rangatira retaining their tino rangatiratanga (chiefly authority). Since Māori political governance was centred on kinship groups and decision-making authority was held locally and centred on a rangatira being able to convince their kin (who comprised their hapū) to agree to their decisions and acknowledge their mana (power), Te Tiriti's arrangement of shared authority between Māori and the Crown was consistent with Māori governance structures. Tikanga and the concept of tino rangatiratanga allowed for the acknowledgement and the assertion of shared authority (alike to the contemporary co-governance and co-management arrangements discussed in chapters 5 and 6) between multiple groups (whanau, hapū and iwi) and therefore it was not necessary under Māori tikanga that many rangatira cede their mana (their sovereignty) to Britain (or anyone else).

SECOND ARTICLE

2

Article II, in both versions, acknowledged Māori decision-making authority and guaranteed Maori would retain possession of their land, waterways, forests, and other resources. In Te Tiriti, the second article guaranteed that rangatira and hapū would retain their "tino rangatiratanga" and the protection of "o rātou taonga katoa" by the British . The term rangatiratanga assured Māori that they retained conrol over their own affairs as well as all their taonga (including whenua/land, awa/rivers). Recognition of rangatirangta seemed to Māori to mean that rangatira and the Crown would govern collaboratively and continue to work together to build a nation and extend the long-standing history of alliance and friendship between the Crown and northern Māori. The English version refers to Māori abilities to retain only their property rights ("their full exclusive and undisturbed possession of their Lands and Estates Forests Fisheries and other properties"). However, the Māori word taonga extended beyond the ownership of land and other natural resources and extended to include socio-cultural and spiritual possessions. Under Article II, the Crown also gained the exclusive right (pre-emptive) to purchase property directly from Māori (with no land sales to other parties allowed). It is unlikely in 1840 that Māori who signed Te Tiriti understood what this section of agreement meant. Within Māori tikanga, it was not possible to permanently alienate land because land was not owned, and was tied to people's identities. Rather it was more likely that Māori understood the provision to be in regard to the allocation of usage rights (which were also shared and conditional rather than absolute and inalienable) and not a permanent severing of the connection of a hapū to its whenua.

THIRD ARTICLE

3

Article III extended Māori the same rights and protections as British citizens. One of the key aspects of the Treaty for Māori was that it created a reciprocal and ongoing relationship between Māori and British; Article II assured Māori ongoing capacities to access European knowledge and technology and to (supposedly) afford them protection from scrupulous Pākehā settlers.

Sources: Jackson, 1993; Mutu, 2011; Healy et al., 2012; Orange, 2015; Waitangi Tribunal, 2014; Barnes & McCreanor, 2019; Mead, 2016; Feint, 2017; Jackson, 1993; O'Malley, 2013b; Parsonson, 2017.

Fig. 3.6 Articles of the Treaty

As pointed out by legal scholars, since the British Crown provided both English and Māori versions of the Treaty/Te Tiriti then the legal doctrine of *contra proferentem* (in use in 1840 and to the present-day) should be applied (Orange 2015; Suter 2014). The doctrine *Contra proferentem* (also known as the "interpretation against the draftsman") is a contractual interpretation which considers that, where an agreement, promise or term is ambiguous, the preferred meaning is the one that works against the interests of the party (the British Crown) who provided the vague wording (Chan and Fan 2018; Grammond 1994; Suter 2014). Accordingly, the correct meaning of Te Tiriti/The Treaty would be the Māori rather than British interpretations.

The Treaty presented to rangatira at Waitangi laid out a general statement of intentions and left the exact details of the nature of the partnership arrangements between Māori and the Crown undefined. The records of the oral discussions that took place during negotiations are few and far between. Since Māori society was traditionally a verbal rather than written culture, and important decisions were made by debates and discussions amongst rangatira (and hapū members), the absence of accurate records of what was said by rangatira about the Treaty and how Hobson and other British responded means we lack a full understanding of how Māori interpreted the Treaty. The limited, highly incomplete, records (written by missionaries, officials and local Pākehā settlers) indicate that Māori rangatira who met on 5 February clearly understood that they were faced with an important decision. Many rangatira spoke of their fears that they would lose their independence and be nothing more than slaves of the Governor (some reportedly stated that they did not want to end up like Australian Aboriginal peoples "having to break stones on the road") (Anderson et al. 2015, p. 203). Other speakers raised the issue of land dealing. They demanded that British Resident Busby and missionaries in the Bay of Islands return to the land they had initially purchased from Māori as they never intended to lose their land rights permanently.

While growing debates occurred within and between Māori and Pākehā communities about how and where British law and order was to be introduced, how it reconciled with "rangatiratanga", and what the new arrangements for the government were to be, for many Māori such debates were removed from their daily lives. Within the Waikato and Waipā catchments, Māori communities enjoyed a period of growing prosperity and productive trade with Pākehā. Indeed, for much of the 1840s and early-to-mid 1850s Pākehā settlers (notably the growing settlement of Auckland) depended heavily on iwi from the Waikato and Waipā river catchments for their necessary supplies.

Historical Context: The Invasion, Raupatu (Confiscation) and Alienation of Whenua 1863–1885

While the Te Tiriti o Waitangi/Treaty of Waitangi, signed by representatives of various Māori iwi and the British government in 1840, was meant to guarantee the protection of Māori land and natural resource rights, colonial officials did not honour the terms of the Treaty. The settler-led colonial government actively sought to appropriate Māori land and limit the abilities of Māori to exercise rangatiratanga over natural resources using the military, financial, and legal mechanisms (see Fig. 3.7: Timeline of key historical events). Colonial officials and settlers repeatedly expressed their desires to acquire the whole of the floodplains of the Waikato and Waipā rivers for Pākehā settlement at the same time as various Waikato Māori iwi and hapū were forming a pan-tribal alliance (the King Movement—Kīngitanga) that opposed selling land to Pākehā and sought to retain Māori political authority (rangatiratanga) over their whenua, awa, and other taonga including the wetlands. The Crown considered such efforts as acts of rebellion against the British Crown and used military and later legal mechanisms to undermine Māori political sovereignty, land tenure arrangements, and relationships with their whenua and awa (New Zealand Government 1865; Unknown Author 1864).

In the Waikato district, the appropriation of Māori land and waterways came first through military actions. Between July 1863 and April 1864, British military forces, (consisting of 12,000 imperial troops, 4000

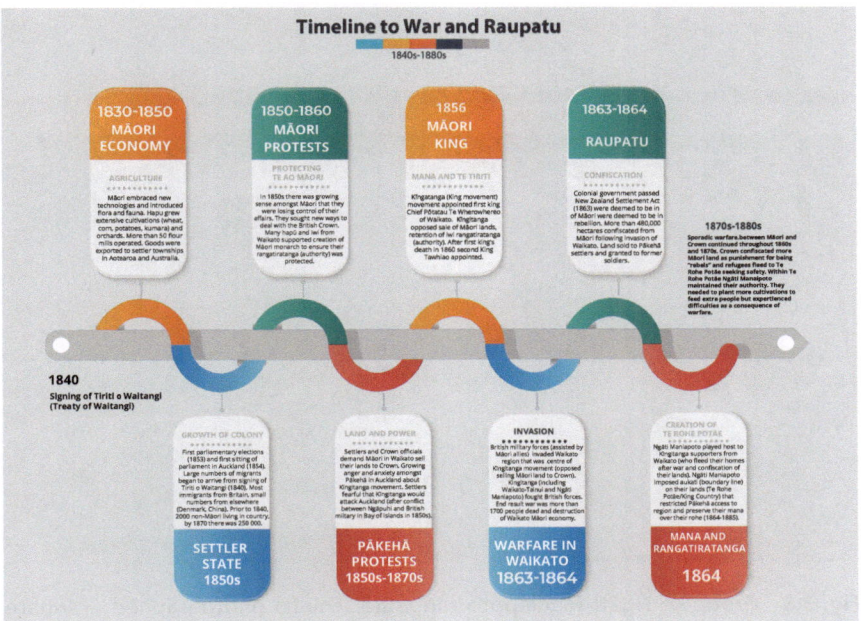

Fig. 3.7 Timeline of key events in Waipā 1860s–1880s

colonial soldiers and several hundred Māori tribal allies fighting against 5000 part-time Kingite soldiers), invaded and occupied the Waikato district (Belich 2015; O'Malley 2016). The result of the Waikato War was the death of approximately 1700 people, as well as the destruction of Māori economies. After the war, villages and crops were ruined, flour mills were burnt, and livestock looted. In 1865, the colonial government announced the confiscation of 485,000 hectares of land from Māori communities in the Waikato, including a large portion of the lower reaches of the Waipā River. The confiscated lands were allocated to Pākehā settlers. Members of Waikato-Tainui iwi (who supported the Kīngitanga movement) sought refuge with their kin, Ngāti Maniapoto, in the area south of the Pūniu River, which became known as Te Rohe Pōtae. Te Rohe Pōtae remained under the rangatiratanga of Ngāti Maniapoto until 1882 when iwi rangatira (leaders) (see Fig. 3.8: Photograph of Ngāti Maniapoto chiefs) agreed to allow the colonial government to begin to

Fig. 3.8 Group of Ngāti Maniapoto rangatira (chiefs) photographed at whare komiti (committee meeting) at Haerehuka, Rohe Pōtae (King Country) on 4 June 1885 by Alfred Burton. The meeting was to discuss whether to allow the Crown to construct a railway through their territory. Back row, from left: Rewi Maniapoto, Tawhana Tikaokao, Taonui Hikaka, Hone Wetere Te Rerenga. Front row, from left: Te Rangituataka, Te Naunau Hikaka. Source: Alfred Henry Burton (photographer), PA7-36-30. Alexander Turnbull Library, Wellington, New Zealand

construct a railway line (the "Main Trunk Line") through the area, which resulted in the area being 'opened up' to Pākehā settlers (Belgrave 2017; Waitangi Tribunal 2018).

With the commencement of works to construct the railway in the middle and upper reaches of the Waipā River, the loss of Māori land, access to resources, and accompanying decision-making authority occurred insidiously and cumulatively through the mechanisms of the settler-colonial law (legislation and the judiciary) rather than through acts of war (see Table 3.1: Land loss within Te Rohe Pōtae district) (Belgrave 2017; Williams 1999). Both military and legal means achieved the same ends: to sever the links between the Indigenous people and their rohe. One adopted abrupt and apparent methods to obtain it, the

Table 3.1 Land loss within Te Rohe Pōtae district

Year	Percentage of land area held by Māori	Notes about alienation
1865	93.1	Government acquisition of Māori land prior to 1865 was approximately 6.7 per cent of total land area (129 181 acres) within Rohe Potāe district (as defined by Waitangi Tribunal Inquiry). The total area of land alienated from Māori was 7.9 per cent.
1889	93	The Native Land Court's entry into the district in 1886 was a major factor in establishing and accelerating Māori land alienation. More than 40 per cent of land within the district was alienated between 1889–1910. Overall, the majority of the alienations was from government purchasing (97 per cent).
1910	49	Between 1910–1931 the pace of land alienation slowed, with Māori land holdings reduced to just under 50 percent of pre-1840 levels in 1910. Private acquisition of Māori land was significant feature of the time period.
1931	24	Pace of land alienation continued to slow, with Māori seeking to retain what little land that remained within the district. During the 1950s the local government compulsory acquired Māori land (as well as non-Māori land) along the Waipā River and its tributaries as well as other river systems in the Waikato Region for flood control works. The land was acquired ("taken") under the Public Works Act.
1966	18	Most land alienation in the second half of the twentieth century in Rohe Potāe is a consequence of Māori seeking to convert (Europeanise) their Māori land title (whereby each parcel of land is divided into shares held by different members of a whānau) to freehold title (individual owners).
1975	15	

other used slow insidious but still violent means. In particular, the operations of the Native Land Court (which commenced operations in Ngāti Maniapoto territory in 1886) involved blocks of Māori land going through a legal process to convert them from communally held (wherein usage rights were shared between whānau and hapū) into individualised land titles. The activities of the Native Land Court effectively continued where the land confiscations of 1863 had left off; for more details about the Native Land Court (defined by one scholar as an "act of war") see the works of historians including Judith Binney, Richard Boast, and David Williams (Binney 2001; R. Boast 2008; R. P. Boast 2017; Williams 1999). Once converted to individualised titles, Māori were frequently forced to sell their land due to financial necessity (to pay the high prices of court fees, government taxes, and costs of living). The central government acquired the majority of Rohe Pōtae land in this manner; the government were able to purchase Māori land heavily reduced prices and then on-sell the land to settlers at far higher prices.

On top of that, successive acts of parliament gave central government departments and local government bodies broad-ranging powers to acquire or fundamentally alter the characteristics of land and waterways in the name of 'development' and 'improvement', which included the drainage and destruction of the extensive wetlands of the Waipā River and its tributaries. Local government bodies (empowered by central government) authorised water- and land-scape remaking activities (draining wetlands, realigning rivers, building railways and roads, and removal of native vegetation). In doing so, government officials (deliberately and incidentally) ignored the rangatiratanga of iwi over their awa and whenua, and their responsibilities as kaitiaki (environmental guardians) to care for and maintain the mauri of the river, land, biota and their people. The concerns of iwi and their reflections back to the invasion, confiscations, and breaches of the Treaty illustrates a critical aspect of the Indigenous environmental justice theorising, which recognises the connection between past and present injustices.

Conclusion

In this chapter, we provided an overview of the histories and values of iwi and hapū living within the Waipā River catchment before and immediately following colonisation. The chapter presents the building blocks for readers to later understand the consequences of the settler efforts to radically remake Māori waterscapes and landscapes, and the negative impacts of those changes for iwi, hapū and whānau. Such historical analyses are often engaged to explain the emergence and significance of environmental injustices on Indigenous peoples and their traditional lands and waters (Clark 2002; Hooks and Smith 2004; Harris and Harper 2011) (Boone and Buckley 2017; S. Harris and Harper 2011; Hooks and Smith 2004; Vickery and Hunter 2016; Wakild 2013). Such historical perspectives, as we demonstrate through the next examples of wetland drainage, flooding and water pollution, offer essential insights into how environmental injustices accumulated over time and that historical injustices are in fact at the heart of the present-day environmental justice issues facing iwi in Aotearoa. Moreover, some scholars argue that without an appreciation of histories, contemporary issues (be it freshwater degradation, climate change, biodiversity, poverty, discriminatory policies) are open to misinterpretation. As Lord and Shutkin wrote in 1994: "To approach history casually and complacently is to evade history's inevitably multiplications facts and to mask the many meanings the facts could support" (Lord and Shutkin 1994, 5) (Lord and Shutkin 1994, p. 5). The deliberate or incidental disregard of histories presents significant implications for not only how people perceive Māori or Aotearoa New Zealand histories but also their understandings of environmental justice (Lord and Shutkin 1994).

In the specific context of Indigenous environmental justice, an ever-expanding body of Indigenous and decolonial scholarship demonstrates the interwoven relationships between colonialism and environmental injustice (Curley 2019; Dhillon 2020; McGregor 2014; Whyte 2018; Winter 2019). In the next three chapters, we highlight how settler colonialism is not confined to the past and continues to be acted on and evolve; it is thus, critical to consider settler colonialism as an ongoing process that continues to shape Aotearoa New Zealand in the twentieth century. In the next chapter (Chap. 4) we provide an in-depth exploration of settler-led efforts to remove the wetlands of the Waipā River, and how those ecologically transformative acts resulted in a litany of environmental injustices for generations of Maori who whakapapa to the waterscapes of Te Awa o Waipā.

References

Adams, P. (2013). *Fatal Necessity: British Intervention in New Zealand, 1830–1847*. Wellington: Bridget Williams Books.

Anderson, A. (2002). A Fragile Plenty: Pre-European Maori and the New Zealand Environment. In E. Pawson & T. Brooking (Eds.), *Environmental Histories of New Zealand* (pp. 19–34). Auckland: Oxford University Press.

Anderson, A. (2003). *Prodigious Birds: Moas and Moa-Hunting in New Zealand*. Cambridge: Cambridge University Press.

Anderson, A. (2016). The Making of the Maori Middle Ages. *Journal of New Zealand Studies, 23*, 2.

Anderson, A., Binney, J., & Harris, A. (2012). *Tangata Whenua: An Illustrated History*. Wellington: Bridget Williams Books.

Anderson, A., Binney, J., & Harris, A. (2014). *Tangata Whenua: An Illustrated History*. Wellington: Bridget Williams Books.

Anderson, A., Binney, J., & Harris, A. (2015). *Tangata Whenua: A History*. Wellington: Bridget Williams Books.

Ataria, J., Mark-Shadbolt, M., Mead, A. T. P., Prime, K., Doherty, J., Waiwai, J., et al. (2018). Whakamanahia Te mātauranga o te Māori: Empowering Māori Knowledge to Support Aotearoa's Aquatic Biological Heritage. *New Zealand Journal of Marine and Freshwater Research, 52*(4), 467–486.

Attwood, B. (2001). Learning about the Truth: The Stolen Generations Narrative. *Telling Stories: Indigenous History and Memory in Australia and New Zealand*, 183–212.

Attwood, B. (2014). Law, History and Power: The British Treatment of Aboriginal Rights in Land in New South Wales. *The Journal of Imperial and Commonwealth History, 42*(1), 171–192.

Ballantyne, T. (2012). *Webs of Empire: Locating New Zealand's Colonial Past*. Wellington: Bridget Williams Books. Retrieved April 21, 2017, from https://books.google.co.nz/books?hl=en&lr=&id=o7ipY4SmDqoC&oi=fnd&pg=PR2&dq=related:rMTvJd8z9nQJ:scholar.google.com/&ots=9MSbl8yiyI&sig=vt6yqzedN4BM9d9ozDdAX8pKrz8.

Ballara, A. (1998). *Iwi: The Dynamics of Māori Tribal Organisation from C.1769 to C.1945*. Wellington: Victoria University Press.

Bambridge, T. (2013). *The Rahui: Legal Pluralism, Environment, and Land and Marine Tenure in Polynesia*. Canberra: ANUE Press.

Ban, C. (1998). Wetland Archaeological Sites in Aotearoa (New Zealand) Prehistory. *Hidden Dimensions: The Cultural Significance of Wetland Archaeology, 1*, 47.

Barber, L. (1978). *The View from Pirongia: The History of Waipa County.* Te Awamutu: Richards Publishing in association with Waipa County Council.

Barrett, G. (2012). *Wai 898 A109 Oral and Traditional History Volume Ngāti Tamainupō, Kōtara and Te Huaki* (Waitangi Tribunal). Wellington: Crown Forestry Rental Trust. Retrieved May 26, 2020, from https://forms.justice. govt.nz/search/Documents/WT/wt_DOC_42325213/Wai%20 898%2C%20A109.pdf.

Bassett, K. N., Gordon, H. W., Nobes, D. C., & Jacomb, C. (2004). Gardening at the Edge: Documenting the Limits of Tropical Polynesian Kumara Horticulture in Southern New Zealand. *Geoarchaeology, 19*(3), 185–218.

Belgrave, M. (2017). *Dancing with the King: The Rise and Fall of the King Country, 1864–1885.* Auckland University Press.

Belgrave, M., Belgrave, D., Proctor, J., Joy, M., Togher, S., Young, G., et al. (2011). *Te Rohe Pōtae Environmental and Wāhi Tahu Report (A Report Commissioned by the Crown Forestry Rental Trust for the Waitangi Tribunal's Te Rohe Pōtae District Inquiry No. A/76).* Wellington: Crown Forestry Rental Trust.

Belich, J. (1996). *Making Peoples: A History of the New Zealanders, from Polynesian Settlement to the End of the Nineteenth Century.* Auckland: Penguin Press.

Belich, J. (2009). *Replenishing the Earth: The Settler Revolution and the Rise of the Angloworld.* Oxford: Oxford University Press.

Belich, J. (2015). *The New Zealand Wars and the Victorian Interpretation of Racial Conflict.* Auckland: Auckland University Press.

Best, E. (1904). Notes on the Custom of Rahui. Its Application and Manipulation, as Also Its Supposed Powers, Its Rites, Invocations and Superstitions. *The Journal of the Polynesian Society, 13* 2(50), 83–88.

Best, E. (1930). Maori Agriculture. Cultivated Food-Plants of the Maori and Native Methods of Agriculture. *The Journal of the Polynesian Society, 39* 4(156), 346–380.

Best, E. (1931). Maori Agriculture. Cultivated Food-Plants of the Maori and Native Methods of Agriculture. (Continued). *The Journal of the Polynesian Society, 40* 1(157), 1–22.

Binney, J. (2001). The Native Land Court and the Maori Communities. In J. Binney, J. Bassett, & E. Olssen (Eds.), *The Shaping of History: Essays from* the *New Zealand Journal of History* (p. 143). Wellington: Bridget Williams Books.

Boast, R. (2008). *Buying the Land, Selling the Land: Governments and Māori Land in the North Island 1865–1921*. Wellington: Victoria University Press.

Boast, R. P. (2017). The Native Land Court and the Writing of New Zealand History. *Law & History, 4*, 145.

Boone, C. G., & Buckley, G. L. (2017). Historical Approaches to Environmental Justice. In R. Holifield, J. Chakraborty, & G. Walker (Eds.), *The Routledge Handbook of Environmental Justice* (pp. 222–230). New York and London: Taylor & Francis.

Borell, M., & Joseph, R. (2012). *Ngāti Ti Apakura Te Iwi Ngāti Apakura Mana Motuake. Report for Ngāti Apakura Claimants and the Waitangi Tribunal. Wai 898 A/097. Prepared for the Ngāti Apakura claimants for Te Rohe Pōtae Inquiry (Wai 898) and Commissioned by the Crown Forestry Rental Trust* (No. Wai 898 A/097). Wellington: Crown Forestry Rental Trust.

Boswijk, G., & Johns, D. (2018). Assessing the Potential to Calendar Date Māori Waka (Canoes) Using Dendrochronology. *Journal of Archaeological Science: Reports, 17*, 442–448.

Boulton, L. (2007). *Hapu and Iwi Land Transactions with the Crown and Europeans in Te Rohe Potae Inquiry District, c.1840–1865* (Waitangi Tribunal No. A70). Wellington: Waitangi Tribunal.

Brookes, B. (2016). *A History of New Zealand Women*. Wellington: Bridget Williams Books.

Buck, P. (1950). *The Coming of the Māori*. Wellington: Whitcombe & Tombs.

Cassels, R. (1972a). Locational Analysis of Prehistoric Settlement in New Zealand. *The Australian Journal of Anthropology, 8*(3), 212.

Cassels, R. (1972b). Human Ecology in the Prehistoric Waikato. *The Journal of the Polynesian Society, 81*(2), 196–247.

Chan, K., & Fan, C. (2018). *The Principle of Contra Proferentem and the Interpretation of Arbitration Agreements* (SSRN Scholarly Paper No. ID 3357912). Rochester, NY: Social Science Research Network.

Collins, A., Turner, K., & Te Huia, M. K.-H. (2012). *Wai 898 A94 Te Kurutao a Maahanga Te Pū o te Tao Te Pū Kotahitanga Oral and Traditional History Volume of Ngāti Maahanga* (Waitangi Tribunal). Wellington: Crown Forestry Rental Trust. Retrieved April 25, 2020, from https://forms.justice.govt.nz/search/Documents/WT/wt_DOC_42292535/Wai%20898%2C%20 A094.pdf.

Cowan, J. (1901). The Romance of the Rohepotae. *New Zealand Illustrated Magazine, IV*, 32.

Cromarty, P., & Scott, D. A. (1995). *A Directory of Wetlands in New Zealand*. Wellington: Department of Conservation.

Cummins, P., Ward, C., & Museum, K. (2004). *A History of Kawhia & Its District*. Kawhia, NZ: Kawhia Museum.

Cunningham, M. (2014). *The Environmental Management of the Waipa River and Its Tributaries. Case-Study Commissioned by the Waitangi Tribunal for Te Rohe Potae district inquiry (Wai 898)* (District Inquiry Research Report No. A150 (Wai 868_). Wellington: Waitangi Tribunal.

Curley, A. (2019). "Our Winters' Rights": Challenging Colonial Water Laws. *Global Environmental Politics, 19*(3), 57–76.

Davidson, J., Findlater, A., Fyfe, R., MacDonald, J., & Marshall, B. (Eds.). (2011). Connections with Hawaiki: The Evidence of a Shell Tool from Wairau Bar, Marlborough, New Zealand. *Research Report, 2*(2), 10.

Dhillon, C. M. (2020). Indigenous Feminisms: Disturbing Colonialism in Environmental Science Partnerships. *Sociology of Race and Ethnicity*, 2332649220908608. https://doi.org/10.1177/2332649220908608.

Downes, T. W. (1918). Notes on Eels and Eel-weirs (Tuna and Pa-tuna). *Transactions and Proceedings of the New Zealand Institute, 50*, 296–316.

Durie, E. T. (1994). Custom Law: Address to the New Zealand Society for Legal and Social Philosophy. *Victoria University of Wellington Law Review, 24*, 325.

Ellison, S., Greenshill, A., Hamilton, M., Te Kanawa, M., & Rickard, J. (2012). *Wai 898 A99 Tainui Oral and Traditional Historical Report* (Waitangi Tribunal). Wellington: Crown Forestry Rental Trust. Retrieved May 1, 2020, from https://forms.justice.govt.nz/search/Documents/WT/wt_DOC_4231 7979/Wai%20898%2C%20A099.pdf.

Firth, R. (2012). *Primitive Economics of the New Zealand Maori (Routledge Revivals)*. London: Routledge.

Forster, M. E. (2012). *Hei Whenua Papatipu: Kaitiakitanga and the Politics of Enhancing the Mauri of Wetlands* (Thesis). Massey University. Retrieved May 4, 2020, from https://mro.massey.ac.nz/bitstream/handle/10179/3336/02_whole.pdf.

Francis, A. (2011). *The Rohe Potae Commerical Economy in the Mid-Nineteenth Centuryl, c. 1830–1886* (Report commissioned by the Waitangi Tribunal for Te Rohe Potae District Inquiry (Wai 898) No. Wai 898 A/26). Wellington: Waitangi Tribunal.

Furey, L. (n.d.). *Maori Gardening: An Archaeological Perspective.* Wellington: Science & Technical Publishing Department of Conservation.

Grammond, S. (1994). Aboriginal Treaties and Canadian Law. *Queen's Law Journal, 20*, 57.

Grimshaw, P. (2002). Interracial Marriages and Colonial Regimes in Victoria and Aotearoa/New Zealand. *Frontiers: A Journal of Women Studies, 23*(3), 12–28.

Gumbley, W., Higham, T. F. G., & Low, D. J. (2004). Prehistoric Horticultural Adaptation of Soils in the Middle Waikato Basin: Review and Evidence from S14/201 and S14/185, Hamilton. *New Zealand Journal of Archaeology, 25*(2003), 5–30.

Haar, J. (2009). Entrepreneurship and Maori Cultural Values: Using 'Whanaungatanga' to Understanding Maori Business. *NZJABR, 7*(1), 16.

Hammer, G. E. J. (1991). *A Pioneer Missionary: Raglan to Mokau 1844–1880: Cort Henry Schnackenberg.* Wesley Historical Society (New Zealand): Auckland.

Hargreaves, R. P. (1959). The Maori Agriculture of the Auckland Province in the Mid-Nineteenth Century. *The Journal of the Polynesian Society, 68*(2), 61–79.

Harmsworth, G., Awatere, S., & Robb, M. (2016). Indigenous Māori Values and Perspectives to inform Freshwater Management in Aotearoa-New Zealand. *Ecology and Society, 21*(4).

Harris, S., & Harper, B. (2011). A Method for Tribal Environmental Justice Analysis. *Environmental Justice, 4*(4), 231–237.

Healy, S., Huygens, I., & Murphy, T. (2012). *Ngapuhi Speaks: He Whakaputanga and Te Tiriti o Waitangi: Independent Report on Ngapuhi Nui Tonu Claim.* Te Kawariki & Network Waitangi Whangarei.

Hikuroa, D. (2016). Mātauranga Māori—The ūkaipō of Knowledge in New Zealand. *Journal of the Royal Society of New Zealand.*

Hogg, A., Gumbley, W., Boswijk, G., Petchey, F., Southon, J., Anderson, A., et al. (2017). The First Accurate and Precise Calendar Dating of New Zealand Māori Pā, using Otāhau Pā as a Case Study. *Journal of Archaeological Science: Reports, 12*, 124–133.

Hooks, G., & Smith, C. L. (2004). The Treadmill of Destruction: National Sacrifice Areas and Native Americans. *American Sociological Review, 69*(4), 558–575.

Horrocks, M., Smith, I. W., Nichol, S. L., & Wallace, R. (2008). Sediment, Soil and Plant Microfossil Analysis of Maori Gardens at Anaura Bay, Eastern North Island, New Zealand: Comparison with Descriptions made in 1769 by Captain Cook's Expedition. *Journal of Archaeological Science, 35*(9), 2446–2464.

Internal Affairs. (Undated). Internal Affairs Department of Wildlife Service, "Lake Ngaroto: Inventory and Management Plan", Draft, Undated, AANS W3832 18, 30/1/21, Archives New Zealand, Wellington. Unpublished.

Irwin, G. (2013). Wetland Archaeology and the Study of Late Māori Settlement Patterns and Social Organisation in Northern New Zealand. *The Journal of the Polynesian Society, 4*(122), 311–332.

Jackson, M. (1993). Land Loss and the Treaty of Waitangi. *Te Ao Märama: Regaining Aotearoa. Mäori Writers Speak Out, 2.*

Jones, C. (2016). *New Treaty, New Tradition: Reconciling New Zealand and Maori Law.* Toronto: University of British Columbia. Retrieved June 12, 2019, from https://books.google.co.nz/books?hl=en&lr=&id=DSLCDAAAQBAJ &oi=fnd&pg=PT5&dq=Jones+2016+New+Treaty&ots=09dWY_ fMZ0&sig=vEDmsW4b2_KETAJpQfJWLcrDjRg#v=onepage&q=Jo nes%202016%20New%20Treaty&f=false.

Kennedy, N. C. K. (2017). *Kaitiakitanga o te Taiao-Reconciling Legislative Provisions and Outcomes for Māori* (Thesis). University of Waikato. Retrieved May 4, 2019, from https://researchcommons.waikato.ac.nz/handle/10 289/11121.

Leach, H. (2005). Gardens without Weeds? Pre-European Maori Gardens and Inadvertent Introductions. *New Zealand Journal of Botany, 43*(1), 271–284.

Lord, C. P., & Shutkin, W. A. (1994). Environmental Justice and the Use of History. *Boston College Environmental Affairs Law Review, 22*(1), 1–26.

Lorrey, A. M., Boswijk, G., Hogg, A., Palmer, J. G., Turney, C. S. M., Fowler, A. M., et al. (2018). The Scientific Value and Potential of New Zealand Swamp Kauri. *Quaternary Science Reviews, 183*, 124–139.

Luiten, J. (2011). *Local Government in Te Rohe Potae (A Report Commissioned by the Waitangi Tribunal for the Te Rohe Potae Casebook Research Program).* Wellington: Waitangi Tribunal.

Marsden, M. (2003). *The Woven Universe: Selected Writings of Rev. Māori Marsden.* Otaki: Estate of Rev. Māori Marsden.

Matthews, P. (1985). Nga taro o Aotearoa. *The Journal of the Polynesian Society, 94*(3), 253–272.

McGregor, D. (2014). Traditional Knowledge and Water Governance: The Ethic of Responsibility. *AlterNative: An International Journal of Indigenous Peoples, 10*(5), 493–507.

Mead, H. M. (2016). *Tikanga Maori (Revised Edition): Living by Maori Values.* Wellington: Huia Publishers.

New Zealand Government. (1865). Proclamation of Native Lands Under the New Zealand Settlements Act. *Daily Southern Cross*, p. 5.

Ngata, S. A. T. (2004). *Ngā Mōteatea*. Auckland: Auckland University Press.

Lord Normanby. (1908). Lord Normanby to Captain Hobson, R.N. Downing Street, 15 August 1838. In R. McNab (Ed.), *Historical Records of New Zealand, Vol. I*. Wellington: John Mackey. Retrieved March 19, 2020, from http://nzetc.victoria.ac.nz/tm/scholarly/tei-McN01Hist-t1-b10-d135.html.

O'Malley, V. (2013). *The Meeting Place: Maori and Pakeha Encounters, 1642–1840*. Auckland: Auckland University Press.

O'Malley, V. (2016). *The Great War for New Zealand: Waikato 1800–2000*. Wellington: Bridget Williams Books.

O'Malley, V. (2017). *He Whakaputanga: The Declaration of Independence, 1835*. Wellington: Bridget Williams Books.

Orange, C. (2015). *The Treaty of Waitangi*. Wellington: Bridget Williams Books.

Parsons, M., & Fisher, K. (2020). Indigenous Peoples and Transformations in Freshwater Governance and Management. *Current Opinion in Environmental Sustainability*. https://doi.org/10.1016/j.cosust.2020.03.006.

Parsons, M., & Nalau, J. (2016). Historical Analogies as Tools in Understanding Transformation. *Global Environmental Change, 38*, 82–96.

Parsons, M., Nalau, J., Fisher, K., & Brown, C. (2019). Disrupting Path Dependency: Making Room for Indigenous Knowledge in River Management. *Global Environmental Change, 56*, 95–113.

Parsonson, A. (2017). *The Fate of Maori Land Rights in Early Colonial New Zealand: The Limits of the Treaty of Waitangi and the Doctrine of Aboriginal Title. Law, History, Colonialism*. Manchester University Press. Retrieved March 19, 2020, from https://www.manchesterhive.com/view/97815261 19704/9781526119704.00021.xml.

Paterson, L. (2010). Hawhekaihe: Maori Voices on the Position of "Half-castes" Within Maori Society. *The Journal of New Zealand Studies, 9*. https://doi.org/10.26686/jnzs.v0i9.121.

Pauling, C., & Ataria, J. (2010). *Tiaki Para: A Study of Ngāi Tahu Values and Issues Regarding Waste*. Lincoln: Manaaki Whenua Press.

Petrie, H. (2013). *Chiefs of Industry: Maori Tribal Enterprise in Early Colonial New Zealand*. Auckland: Auckland University Press.

Pick, R. D. (1967). An Island Occupation Site on Lake Ngaroto on Pierce's Farm at the Northern End of the Lake. *The Journal of the Te Awamutu Historical Society, 2*(1), 19–21.

Pick, R. D. (1968). Waikato Swamp and Island Pa. *New Zealand Archaeological Association Newsletter, 11*(1), 30–35.

Pond, W. (1997). *The Land with All Woods and Water, Rangahaua Whanui National Theme U* (p. 123). Wellington: Legislation Direct.

Pool, I. (2015). Maori Demography and the Economy to 1840. In *Colonization and Development in New Zealand between 1769 and 1900* (pp. 151–176). Springer.

Robb, M. J. G. (2014). *When Two Worlds Collide: Mātauranga Māori, Science and Health of the Toreparu Wetland* (Thesis). University of Waikato. Retrieved May 4, 2019, from https://researchcommons.waikato.ac.nz/handle/10289/8776.

Salmond, A. (2017). *Tears of Rangi: Experiments Across Worlds*. Auckland: Auckland University Press.

Salmond, A. (2018). *Two Worlds: First Meetings Between Maori and Europeans 1642–1772*. Auckland: Penguin Group New Zealand, Limited.

Schnackenberg, E. H. (1935). *The Pohutukawas of Kawhia: Tales, Traditions & Legends Relating to Kawhia's Famous Christmas Trees*. Kawhia: Kawhia Settler.

Shawcross, K. (1967). Fern-Root, and the Total Scheme of 18th Century Maori Food Production in Agricultural Areas. *The Journal of the Polynesian Society, 76*(3), 330–352.

Shawcross, W. (1968). The Ngaroto Site. *New Zealand Archaeological Association Newsletter, 11*(1), 2–29.

Smith, S. P. (2011). *Hawaiki: The Original Home of the Maori: With a Sketch of Polynesian History*. Christchurch: Cambridge University Press.

Stevens, K., & Wanhalla, A. (2017). Intimate Relations: Kinship and the Economics of Shore Whaling in Southern New Zealand, 1820–1860. *The Journal of Pacific History, 52*(2), 135–155.

Stirling, B. (2016). *From Busby to Bledisloe: A History of the Waitangi Lands. A Report Commissioned by the Waitangi Marae Trustees and the James Henare Maori Research Centre* (Waitangi Tribunal No. Wai 1040 W5). Unpublished: Waitangi Marae Trustess and the James Henare Maori Research Centre.

Stokes, E. (1988). *Mokau: Māori Cultural and Historical Perspectives*. Hamilton: University of Waikato.

Stokes, E. (2013). Contesting Resources: Māori, Pākehā and a Tenurial Revolution. In E. Pawson, & T. Brooking (Eds.), *Making a New Land: Environmental Histories of New Zealand* (pp. 52–69). University of Otago Press: Dunedin.

Suter, B. (2014). *The Contra Proferentem Rule in the Reports of the Waitangi Tribunal*. Retrieved March 19, 2020, from http://researcharchive.vuw.ac.nz/handle/10063/4404.

Tāne, W. (2017). *Cultural Impact Assessment: An Assessment of Cultural Impacts of the Proposed Happy Valley Milk Ltd Dairy Factory on Redlands Road, Otorohanga, July 2017. Report commissioned by Nehenehenui Regional Management Committee*. Ōtorohanga: Nehenehenui Regional Management Committee.

Taumoefolau, M. (1996). FROM * SAU 'ARIKI TO HAWAIKI. *The Journal of the Polynesian Society, 105*(4), 385–410.

Tauriki, M., Ngaia, T. I., Roa, T., Maniapoto-Anderson, R., Barrett, A., Douglas, T., et al. (2012). *Ngāti Maniapoto Mana Motuhake Report for Ngāti Maniapoto Claimants and the Waitangi Tribunal. A/110 (Waitangi Tribunal No. Wai 898 A/110)*. Hamilton: Crown Forestry Rental Trust.

Tawhai, V., & Gray-Sharp, K. (2011). *Always Speaking: The Treaty of Waitangi and Public Policy*. Wellington: Huia Publishers.

Te Hurinui Jones, P. (1995). *Ngā iwi o Tainui: The Traditional History of the Tainui People: ngā koorero tuku iho a ngā tūpuna*. Auckland: Auckland University Press.

Thorne, F. (2011). *Te Maru-o-Hikairo: Oral and Traditional History Report of Ngati Hikairo (Wai 898 A98)*. Wellington: Waitangi Tribunal.

Unknown Author. (1864). The Native Rebellion. *New Zealand Herald*, 25 June, p. 5. Auckland.

Unknown Author. (1902). Art. Ii.—The Cultivation and Treatment of the Kumara by the Primitive Maoris. *Transactions and Proceedings of the Royal Society of New Zealand*, 42.

Vickery, J., & Hunter, L. M. (2016). Native Americans: Where In Environmental Justice Research? *Society & Natural Resources, 29*(1), 36–52.

Waitangi Tribunal. (1993). *The Waitangi Tribunal and the Motunui-Waitara Claim*. Wellington: Waitangi Tribunal.

Waitangi Tribunal. (2014). *He Whakaputanga me te Tiriti the Declaration and the Treaty: The Report on Stage 1 of the Te Paparahi o Te Raki Inquiry*. Wellington: Legislation Direct.

Waitangi Tribunal. (2018). *Te Mana Whatu Ahuru: Report on Te Rohe Pōtae Claims Pre-Publication Version Parts I and II*. Wellington: Unpublished.

Wakild, E. (2013). Environmental Justice, Environmentalism, and Environmental History in Twentieth-Century Latin America. *History Compass, 11*(2), 163–176. https://doi.org/10.1111/hic3.12027.

Wanhalla, A. (2015). *In/visible Sight: The Mixed-Descent Families of Southern New Zealand.* Wellington: Bridget Williams Books.

Wanhalla, A. (2017). Intimate Connections: Governing Cross-Cultural Intimacy on New Zealand's Colonial Frontier. *Law & History, 4,* 45.

Whyte, K. (2018). Settler Colonialism, Ecology, and Environmental Injustice. *Environment and Society, 9*(1), 125–144. https://doi.org/10.3167/ares.2018.090109.

Williams, D. V. (1999). *"Te Kooti Tango Whenua": The Native Land Court 1864–1909.* Wellington: Huia Publishers.

Winter, C. J. (2019). Does Time Colonise Intergenerational Environmental Justice Theory? *Environmental Politics,* 1–19. https://doi.org/10.108 0/09644016.2019.1569745.

Worthy, T. H., & Swabey, S. E. J. (2002). Avifaunal Changes Revealed in Quaternary Deposits Near Waitomo Caves, North Island, New Zealand. *Journal of the Royal Society of New Zealand, 32*(2), 293–325.

4

Remaking Muddy Blue Spaces: Histories of Human-Wetlands Interactions in the Waipā River and the Creation of Environmental Injustices

There are a diversity of wetlands within Aotearoa New Zealand and a variety of ways of classifying them. Throughout the nineteenth and early twentieth centuries, Europeans used the term "swamp" to refer to all Aotearoa's wetlands. However, technically the term swamp relates only to an area consisting of pooled water and some vegetation cover (Park 2002). Wetlands, from a western scientific perspective, are now broadly defined as "lands transitional between terrestrial and aquatic systems where an oversupply of water for all or part of the year results in distinct wetland communities" (Ausseil et al. 2015; Clarkson et al. 2013, p. 193). Scientists classify wetlands in different ways, including freshwater areas with emergent plants (palustrine), saltwater estuaries and lagoons (estuarine), and freshwater lakes (lacustrine); as well as coastal, interior and riverine; and swamps, bogs, and mires (Clarkson et al. 2013; Parsons 2019). Mātauranga Māori (Māori knowledge) classifies wetlands using different terms, including roto (lake) and moana (lacustrine), poharu (palustrine), manga (creek) and awa (riverine), and muriwai, wahapū and hāpua (estuarine). All the main types of wetlands are found in Aotearoa, in this chapter, we adopt the terms wetlands and repo (wetlands) to refer to the

© The Author(s) 2021
M. Parsons et al., *Decolonising Blue Spaces in the Anthropocene*, Palgrave Studies in Natural Resource Management, https://doi.org/10.1007/978-3-030-61071-5_4

diversity of wetland types within the Waipā catchment (Denyer and Robertson 2016; Phillips et al. 2002).

For many commentators, this loss of wetlands is simply a footnote in the broader story of colonial settlement and modernising development; an inevitable consequence of economic and social progress that all societies undertook. Yet Aotearoa's figures are amongst the greatest extent of wetland reduction in the developed world (Clarkson et al. 2013; Park 2001). Less than 10 per cent of its original (pre-human) wetlands remain (as of 2015 figures), with 16 per cent of wetlands retained in the South Island, while less than five per cent remains in the North Island (Ausseil et al. 2015; Clarkson et al. 2013). In the Waikato region, only 8.9 per cent of the wetlands remain (see Figs. 4.1 and 4.2). In comparison, since European colonisation commenced, wetland loss in other settler-colonial states range from a 50 per cent loss in Australia to 53 per cent in the United States of America, and between 65 to 80 per cent in Canada (Davidson 2014; Denyer and Robertson 2016; Mitsch and Gosselink 2000; Park 2002; Parsons 2019). During the same period, the wetlands of European nations were likewise being reduced. France recorded a small reduction (10 per cent reduction). Whereas Netherlands (famous for its extensively re-engineered waterways and dykes) and Britain (where drainage works began in East Anglia from the sixteenth century) both registered a decrease of 60 per cent. In comparison, since European colonisation, the United States has lost 53 per cent of its wetlands, Canada between 65–80 per cent, and Australia 50 per cent (Davidson 2014; Denyer and Robertson 2016; Mitsch and Gosselink 2000; Park 2002). Thus, while many countries drained wetlands and converted them to grasslands and urban areas, the amount and speed of wetland loss in Aotearoa were particularly pronounced in the nineteenth and twentieth century.

In the present-day, increased concerns are being raised about that the state of rivers, including issues of pollution, degradation, and water scarcity, yet wetlands seldom feature in discussions of river health (Azarnivand et al. 2017; Chetty and Pillay 2019; Flint et al. 2017; Hemming et al. 2017; Kansal 2018; O'Donnell and Macpherson 2019). Indeed, a wealth of scholarship examines the plethora of freshwater issues (including river management, drinking water, water allocation) facing Indigenous

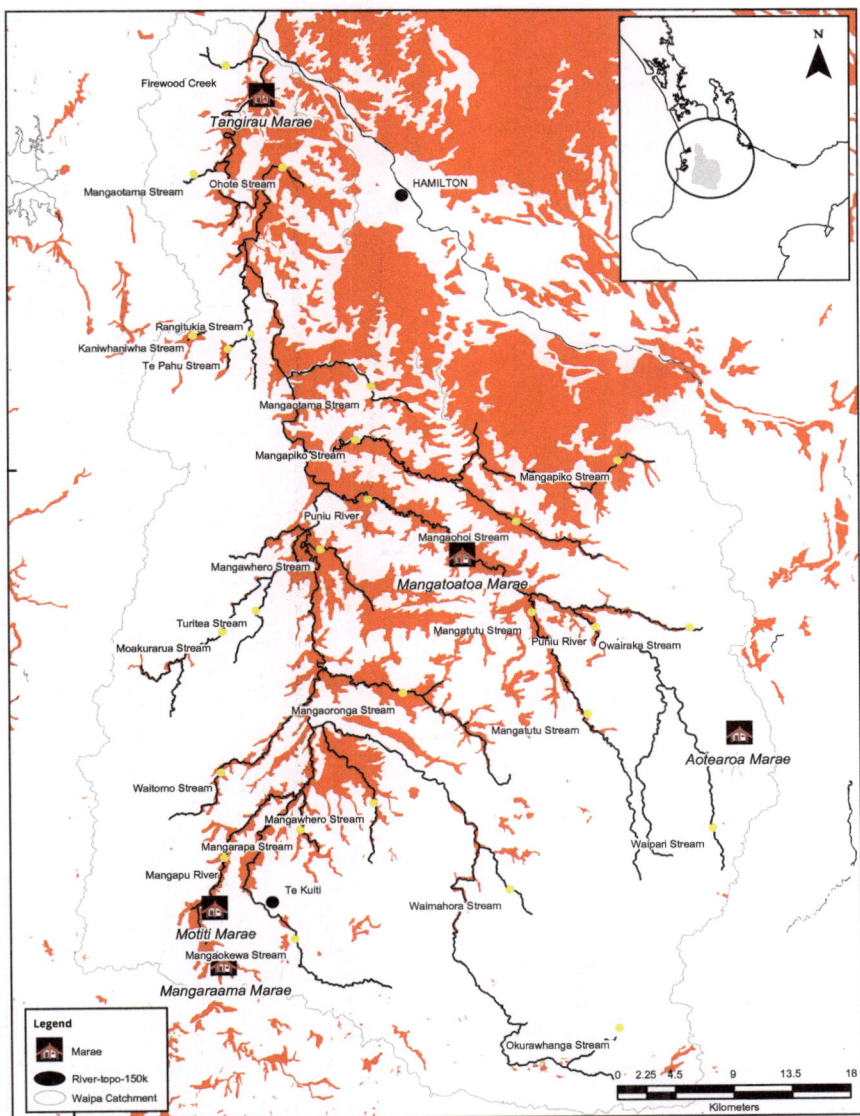

Fig. 4.1   Map of pre-human wetlands in the Waipā catchment

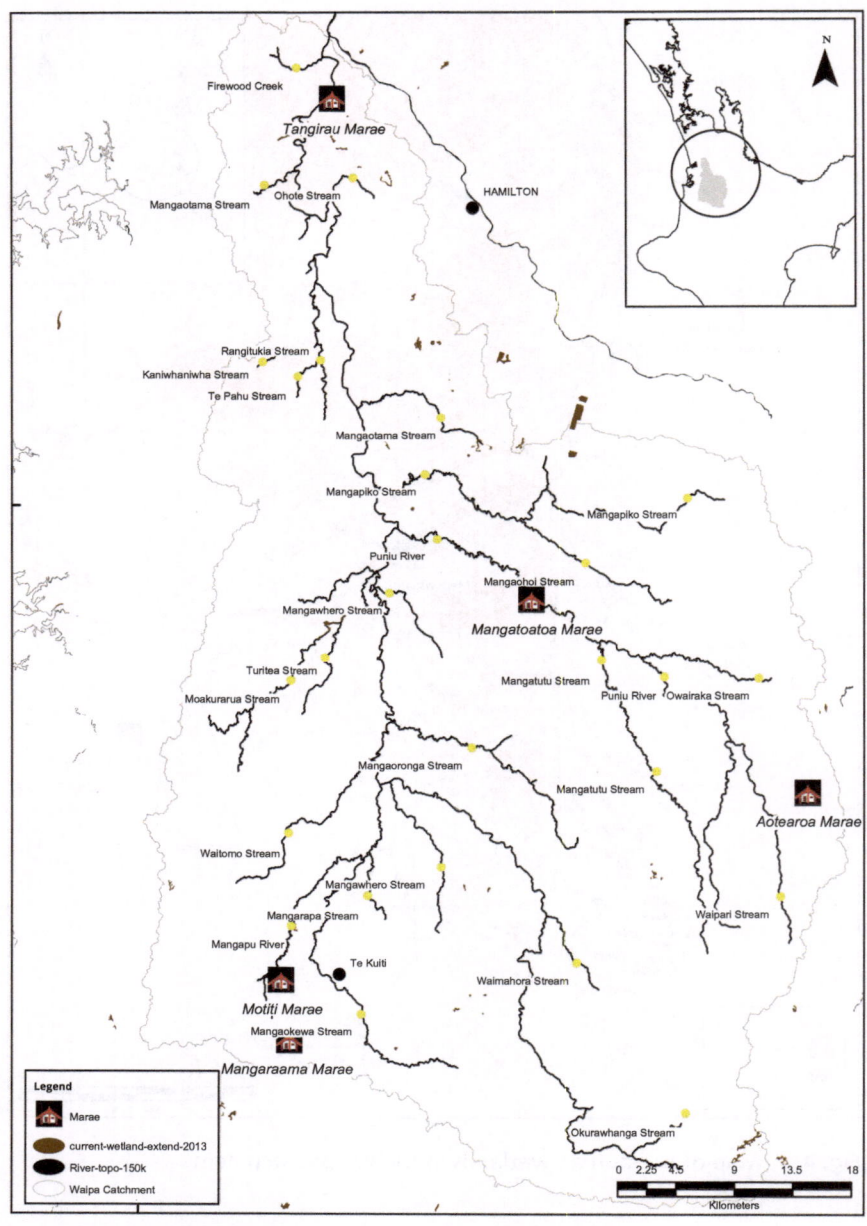

Fig. 4.2 Map of wetlands in Waipā catchment in 2016

peoples; however, wetlands receive only cursory mention (Bradford et al. 2016; Durette et al. 2009; Hanrahan 2003; Hemming et al. 2017; S. Jackson 2018; Morris and Ruru 2010; Patrick 2011; Poelina et al. 2019). The lack of explicit focus on wetlands is reflective, we suggest, of the privileging of settler-colonial histories, knowledge, values, and ways of conceptualising freshwater geographies. In this chapter, we document how the physical and discursive removal of wetlands, from waterscapes and landscapes, were manifestations of settler-colonial domination. We suggest the need to re-focus (decolonise) freshwater histories to encompass wetlands as an integrated and essential part of historical (and future) healthy and productive waterscapes and landscapes. Accordingly, in this chapter, we challenge the collective amnesia about wetlands in accounts of academic and practitioner dialogue about river health and management and through a historical case study demonstrate how wetland reduction (erasure) was a constitutive part of the mechanisms of settler-colonial domination. We argue that wetland loss was (and is still) an environmental injustice that had specific implications for Indigenous peoples due to their material, socio-cultural, and spiritual connections.

The incremental actions taken by individuals and institutions over decades beginning with the invasion of the Waikato in 1863 and continuing into the first three decades of the twentieth century amounted to ecological transformations of Māori waterscapes (outlined in Chap. 3). Colonialism and capitalist exploitation were tightly bound together, with Māori livelihoods, natural resource management and land-tenure arrangements, and governance structures all radically disrupted by the introduction of new materials, biota, technologies, peoples, policies and institutional structures (Whyte 2018, p. 135). By seeking to establish ongoing environmental injustices for Māori communities (distributive, procedural, recognitional, and cosmological injustices). We focus on the context of settler colonialism within Aotearoa New Zealand and how wetland drainage works strategically undermined Māori resilience, which encompassed the health and wellbeing of Māori communities as well as their more-than-human relatives (including their rivers, wetlands, lakes, and biota). Countless relationships that link settler colonialism, environmental injustice, and wetland drainage (as acts of environmental violence). In the previous chapter, we explored how Māori principles of mauri (life force) and whakapapa (genealogy) were essential parts of how Māori conceptualised and interacted with their waterscapes and constitutive elements of Māori collective continuance across generations of occupation. In this chapter, we examine the disruption and destruction wrought by settler colonialism on Māori communities and waterscapes within the Waipā. One avenue by which settler-colonial violence committed environmental injustice against Māori communities was by strategically weakening Māori health and wellbeing and their collective continuance through the destruction of Māori relationships with their awa (rivers) and repo. At least two forms of environmental injustices (misrecognition and procedural) were a product of settler colonialism and the eco-violence that occurred as a consequence of colonisation. In particular, we highlight how specific government policies, decisions of the courts, and government departments and settler society's failure to recognise Māori relationships with, values attached to, and rights to use and practice rangatiratanga over wetlands, waterways, and lands (supposed guaranteed to Māori under Article Two of the Treaty of Waitangi) formed the basis of cumulative and ongoing environmental injustices.

The structure of the chapter is as follows; we first provide a brief of historical events that coincided with and in some instances directly feed into settler-led operations to drain the wetlands. Next, we explore white (Pākehā) settlers' perceptions of the wetlands of the Waipā River and neighbouring river systems as hazardous and unhealthy spaces. We locate settler conceptualisations of wetlands as hazards as part of broader settler-colonial narratives about what was considered socially, economically, and morally acceptable waterscapes and landscapes for settlers to live and work in (as opposed to Indigenous and non-Indigenous non-white peoples). We then move onto discussing the strategies that were employed to reduce and replace wetlands by the settler state as well as individual settlers; this includes the extensive use of poorly paid non-white workers (both Māori and non-Māori non-Pākehā) who undertook the physical labour of draining the wetlands. Lastly, we highlight how Māori communities (as individuals and members of whānau, hapū and iwi) sought to challenge settler-colonial domination through spontaneous protests, legal cases, and political negotiations to conserve wetlands.

Settler Imaginative Geographies of the Waipā: 1850s–1860s

In the decade before the invasion of the Waikato, colonial officials and settlers repeatedly proclaimed their desires to acquire the whole Waikato and Waipā River floodplains for Pākehā settlement; this aim conflict with the aims of the Kīngitanga movement (the King movement) and individual iwi to retain their rangatiratanga (authority) over their whenua (land) and retain their mode of life centred on their tikanga (laws) and mātauranga Māori (Māori knowledge). Indeed, settler imaginative geographies of the Waikato discursively dislodged the presence of Māori and their repo in the district even before the actions of the armed forces. Throughout the 1850s and early 1860s, colonial media and government officials' writings depicted the environs south of the growing Pākehā township of Auckland as a wild and fertile landscape, home to the rapidly depopulating and soon to be erased native population. The Waikato delta was, one journalist reported in 1859, dotted with the "remains of Māori villages" in a state of "fast hastening … decay", which imparted an "air of picturesque and rustic beauty" to the landscape that was in desperate need to Pākehā settlement (Unknown Author 1859). The Waikato region was imagined by Pākehā commentators to be lying in a state of unuse awaiting the "Anglo-Saxon race" to transform it into a thriving and prosperous region, despite the operations of extensive Māori horticultural operations.

A central part of the settler-colonial project involved the discursive erasure of Māori agricultural success and economic development in the Waikato before and after the 1863 invasion. With the coming of the Pākehā, Māori hapū and iwi within the Waikato and Waipā catchment quickly embraced new knowledges, technologies, plants and animals (Unknown Author 1846) and by the 1840s iwi and hapū were growing diverse cultivations, which included kumara, potatoes, taro, corn, wheat, peaches and apples, as well as other goods (flax and timber) for their use as well as to exchange and trade with other groups (see Fig. 4.3: Illustration of peach orchards along the Waipā River) (Unknown Author 1859, 1864b). In addition to food, hapū (sub-tribes) along the Waipā and

Fig. 4.3 The drawing shows houses at the village of Whatawhata on the banks of the Waipā River (located 12 kilometres west of the present-day city of Hamilton). On the left of the drawing, several houses are visibly located in front of peach trees in blossom. Figures are shown descending the hill, and a Māori man is in a waka (canoe) is going across the river. British soldier Joseph Osbertus Hamley made the drawing in 1864 (many of which were copied from his superior officer) during the British military invasion of the Waikato. (Source: Ref: E-047-q-013. Alexander Turnbull Library, Wellington, New Zealand)

Waikato rivers operated 50 flour mills, with individual hapū constructing their mills to grind their wheat into flour that was exported to settlers living in rapidly expanding urban areas in Aotearoa (Auckland, New Plymouth, Wellington) and Australia (Sydney and Melbourne) (Grey 1849; Hursthouse 1861; Morgan 1849; Unknown Author 1854, 1859). As a consequence of the success of Māori agricultural and economic activities in the 1840s and 1850s, Pākehā journalists dub the Waikato-Waipā Delta "the Garden of New Zealand" (Unknown Author 1854) and the "Granary of North Island" (Hargreaves 1961, p. 72). Yet, rather than acknowledge Māori agricultural operations of Waikato Riverine Māori as evidence of Māori skills as farmers, settler-colonial government officials, travel writers, and journalists chose to ascribe agricultural development to the region's fertile soils and good climatic conditions (R. P. Hargreaves 1959; Hargreaves 1961).

By the 1850s settler-colonial discourse narrated that Māori communities living within the Waipā River catchment (and elsewhere) were making poor use of their lands. As one settler journalist wrote regarding the Waikato:

> We have been accused at home 'of coveting Naboth's vineyard,' when we have looked with [a] wistful eye on the extensive plains and rich alluvial flats retained by the natives, which they have never cultivated nor ever will! Surely there is no analogy! Naboth's vineyard[1] was cultivated, or it would not have been a vineyard at all; and I deny that any right-minded man covets the miserably cultivated clearings of the natives. (Unknown Author 1863)

Such narratives—that Māori did not make productive use of their lands and was doomed to extinction due to their racial defects—were used to justify government actions to dispossess Māori of their whenua (through military, legal, and economic measures) (New Zealand Parliament 1858). The writings of one journalist from September 1863 offers insights into how settlers perceived the Waikato-Waipā delta and the ways in which the discursive erasure of Māori economic activities and physical presence in the district served to justify settler state interventions to dispossess Māori:

> The most painful feeling is produced by the sight of rich plains of immense extent, capable of profitable cultivation, with comparatively little trouble and expense, lying waste and unoccupied, while thousands of ready hearts and hands in the old country [Britain] are wanting the means of subsistence. (Unknown Author 1863)

He argued that mass migration of "thousands, and tens of thousands, of industrious and necessitous [English] country-men into the fertile

---

[1] Naboth's vineyard is a biblical reference to the Old Testament (Book of Kings), the bible story tells of Naboth was a citizen of Jezreel who was executed by the Queen so that her husband could take possession of Naboth's vineyard.

plains of the Waikato" would ensure that the remnant Māori population would soon be "swamp[ed]" and the lands "rapidly brought into cultivation" (Unknown Author 1863). Such narratives, which linked together race-based theories (that Māori and other Indigenous peoples were "doomed" to extinction) with the socio-political and economic agendas of settler-colonial capitalism, positioned the land as lying unused awaiting the arrival of hardworking white settlers (specifically white men) to prune, plant, and develop the land, waters, and resources. Such settler-colonial stories of undeveloped landscapes, told and re-told in settler societies (including Aotearoa Australia, Canada, the United States), were notable for what they were missing, with the presence of Indigenous inhabitants often only mentioned in regards to their former presence, their depopulation, and failures (be it to cope with civilisation, to manage and develop resources) (Edmundson 2019; Ellinghaus 2006; Kelm 2005; McGregor 1997; Wolfe 2006).

Colonial media and government officials' writings increasingly imagined the Waikato lowlands as a future agricultural arcadia for settlers (not Māori), a so-called "Britain of the South", with no room in these accounts for mention of wetlands (Hursthouse 1861). Prior to the invasion published or unpublished records of settlers, missionaries, or colonial officials did not prominently (if at all) make mentions of wetlands (peat, mire, swamp, bog) (Hammer 1991; Howe 1970; Morgan 1862, 1864; Unknown Author 1859, 1867). Sometimes Pākehā travellers recounted how during their journeys along the Waikato and Waipā rivers how their canoes (waka supplied by their Māori guides) became stuck in muddy areas and they (following the lead of their Māori guides) were forced to wade through water and mud (Unknown Author 1859, 1862; Von Hochstetter 1867). However, these were brief interludes to narratives that focused on the ample natural resources that awaited Pākehā occupation and the seemingly limitless potential for agricultural development. As one visitor wrote the land around Te Awamutu reminded him more of "England that any part of the colony I have seen: the level plains and gently sloping rises, … add beauty to the would, if all under cultivation, be a second Leicestershire" (Unknown Author 1864b). Such descriptions were often tantamount to wishful (one cannot say hopeful) imaginative geographies. They were (and are) a mode of envisioning landscapes and waterscapes that underpinned the entire history of settler-colonial projects (S. Dench 2011).

The representations of the Waikato and Waipā floodplains as being unoccupied and unproductively used land epitomises the settler-colonial gaze. It was (and is) a vision, on the one hand, that dismissed the legitimacy of Māori occupation, livelihoods, and way of life (see Fig. 4.3). On the other hand, they saw an entire settler-colonial body politic to come, which included the establishment of townships, infrastructure, and institutional arrangements, and everyday practices associated with colonial governance and the built environment. Thus, a settler-colonial polity to come was being imagined and projected onto the visual field of the Waikato in 1863–1864 as warfare in the Waikato was still ongoing. The map of Queenstown (later renamed Newcastle before reverting once again to its original Māori name of Ngāruawāhia) attests to how the settler gaze first erased. Indigenous wetlands, places, names, and bodies on paper even before the physical process of removal was completed (via the military invasion, confiscation of Māori land, drainage of wetlands, and construction of new township).

The surveying, naming, and roading layout of the township of Ngāruawāhia, located where the Waipā River joins the Waikato River, is a case in point. The location was (and remains still) the centre of Kīngitanga movement and is the traditional lands (rōhe) of Waikato-Tainui iwi (see Chap. 3, Fig. 3.1: Map of the location of hapū and iwi). During the invasion, the site was used as a military encampment, and later the area was carefully laid out by surveyors in the shape of the Union Jack (see Fig. 4.4: Map of Queenstown). The inscription one of the most obvious symbols of British sovereignty on an area of profound cultural, political, and spiritual significance for Waikato-Tainui Māori and other groups affiliated with the Kīngitanga movement was an assertion of colonial power. It was part of wider efforts to erase Māori history, values, and rights, and more specifically to erase Waikato-Tainui rangatiratanga (which breached article two of the Treaty/Te Tiriti). The roads were laid out in neat, straight lines and given names that reflected British (in keeping with Anglo-Saxonism discourse) colonial power. On the map there is no sign of wetlands nor Māori (past, present, or future) presence; just as Māori was rendered out of place in the settler-colonial order of things, so

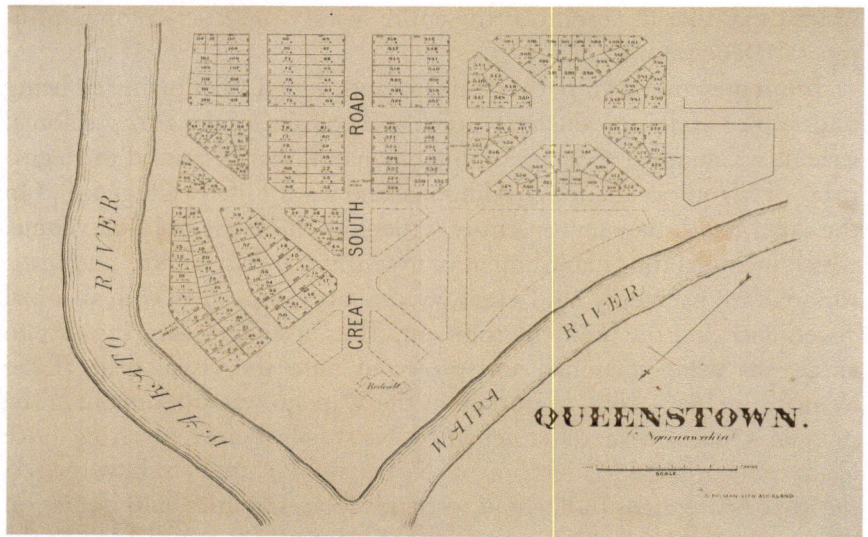

Fig. 4.4 Map of Queenstown (Ngāruawāhia) showing allotments for sale in 1865. (Source: Auckland Libraries Heritage Collections NZ Map 4498-22)

too were wetlands. Yet, the straight, fixed, and well-defined lines on a map did not accord to the reality of life on the muddy, swampy, unstable, and porous ground of the newly confiscated lands. Just as the wetlands did not simply disappear because of colonial displeasure so too Māori presence within the area was not erased despite colonial violence and confiscation, both the continuation of mud and Māori created continued anxieties amongst Pākehā. Such maps both were a product of and reinforced settlers' imaginative geographies of the Waikato. Maps and surveys, as the works of historians' and geographers' demonstrates (Brealey 1995; Byrnes 2001; Cameron 2011; S. J. Dench 2018), were (and still are) modes of envisioning landscapes and waterscapes that formed a critical part of settler colonialism. While wetlands were absence from pre-invasion accounts of the plains, British military officials did note the presence of wetlands in the Waikato-Waipā Delta posed difficulties to soldiers and drew maps that indicated the presence of wetlands (see Figs. 4.5 and 4.6).

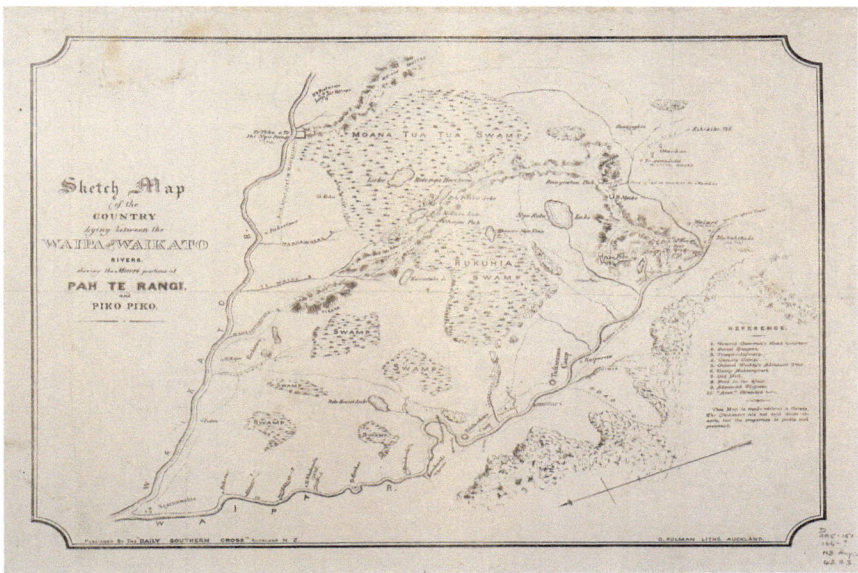

Fig. 4.5 Sketch Map of Waipā and Waikato Rivers showings some of the wetlands in the lower section of the Waipā River (note Ngāruawahia at the meeting point of the Waipā and Waikato Rivers). (Source: Auckland Libraries Heritage Collections NZ Map 4273)

Post-Invasion Realities: Life on/in the Wetlands

In the aftermath of military invasion of the Waikato (1864 onwards), the land confiscated from iwi was surveyed and subdivided into freehold blocks for lease or purchase by Pākehā settlers (either single or married men). And the extensive wetlands could no longer be overlooked as surveyors and settlers walked on and waded through the water and mud (Cowan 1928; Unknown Author 1865a, 1865b). British imperial soldiers were given the option of taking up a land grant (with the amount of land awarded based on rank and length of service); with three years of "good" service allowing rank and file soldiers (privates) to obtain 50 acres of land in the Waikato (Hamilton 1968; McLellan 2017). Approximately 60,000 hectares were subdivided into land in and around the military settlements of Alexandra (Pirongia), Kihikihi, Whatawhata, Cambridge,

Fig. 4.6 Showing details of Map 5 including part of Ruakuhia wetland (left), Lake Ngāroto (centre-left), Piko Piko as well as other Māori pā, and Otawhao mission is also recorded (centre-left). (Source: Auckland Libraries Heritage Collections NZ Map 4273)

and Kirikiriroa (Hamilton); other parcels of land was also sold to former soldiers. The newly surveyed land, however, did not always accord to the previous agrarian utopian descriptions of missionaries, government officials, and travel writers, and most soldiers walked off their land declaring too swampy (McLellan 2017). In her memoir, Bernice Monrath Johnstone, for instance, recounted how her father purchased their family's 50-acre farm, situated alongside the Mangapiko Stream between Te Awamutu and Pirongia, in the late 1860s from former soldier "who was eager to get away from the close vicinity to Maori camps, settlements, and the King Country, and glad to pass it onto someone more courageous than himself" (Johnstone and Roberts 2004, p. 9). She recalled that both she and her mother found serious fault with the farm due to its isolated position (with few Pākehā farms in the vicinity) and it was "more waste land than farmed land—acres of swampy bush country, acres of fern, and acres of scrub". The "sandy piece of land" was broken up by the "Mangapiko Stream [which] twisted and turned forming boundaries of narrow points and sharp bends ... [with] more than half the farm ... broken into by many swamps" (Johnstone and Roberts 2004, p. 14).

Many settlers similarly complained about their newly acquired lands with a litany of objections listed in letters to newspapers and politicians about the deficiencies of sections they obtained (by grant, lease or purchase) from the central government. In contrast to the earlier depictions of fertile and prosperous lands by travel writers and members of the media, Pākehā settlers found "swamp or ... springy flax flats", which were frequently impassable due to heavy rains (see Fig. 4.7) (Unknown Author

Fig. 4.7 Horse and cart journey across the wetlands of the Waikato-Waipā delta (circa the 1880s). The area is probably located of the property known as the Broadlands owned by Hubert Valie (along Waikato River rather than Waipā). Still, it does give a good indication of what the landscapes/waterscapes looked like before drainage work. (Source: Auckland Libraries Heritage Collections 2-V421)

1864a, 1864c, 1865b, 1865c). Some declared that they were being forced into "bastard system of farming" that was not "convenient nor healthy"; they narrated their efforts to farm on the swamp-filled grounds of the Waipā and Waikato rivers as alike to a battle against inevitable financial and physical ruin (Unknown Author 1865c). The task of digging drains and removing the wastelands of "black, cold swamps" was something that was often declared by experts and settlers alike both as tiresome and potentially deadly and undesirable work for people (specifically white men) to perform (Unknown Author 1887, 1909). One journalist declared only a special type of man, a "swamper", was capable of such work (see Fig. 4.8). Swampers, the writer declared, possessed the necessarily care-lessness with his health to be able to withstand the process of transform-ing the wetlands into farmlands. Daily activities of "walking through the damp rushes" in "stinking water", "cutting and lifting masses of dripping peat" were all supposedly hazardous to one's health. The men who worked to drain the wetlands were supposedly left physically drained of strength; their health eroded from working in the unhealthy (potentially malaria-filled and miasmatic) atmosphere that left them with "rheumatism and sciatica" (Unknown Author 1887). The smells of wetlands of decompos-ing vegetation, of sulphur (which indicates anaerobic processes), were not only off-putting for Pākehā but due to their sanitary and miasmatic understandings of disease-causation directly linked to ill-health and potentially death. Despite the discovery of bacteria and the emergence of bacteriological science in the mid-nineteenth century, most doctors and settlers in Aotearoa continued to believe that miasmas (bad airs) caused ill-health. Miasmas abounded in diverse and highly uncertain ways (through decomposing matter, foul air, strong odours, human and ani-mal waste, stagnant water, factories, particular environments) (J. Beattie 2008; Chiang 2008; Flikke 2016; Halliday 2001; Kiechle 2017; Parsons 2019; Worboys 1994). The stench of composting vegetation and animal matter, and peaty soils, all of which were found in abundance within the Waipā wetlands were read as persistent threats to one's health. Accordingly, settlers (or their descendants) increasingly sought to employ non-white people ("the Other" if we employ the theoretical concept from Edward Said's seminal work *Orientalism*) most commonly Māori but also some-times non-Māori non-British people (including Croatian and Chinese)

Fig. 4.8 Swampers digging a drainage canal through a wetland. (Source: Auckland Library Heritage Collections NZG-19100209-28-2)

to dig the drains and farm the swampy lands (Said 1978); as they (Other "races") showed an "aptitude for horticulture" (St. John 1873), was cheaper to employ, was unconcerned about the dangers of miasmas, and were supposed gravitated towards nomadic lifestyles (Johnstone and Roberts 2004; Jones 1996; McGovern 1986; Meyer 1996). Indeed, the canals, flood levees and roads that now define the Waipā floodplains were constructed on land confiscated and/or acquired from Māori and built by Māori labourers; yet, Māori also actively opposed and mounted protests against such works.

At the same time as Māori were being employed to drain wetlands to facilitate settler projects to transform swampy grounds into pastures, doctors blamed Māori ill-health (with mortality and morbidity rates far higher from Māori than Pākehā) on Māori unhealthy practices of living and working on or near wetlands (AJHR 1875, 1885a). These ideas, as we outline in our other works, were infused socioeconomic, medical, and race-based theories that justified Indigenous dispossession and drainage works. Such theories reinforced the longstanding prejudices that many settlers held about both Indigenous peoples and wetlands as being need to undesirable and an impediment to settler-colonial progress (Carlson 2010; Parsons and Nalau 2016; Parsons 2019) (Fig. 4.9).

The abundance and vagueness of miasmas (with no clear boundaries between airs, lands, waters, and bodies) meant that miasmatic thinking about disease-causation continued to influence settler-colonial understandings of wetlands throughout Aotearoa. In the Waikato, Pākehā doctors and government officials in the late nineteenth century sought to reassure the Pākehā public that their health and wellbeing was secure, despite the continued presence of large wetlands areas and neighbouring Māori communities within Rohe Pōtae (Unknown 1876). Race and environmental health were interwoven together, with Nevertheless, outbreaks of "typhoid fever and colonial fever" in the district in 1876 were reported by the *Waikato Times* newspaper as "due doubtless to the heavy rains of early summer, followed by great heat, generating miasma", with similar climatic conditions and fevers recorded in 1860 (Unknown 1876). Elsewhere in the country, high mortality rates were ascribed to building houses out of green timber, on undrained land, and climatic changes in the area. The omnipresent dangers of "damp vapour arising"

HOW SETTLERS IN THE KING COUNTRY GET ON WITHOUT ROADS: LAYING A "CORDUROY" TRACK OVER SWAMPY COUNTRY FOR THE PACKING OF SUP-PLIES INTO THE BACK-BLOCKS.   P A. Mussen, Photo.

Fig. 4.9 Two Māori workers and a Pākehā boy constructing a road in the King Country/Rohe Pōtae (circa 1910). (Source: Auckland Libraries Heritage Collection AWNS-19101124-6-5)

were deemed "highly prejudicial to the health" of residents (AJHR 1875, p. 20). In 1875, for instance, the population of the township of Napier in the Hawke's Bay (North Island) experienced an outbreak of a "fever" (incorrectly labelled malaria) which was blamed by officials on the "noisome emanations from the swamp" beside the town (Unknown Author 1878a). The "tepid swamps", it was reported, poisoned the "otherwise pure air" and ruined "health, retard[ed] settlement, and br[ought] hundreds to an early grave" (Unknown Author 1879). In total, the deaths of 140 people in Napier (out of a total population of 3000) were linked to the 1875 fever epidemic. In 1875, the New Zealand Parliament introduced the first legislation that specifically authorised wetland drainage. The Napier Swamp Nuisance Act enabled local government officials to "fill in" (meaning to drain, establish levees, and build up the soil) any parcel of land deemed to be a muddy watery "nuisance" without the consent of the landowners (New Zealand Parliament 1875). While the 1875 legislation did not mention a specific disease, the legislation and discussions surrounding it (in parliament and newspapers) highlighted Pakehā perceptions of wetlands as spaces of noxious airs, unruly waters, and unproductive flora and flora (Unknown Author 1879). The Napier experience similarly influenced perceptions of the wetlands of the Waikato-Waipā Delta and saw more calls from Pākehā for specific local and central government support for wetland drainage works. Wetland drainage was not only meant to provide Pākehā with economic security but also with significant health benefits through the eradication of dangerous disease-inducing miasmas (J. Beattie 2008; J. J. Beattie 2005).

Exotic tree plantings were, alongside digging of drainage canals, essential steps that settlers needed to take to eradicate the hazards posed by wetland living. In April 1876 a correspondent for the newspaper *Waikato Times* to praised the planting of "double rows of pinus insignus and eucalyptus" in Ngāruawahia (see Fig. 4.10). The exotic trees, the unnamed commentator wrote, were not only aesthetically pleasing but also vital preventive health aids to purify the air from dangerous odours (Unknown Author 1876a). The newspaper published follow up articles that promoted the planting of eucalyptus trees as to act as an "active purifier neutralising the miasma of the atmosphere and absorbing the superabundant moisture below the surface" (Unknown Author 1876b). These newspaper articles sought to encourage other Town Boards in Waikato to follow the example of Ngāruawahia and plant trees to improve the health. Women in the Waikato were likewise encouraged plant "the most odorous flowers" (Unknown Author 1875) including "lavender, musk, cherry laurel, clove, fennel, narcissus, heliotrope, hyacinth, and mignonette" that were supposed "endowed with health-preserving properties" that could remove miasmas from the air (Unknown Author 1873). Accordingly,

Fig. 4.10 View of Aikens Street in Ngāruawahia (1910) showing a row of mature eucalyptus trees in front of the houses on the right side of the road. (Source: Green & Colebrook (Firm). Aikens Street, Ngāruawahia, 1910—Photograph was taken by G & C Ltd. Ref: 1/2-000256-G. Alexander Turnbull Library, Wellington, New Zealand)

the cultivation of flowers was deemed critical work that served to be "delightful and humanising in itself, but one which in a way ... confers a positive benefit on society so great that it can hardly be overrated" especially in areas where there were much poisonous (miasma-filled) airs (such as in large towns and muddy swamps) (Unknown Author 1875). Every Pākehā woman, one author declared in 1873 (paralleling the sentiments of other writers), should surround her dwelling with "sweet-scented flowers and plants", and who ensured "her rooms fragrant with their essences, [was] an angel of health to her family" (Unknown Author 1873).

While Pākehā men were (supposedly even if others overlooked who actually undertook the labour) responsible for ensuring that lands were cleared of Indigenous biota, drained, and remade into neo-British farmlands, Pākehā women were charged with the tasks of ensuring that homes and families were kept healthy through domestic hygiene practices (cleaning, cooking) and planting sweet-smelling home gardens (Unknown Author 1873, 1875, 1885). Such individuals actions to transform spaces from hazardous to healthy were, thus, narrated as an essential part of raced and gendered discourses that circulated within Aotearoa and other settler-colonial societies that emphasised that the Indigenous landscapes, waterscapes, modes of life, dress codes, and even bodies were deficient and needed to be fundamentally changed. One journalist in 1878, during a boat journey up the Waipā River, described his impressions of the Māori and Pākehā who lived at the small village of Whatawhata in highly racialised and gender terms:

> Here ... the two races dwell together in brotherly and sisterly love. Maori and half-caste gamins and 'young colonials' mingle together on the sand of the riverbank and vie with each other in [the] exercise of sound lungs. There a wahine [Māori woman] squats on the extreme verge of the precipice with her pickaninny [sic] carried on her back in a blanket watching her other 'young barbarians all at play;' and just far enough off to preserve pakeha [sic] feminine dignity, is a settler's wife attired in something that looks like silk, and sheltered under a parasol. ... conveys a whole volume of testimony to its civilisation. (Unknown Author 1878b)

The diaries, letters, oral histories, and published accounts (in newspapers, memoirs, and magazines) of Pākehā men and women, who visited and lived in the floodplains of the region, often recounted feelings of fear and despair; with concerns about their new wetland-filled homes intermingling racial, social, medical, and economic anxieties. They wrote accounts—in diaries, letters, memoirs, reports to government, letters to editors—filled with their fears of miasmas, the imagined threats posed by Māori (their resistance, return, violence), the economic difficulties of land conversion, and the desperate need to progress the land and themselves onwards and forwards through acts of 'unwatering' the land and bringing civilisation (defined as "life" itself) to the stagnant lands and waters that surrounded the Waikato and Waipā Rivers (Henderson et al. 1994; Johnstone and Roberts 2004; Jones 1996; Kerry-Nicholls 1884; Savage 1847; Unknown Author 1878c, 1887; Westmacott 1977). As one Pākehā man wrote of the wetlands in 1887:

If I were to paint desolation, if I was to plant as dreary a picture as I could of lifeless gloomy colouring, I would depict the scene before me now. I am in the great swamp region of the Waikato, and before me stretches a wide expanse of stagnant water. (Unknown Author 1887)

Similarly, Margaret Macky, who grew up on a farm beside the Waipā River (near the township of Te Awamutu in the 1870s) recalled her childhood home as one: "with its lowland peat swamp terribly lonely and depressing" and "eerie ... with the stumps of bygone forests covered with ... rushes poking through the rain" (Henderson et al. 1994, p. 85) (See Fig. 4.11). However, she later recalled how the work of her parents and other "pioneers" transformed the "depressing" swamps into the "beautiful district is now ours" (Henderson et al. 1994, p. 85). Macky, like others, about employed the dominant settler-colonial narrative:

Fig. 4.11 Woman standing beside one of the peat lakes within the wetlands of Waipā. (Source: Auckland City Libraries Heritage Collections, 2-V416)

settlers arrived in the wilderness; "battled" against nature (Indigenous); and ultimately created order (modernity, capitalism) out of chaos (nature and subsistence-based livelihoods) and in doing so created economically productive communities (Cumberland 1941; Guthrie-Smith 1969; Levy 1970; Parsons and Nalau 2016; Thomson 1867). Such a narrative was not only linear but also transformative in effect and emphasised the radical reconfiguration of social and ecological communities.

Throughout these earlier accounts (by self-described "settlers" or "pioneers") of wetlands the overwhelming narrative employed by Pākehā settlers was one of detachment (rather than an attachment) to place and the ways in which the existing wetland environment was at odds to their understanding (and preference) of what constituted a healthy, productive, and secure lands and waterways to live, work and play in. Many individual settlers wrote about the enormity of the task they faced transforming "the dreary waste of rushes" into "anything beautiful". Beauty from the gaze of settlers (who often described themselves as "Anglo-Saxon") consisted of "verdant pasture" capable of bearing the "weight horses, cattle, and flocks of sheep" (Unknown Author 1887). Settlers declared that "even this dreary wilderness" of apparently useless wetlands could be remade by hardworking (Pākehā or Anglo-Saxon) settlers "into a scene of smiling pastures and spreading fields" through drainage and planting works (Unknown Author 1887) (see Fig. 4.12). As one Pākehā settler wrote in 1883 that while the "original state" of the environment was a "sombre and disagreeable impression" actions could and were being taken by individuals to remove supposedly unproductive Indigenous vegetation and plant grains and grasses to "impart a pleasing beauty to the landscape" reminiscent of the best of England (Barugil 1883). Such imaginative geographies provided the emergent settler-colonial state in Aotearoa as well as individual settlers' with the belief that their actions (of invasion, violence, occupation and radical changes) were all morally justifiable.

Fig. 4.12 Sketch drawing of wetland "Ta Ringamotu" in Te Rohe Potāe (King Country) 1888. (Source: Ref/A-045-003. Alexander Turnbull Library, Wellington, New Zealand)

Māori Engagements with Wetlands and the Settler-Colonial State

In marked contrast, Ngāti Maniapoto and the Kīngatanga allies they were hosting within their rōhe, in the upper and middle catchment of the Waipā River (Rohe Pōtae), held vastly different understandings of wetlands that centred on their tikanga (customary laws) and mātauranga Māori. Māori conceptualised themselves as the kaitiaki (guardians) of their whenua, awa, repo and sought to practice kaitiakitanga (environmental guardianship) despite the slow violence of settler-colonialism. Local hapū continued to be responsible (under the customary laws of tikanga) for practising kaitiakitanga as a way to ensure the health (hauora) of all who lived there; this included the critical need to maintain the life-force—mauri—and wairua—spiritual integrity—of both human and ecological communities both now and for future generations. In addition to employing mātauranga Māori (Māori knowledge), many Māori sought to engage with Western knowledge and practices and attempted to use the British legal system as a way to practice kaitiakitanga and ensure their modes of life, mauri, wairua and tikanga were maintained despite the onslaught of settler-colonial domination and ecological violence.

In the 1880s, as part of negotiations between the Crown and Ngāti Maniapoto about the proposed railway line, Ngāti Maniapoto rangatira (chief or high rank) sought to ensure specific agreements were reached to protect its wetlands and forests from damage. While both Māori and English versions of the Treaty of Waitangi included a provision (Article Two) that required the Crown to actively protect Māori land, forests, and other taonga (treasures), Ngāti Maniapoto leaders were well-aware by the 1880s that the Crown was not honouring its Treaty promises. Accordingly, its leaders attempted to implement a specific accord that would ensure that the construction of the railway line through Rohe Pōtae would not negatively impact their wetland taonga. Rangatira Hopa Te Rangianini, for instance, demanded reassurance from the Native Minister (John Ballance) in 1885 that the railway would not destroy his mahinga kai (food gathering sites). Ballance reported this back to parliament:

> He [Hopa Te Rangianini] owned a swamp, over which the railway would pass, and he obtained eels, which were his principal food in summer, from this swamp. He said he had heard that in England railways were taken over viaducts, and he asked that this might be done in this case, instead of filling up the swamp. (AJHR 1885b, p. 23)

Likewise, rangatira Aporo Taratutu argued that certain trees also needed to be conserved because of their importance as food sources. He informed government officials that only specific native trees should be logged and used in construction work; they were permitted to use matai trees for railway sleepers, but kahikatea trees needed to be retained "because in summer he used the berries for that tree for food" (AJHR 1885b, p. 23). Minister Ballance informed parliament that the government would be constructing viaducts (see Fig. 4.13: Train over the valley)

Fig. 4.13 Train crossing the Waiteti Viaduct, near Te Kuiti, circa 1890. Note the clearance of vegetation and the construction of roads and buildings below the viaduct, which indicates that the government's promises made to Māori to protect wetlands were completely disregarded. (Source: Auckland Libraries Heritage Collection, 4_1078)

in Te Rohe Pōtae, but for reasons of modern engineering and travel-time efficiency for rail journeys rather than in response to Māori demands.

Minister Balance dismissed iwi concerns about their so-called primitive food sources as nothing more than anxiety that would soon be erased by the money that would roll into the district alongside the trains.

> I agree[d] … that the watercourses should not be interfered with; but … Something has been said about eating berries from trees, and so on; but let me tell you that the money will come to the people through the construction of this railway will be worth all the berries in the world, and the eels, too. (AJHR 1885b, p. 24)

Ballance's comment highlighted the fundamental differences between Māori and Pākehā worldviews and modes of life. The dominant values of Pākehā settler society (for which Ballance was a keen illustration) drew on longstanding European post-Enlightening binary thinking that sought to divide human and non-human and culture and nature and preached the sectarian gospel of Victorian technological advancement and economic development. Ballance, like many other colonial officials in Aotearoa and other settler societies, valued the accumulation of goods (including land and money) above all else. The threats and acts of violence, committed by settlers against Indigenous and other non-European societies, were regularly downplayed as unavoidable consequences of the urgent necessities of 'progress', 'improvement' and 'development'.

Many members of Ngāti Maniapoto expressed their strong desires to maintain their modes of managing their taiao (environment), opposed the activities of the Native Land Court (NLC) to convert their whenua into European-style land titles, and actively challenged efforts to alter their relationships with their whenua, awa and repo. When the first surveys were being conducted for the railway line through Rohe Pōtae in 1884, for instance, a group of Ngāti Maniapoto wahine protested against the Crown's intrusion into their territory by removing all the survey pegs and throwing the surveyors' equipment into the Waipā River (Unknown Author 1884, p. 2). By the 1890s, the NLC and the implementation of Europeanised land tenure arrangements began to cause ongoing tensions amongst extended kin groups (including whānau, hapū and iwi of Ngāti Maniapoto) in the middle and upper stretches of the Waipā River catchment. One Māori wahine and her whānau, after they received individualised land titles (after going through the NLC process) to land at Kakepuku (located between Te Awamutu and Otorohanga and including Te Kawa wetlands), erected fences around their newly surveyed and titled properties. However, other members of Ngāti Maniapoto challenged their claims of exclusive individual Pākehā-style land ownership (rather than Māori collective land rights) and removed the fences. The fence-removers argued that they needed (and indeed possessed the right under Māori laws or tikanga) to access the land (irrespective of who held the title) as they needed to access the waterways to harvest tuna (eels). The owners, once again, chose to use the newly imported settler-colonial laws (rather than Māori laws or tikanga involving parties negotiating face-to-face through hui—meeting) to support their land rights and issued a trespasser notice against other members of their iwi (Wilkinson 1892).

By the 1900s, Māori were already an ethnic minority group in Aotearoa. Decades earlier, in 1840, when rangatira from around Aotearoa met, discussed, debated, and signed the Te Tiriti o Waitangi. In again in 1885 when leaders from Ngāti Maniapoto and other iwi met with and reached an agreement with the government (the NZ Crown) about the construction of the railway, Māori were still numerically dominant within Rohe Pōtae. However, by the early twentieth century the influx of Pākehā settlers, capital, largescale Crown land purchasing activities, the operations of the Native Land Court (NLC), and the extension of the countless institutional apparatuses of the settler state saw this demographical structure change in Rohe Pōtae; as had occurred earlier in other parts of Aotearoa. The world in which Ngāti Maniapoto tikanga, including their mātauranga, practices of kaitiakitanga and rangatiratanga, the cornerstones of life, were shifting on its foundations; if not a metaphorical earthquake, then a definite muddy of waters (to draw on an allusion to the wetlands we are discussing) for Māori communities as they sought to address, discuss, negotiate, engage with and clash over the ontological and epistemological implications of settler colonialism.

Government Responses

In response to Pākehā perceptions of wetlands as problematic, hazardous, and unproductive spaces, a series of acts of parliament were introduced which created and authorised government institutions to systemically drain the country's wetlands. The first legislation was Napier Swamp Nuisance Act (1875), which was followed by the Drainage Act in 1881 (New Zealand Parliament 1881) and the 1893 Land Drainage Act, which allowed for the creation of drainage schemes with government subsidies (New Zealand Parliament 1893). In the first decade of the twentieth century, a suite of different acts of parliament were introduced that established central government-led drainage schemes in the Rangitāiki Plains (Eastern Bay of Plenty) and Hauraki Plains (Waikato) (AJHR 1911, 1913; New Zealand Parliament 1908a, 1910) and local government drainage projects (New Zealand Parliament 1904, 1908b). The Land Drainage Act 1904 allowed for the establishment of local government bodies (drainage boards) specifically responsible for draining wetlands within their local areas. Each drainage board was to be made up of members who were elected by local ratepayers (landowners who paid local government taxes) (New Zealand Parliament 1904). Drainage districts, overseen by separate drainage boards, were quickly established in the Waipā River in the first two decades of the twentieth century. At least twelve operated in the Waipā River and its tributaries; nine in the middle and upper Waipā catchment (Kawa, Kio Kio, Awatene, Mangaorongo, Waipa, Mangawhero, Orahin, Waitomo, Mangapu) and three in the lower catchment (Lower Mangapiko, Upper Mangapiko, and Tua Tua Moana) (New Zealand Parliament 1924; Simmonds 1938; Tua Tua Moana Swamp 1915a, b; Unknown Author 1935, 1970).

Drainage boards were responsible for undertaking extensive works to remove and remake wetlands. Activities included the construction of drainage canals (by hand and by machine) the construction of flood levees (stopbanks), the realignment of watercourses so that each flowed in straighter lines, and the removal of pā tuna (eel weirs) and riparian vegetation. Throughout the late nineteenth and first half of the twentieth century, the government's drainage policy, historian Geoff Park observes, was underpinned by four main ideas: first, that wetlands in their existing state were unproductive wastelands that were only valuable because of their potential to be developed into fertile farmlands; second, wetlands did not hold any scenic value and should not be preserved (unlike certain remnants of Indigenous forests, birds, lakes, and mountains); third, that legally wetlands were future parcels of land (Park 2001). This assumption (of wetlands as potential land) meant that Māori entitlements to wetlands (authority over and rights to access and use resources) were considered by both the courts and the government to transfer with land titles (once wetlands were "unwatered"). Fourthly, the transformation of wetlands into farmlands was declared of national significance that it necessitated both governments (central and local governments) and individuals to intervene and fund it to ensure that the process was a success (Parsons and Nalau 2016; Parsons 2019). These four tenets negatively impacted Māori who's ancestral and livelihoods ties were bound to the Waipā wetland ecosystems, and for whom the existing wetlands were of immense value (socially, culturally, spiritually, and economically) in their watered (undrained) state.

The drainage schemes were essentially cooperative development ventures between the settler state and individuals, which relied on common (European/Pākehā) understandings of how land and water should be used. An essential part of this was the denigration of wetlands, Māori waterscapes (created and maintained by hapū and iwi for generations), and the resources associated them. Drainage and the replacement of wetlands with productive grasslands were positioned as part of the process of creating a civilised society. Māori who lived within and around Waipā River catchment, even if they held land titles, were given limited opportunities to assert different values and understandings of wetlands.

Given wide-ranging powers, drainage boards could acquire private land (specifically targeting Māori land that was undrained), construct drainage works (even if local landowners opposed it), manage watercourses, and impose rates (taxes) on landholders (Parsons and Nalau 2016; Parsons 2019). The various drainage legislation specifically included provisions that targeted Māori land (including section 83 of Drainage Act 1904) and made Māori land eligible for local government rates (section 88(2) of Drainage Act 1904). Since the majority of land in Rohe Potāe was still Māori land in the early twentieth century, Māori landowners were liable for pay rates to whatever drainage board their landholdings were located in (which often included multiple boards). Māori encountered substantive difficulties paying their rates (with few banks willing to give Māori mortgages on their properties or access any other financial assistance), and many were forced to lease or sell their parcels of land out of financial necessity; that is, to ensure they and their whānau were kept feed, clothed, and with a roof over their heads. Also, many took work as manual labourers tasked with clearing forests, draining wetlands, constructing roads, and cultivating lands on behalf of settlers. The operations of the Kawa Drainage Board, which we will briefly discuss, demonstrate the diversity of ways Māori sought to maintain their ties to their whenua and awa, and the different tactics individuals chose to engage with settler colonialism.

Te Kawa Wetlands and the Operations of the Kawa Drainage Board

Te Kawa wetlands (see Fig. 4.14: Map of Railway Line), an area of approximately 6000 acres located between Otorohanga and Te Awamutu, was a major food gathering site (mahinga kai) for local hapū to harvest tuna (freshwater eels) as well as other aquatic biota (Unknown Author 1907b). While at the start of the twentieth century local Māori retained ownership of the area surrounding Kakepuku and Te Kawa wetlands, by 1907 the majority of the 6000 acres was either owned or leased to Pākehā. Media and government reports were filled with descriptions of how Te Kawa wetlands "ha[d] for generations existed as an unprofitable waste", but with Pākehā occupation and proposed drainage works it was "being turned to its legitimate use" Unknown Author 1907a). A local newspaper even wrote a piece that celebrated the success of Pākehā in securing "the necessary signatures" of Māori landholders to gain the leaseholds and proclaimed that actions to drain the wetlands and convert it all to pastures much occur as soon as possible (Unknown Author 1907b). Settlers and politicians, quoted in local media, repeatedly emphasised that the removal of wetlands was vital to their collective goal: the creation of profitable dairy farms. Both explicitly and implicitly, in public and private accounts, the 'settlers' or 'pioneers' (as they described themselves) of the Waipā discussed their beliefs about what was the correct way to live, work, and interact with the land and water, biological and biophysical components of the freshwater system. Māori ways of living, cultivating, and harvesting resources from their whenua, awa (Waipā River), and repo, from the Western/Pākehā worldview, were deemed inappropriate, 'wasteful', and unproductive as the focus was not directed at intensive agriculture and the accumulation of goods by individuals in a way that closely replicated those of Pākehā. Yet, amongst Māori, there was not necessarily a universal agreement about how to engage with settler colonialism, and the newly dominant Western/Pākehā worldview, economic arrangements, legal system, institutions, and ways of interacting with environments.

The "opening up" of the Rohe Potae and the railway:
Ngāti Maniapoto and other members of Kīngtanga rejected initial government proposals to allow the construction of a railway through Rohe Pōtae. Ngāti Maniapoto were determined to retain their rangatiratanga over the area. The question of the railway line prompted conflict amongst Kingitanga supporters, with the Crown initially seeking to negotiate directly with King Tawhaio rather than with Ngāti Maniapoto leadership. After sustained negotiations between government officials, Kīngitanga leaders, and Ngāti Maniapoto chiefs between 1882 and 1885, Ngāti Maniapoto chiefs agreed to allow the survey of Te Rohe Potae and the construction of the railway line through their rohe on the proviso that the government agreed to certain conditions. In return for allowing the railway through Rohe Potae the government made numerous promises to iwi. This included: giving a legal amnesty on Maori 'rebels'; the right of the Kingitanga to continue to govern their own affairs; the prohibiton of liquor from Ngati Maniapoto territory; the provision of a parlimentary seat to a Ngāti Maniapoto cheif (Wahanui); providing schools and hospitals to Māori communities; and that Māori within Rohe Pōtae to continue to exercise exclusive responsibility for the management of their own lands and waters. While the government did honour its promise to grant the amnesty, other promises were not honoured. Most notably the government ignored its guarantees to Ngāti Manaipoto to protect their lands and waters, and did not allow Ngāti Maniapoto to continue to exercise their rangatiratanga and kaitiakitanga over its rohe.

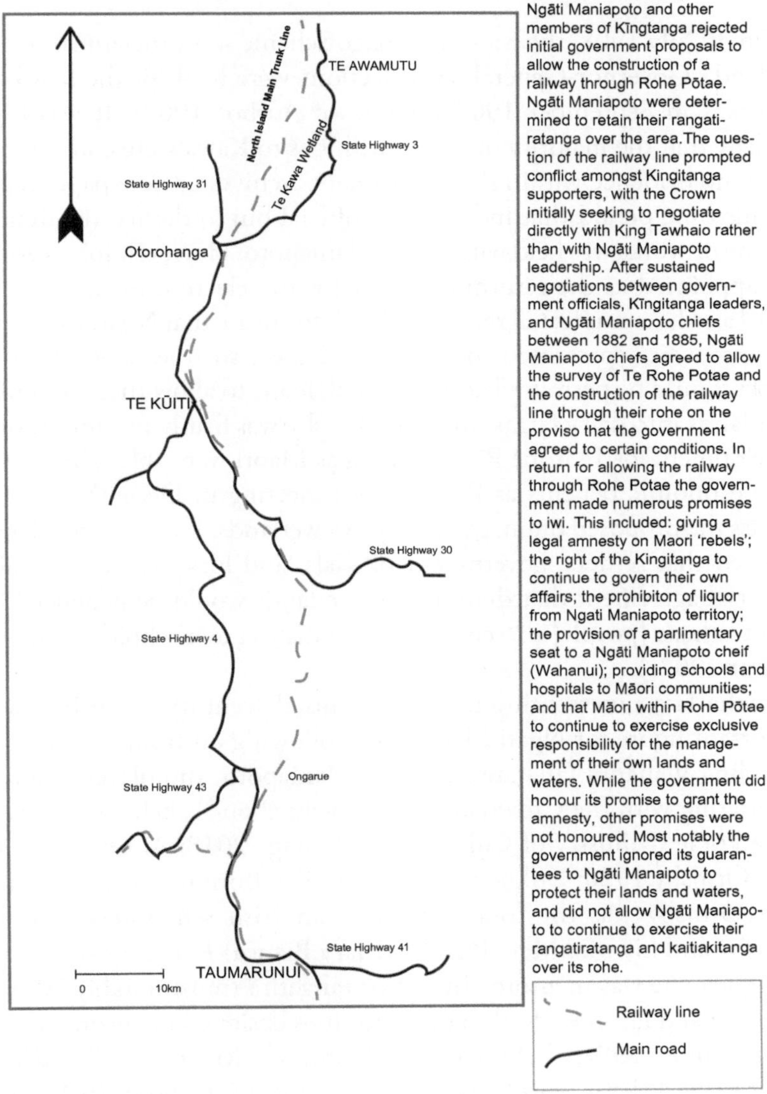

Fig. 4.14 Map showing the central section of the railway line that was constructed through Rohe Pōtae (King Country) following negotiations between Ngāti Maniapoto and the Crown. For a more detailed discussion about Ngāti Maniapoto's decision to allow the railway to be constructed through their territory see recent publications by Michael Belgrave and the Waitangi Tribunal (Belgrave 2017; Waitangi Tribunal 2018)

On 30 July 1908, the Kawa Drainage Scheme was proclaimed by New Zealand's Governor-General, and elections were held for the first board members in September 1908 (Unknown Author 1908). It was highly unlikely that the majority of Māori with ties to Kawa were consulted; we found no evidence within the government archives or newspaper reports for this time period that indicate any official hui to discuss the drainage scheme or works was held with Ngāti Maniapoto. However, John Ormsby was appointed as the returning officer for the election of the Drainage Board and later was its Clerk, and did seek to ensure that Ngāti Maniapoto interests were represented to a limited degree; such as advocating that Māori should be permitted access to bank loans to allow them to develop their lands into dairy farms. Indeed, Ormsby was firmly in support of the 'development' of Te Rohe Pōtae so long as Māori were able to receive the same economic benefits as Pākehā. At a meeting in December 1907 to discuss the proposed drainage of Te Kawa wetlands, for instance, Ormsby declared that settlers, government officials, and local Māori were all in universal agreement that draining the wetlands would be beneficial "not only to those [people] present but to the district as a whole" (Unknown Author 1907c).

In the late nineteenth and early twentieth century, Ormsby was an important figure for Ngāti Maniapoto, following on from earlier leaders including Wahanui Huatare and Rewi Maniapoto, and played a key role negotiating with the Crown and its agencies about Pākehā settlement of Rohe Pōtae (Ministry of Culture and Heritage 2018; A. Ormsby 1907; A. S. Ormsby 1920; Unknown Author 1927). Born in 1854 at Pirongia, John Ormsby was the fourth child of an Irish schoolmaster, Robert Ormsby and his wife Mere Pianika (Mary Bianca) Rangihurihia. From a young age, he was mentored by senior rangatira (most notably Wahanui Huatare) and given leadership opportunities designed to ensure he could act as an intermediary between Te Ao Māori (the Māori world) and Te Ao Pākehā (the Pākehā world) on behalf of Ngāti Maniapoto. In 1883, for instance, Ormsby was appointed to lead the Kawhia Native Committee (the institutional precursor of the local council) and later sought to ensure Māori were given more seats within the committee to try to temper Pākehā settler priorities. Most significantly, in 1884 Ormsby and other successfully petitioned the New Zealand Parliament to allow Wahanui

Huatare to address the House of Representatives on issues relating to Ngāti Maniapoto's lands and the proposed railway line. Wahanui and Ormsby travelled to Wellington and spoke to politicians about Ngāti Maniapoto's commitment to retaining their rangatiratanga over their lands and their willingness to negotiate with the government. The visit bore fruit for Ngāti Maniapoto in that the Native Minister Balance agreed to meet with and form an agreement with representatives of iwi in 1885 (discussed earlier) in which the Crown promised Māori control over their lands, forests and resources, protection from negative effects, and material benefits from the railway (which was subsequently ignored by the settler government as was earlier with Te Tiriti guarantees). Ormsby was later involved in a variety of different administrative and business roles, including as an assessor on behalf of the Native Land Court, the owner of farming ventures, a hotel, butchery, stables, land insurance and bakery; he helped establish the township of Otorohanga and was the chair of the Otorohanga Town Board (Ministry of Culture and Heritage 2018). He even established the first branch of the New Zealand Farmers' Union (which today is known as Federated Farmers) in Rohe Pōtae. He, therefore, came to situate himself in the middle ground or hybrid space between Te Ao Māori and Te Ao Pākehā or Ngāti Maniapoto and settler-colonial worlds. In doing so, Ormsby was able to access financial loans (which most Māori struggled to do at the time), amassed a large portfolio of profitable businesses (in which he employed his Māori relatives), gained political influence (at least at a local government level), and promoted Māori economic development so long as it conformed to the prevailing (Pākehā settler-colonial) capitalist modes of accumulation. He positioned himself as a moderniser who sought to ensure that Ngāti Maniapoto interests were maintained while ensuring that Pākehā settlement and economic development was encouraged. He encouraged Māori to engage in economic activities and employed many of his whānau within his businesses. Ormsby, who straddled multiple worlds—Te Ao Māori and Te Ao Pākehā—like many other Māori leaders in Aotearoa at the time (including Apirana Ngata from the East Coast iwi Ngāti Porou) emphasised the critical need for Māori to embrace Pākehā land development schemes (AJHR 1927; A. Ormsby 1907; A. S. Ormsby 1911). However, Ormsby's views about land, wetland drainage, and economic development stood at odds with other Māori in the area (including his Ngāti Maniapoto kin). The clash in differing ways of engaging with Te Ao Pākehā, maintaining mana (power, prestige, authority), and practising kaitiakitanga was highlighted in disputes surrounding the activities of the Kawa Drainage Board.

Three land parcels under the domain of the Kawa Drainage Board were specifically designated as 'eel reserves' due to the large numbers of pā tuna (eel weirs) that were used by hapū to harvest tuna (freshwater eels) (see Fig. 4.15). The practice of establishing certain eel reserves—Kakepuku 8A (Eel pa) 0.69ha, Kakepuku 8B (Eel pa) 1.38ha, Kakepuku 8C (Eel pa) 2.78ha)—was only, however, ever intended as a temporary measure by the drainage board to appease Māori (who continued to own the majority of land in the region and were only leasing it to Pākehā) (Kawa Drainage Board Clerk 1909). Indeed, pā tuna were of limited value

THE MAORI METHOD OF EEL-CATCHING: A WEIR CONSTRUCTED BY NATIVES ON THE ONGARUE RIVER.
AUCKLAND.   L. Hinge, Photo.

Fig. 4.15 Pā tuna (eel weir) in Ongarue River (located south of the Wāipa River) in 1908. Most pā tuna were comprised of two fences (pā tauremu) that funnelled the tuna (eels) into a hīnaki (eel pot). (Source: Auckland Libraries Heritage Collections AWNS-19080227-16-3)

without the waters that supported tuna, and proposed drainage works threatened to destroy wetlands as mahinga kai for iwi. In response to this threat (to their food harvesting practices, relationships with their repo, and responsibilities as kaitiaki) twelve Māori wrote a letter to the Minister of Native Affairs in 1908 in which they outlined their concerns about the Kawa Drainage Board's proposed works to drain Te Kawa wetlands. They noted that they were not kept informed about the nature of the work, given the opportunity to provide feedback to the board about proposed decisions, or offered any financial compensation for the damage that drainage would cause to their capacities to harvest tuna. In addition, the leaseholder (a Pākehā settler by the name of Walsh) had announced that he planned to implement the drainage works himself, with the support the Kawa Drainage Board, and there was nothing that the Māori owners of the land could do to prevent him. Accordingly, Te Koro and his fellow owners requested that the Minister of Native Affairs:

> protect us in this matter lest our rights [are] wrenched from us by the Pakeha [sic] breaking down and doing away with our eel-weirs without paying compensation. We want, first of all a proper agreement as to payment to us, because the loss of the eel-weirs would deprive us of the support which we gain therefrom, a source of food year after year; and indeed our main source of food supply when the blight destroys our crops. (Te Koro 1908)

Ngawareo Te Koro and the eleven other signatories cited the articles of Te Tiriti o Waitangi that assured Māori continued rangatiratanga over their taonga (including their fisheries) and the Crown's protection of their rights to their resources. They requested that the Minister of Native Affairs intervene to halt the drainage works until an amicable agreement could be reached between the two parties, and also sought financial compensation for the damage to their pā tuna (£2000).

The Under-Secretary of the Native Department requested that government officials based within Rohe Pōtae investigate the matter and provide a report back to the Minister. The Pākehā Judge (connected to the NLC) who inquired into the issue (unsurprisingly) agreed with Kawa Drainage Board's decisions. He declared that wetlands of Te Kawa were "lying unproductive and in its present state is absolutely no value to its Native owners" (President of the Maniapoto-Tuwharetoa District Maori Land Board 1909). Despite its lack of value, the Judge declared, its Māori owners "refused to lease or sell" their lands "except at an exorbitant price". Therefore the Kawa Drainage Board was operating in everyone's best interest. The Kawa Drainage Board was formed by:

> all the persons whose properties will be benefited from the drainage of the swamp, and it appears to me that the Natives are alarmed, because they see, that instead of getting the outrageous price they asked, there is a chance that they will be paid only what is a fair share of the weir. (President of the Maniapoto-Tuwharetoa District Maori Land Board 1909)

After this incredibly limited inquiry, the Minister of Native Affairs essentially washed his hands of the matter and declared it an issue for the courts rather than the government. He informed Te Koro et al. that their only option was to seek legal representation and lodge legal proceedings against Kawa Drainage Board to prevent the drainage works (Fisher 1909). From an iwi Māori perspective, the central government (the Crown) was (as per the Treaty of Waitangi and the 1885 railway agreement) required to protect Māori land and other taonga from damage. So Te Koro and other Māori owners requested that the Crown (and its agency the Native Affairs Department) intervene to stop the drainage works. However, like elsewhere in Aotearoa, the Crown refused to intervene to protect Māori interests and declared that wetland drainage was critical to the progress and development of society; and that the only remedy available to Māori was through the courts.

Later that same year (in November 1909), the Māori owners of the 'eel reserve' Kakepuku 8C (some of whom were signatories to the letter to the Native Minister) were notified by the Kawa Drainage Board that it would construct a drainage canal through their land (Kawa Drainage Board Clerk 1909). The nine owners of Kakepuku 8C vehemently opposed the action as they argued it would destroy their pā tuna and their capacities to harvest tuna. They once again sought to negotiate with the board. They approached local members of parliament for support (which was in keeping with Māori governance protocols that centred on face-to-face meetings and decisions based on the ongoing dialogue between parties until agreements could be reached). However, due to the Minister of Native Affair's refusal to assist Hone Te Anga and his fellow landowners of Kakepuku 8C, they were forced to seek legal representation from a Pākehā legal firm (Bamford & Brown) to help them negotiate with the Kawa Drainage Board. They continued to hope that negotiations with the board (which included Ngāti Maniapoto's John Ormsby) would ensure that they could prevent or at least mitigate the damage caused by drainage works to their pā tuna (Bamford and Brown 1909). Solicitors Bamford & Brown wrote and met with members of the drainage board and outlined their clients' objections to the planned engineering works on their land. On the basis that the said "piece of land [was] an eel pā" that they held "for their benefit and for the benefit of the Ngatingawaero [sic] tribe" (Ngāti Ngāwaero is a hapū of Ngāti Maniapoto), and it was of "great value ... and importance to them" (Bamford and Brown 1909). The proposed drainage canal would, they argued, "destroy the character of the said piece of land as an eel pā", and the owners could not "be adequately compensated for such destruction". Indeed, no loss of money would be sufficient to compensate for the lack of eels. The drainage works, the solicitors warned the drainage board, was both "inequitable" and would "infringe the just legal equitable rights of the objectors to maintain the said piece of land as an eel pā" (Bamford and Brown 1909). The two parties (Bamford & Brown and the Kawa Drainage Board) could not reach an out of court agreement, and in May 1910 the New Zealand Supreme Court heard the case. The Court sided with the Kawa Drainage Board and declared that the Māori landowners could be financially compensated for the loss of the eel weirs. The rights of Māori, the Judge argued, "should not be allowed to stand in the way of draining a large area of the country" (Bamford 1910; Unknown Author 1910).

After the Supreme Court rejected their legal objection of the drainage works, the plaintiffs (Hone Te Anga et al.) then sought to address the issue of compensation, which was allowed under the Land Drainage Act 1908. In May 1910, Harry Bamford, solicitor of the plaintiffs, submitted that since the drainage operations would "put an end to or substantially put an end to the supply of eels in the stream through the said land and will seriously affect the riparian rights of the plaintiffs" they required a large sum—£1500—in compensation (Bamford 1910). To give some sense of how much £1500 was in Aotearoa in May 1910, the figure (according to the New Zealand Reserve Bank inflation calculator) is the equivalent of $260,327.59 New Zealand Dollars (NZD) in December 2019. The Kawa Drainage Board's solicitor (George Kent) declared the amount of money plaintiffs wanted was completely out of the question. Indeed, Kent argued, that the plaintiffs should only receive a far smaller figure because they would receive "certain advantages" from the drainage works, including an increase in their property values and the potential that drains may "increase the number of eels" in the waterways. Moreover, he added that "eel pas were of a diminishing character and that on the admission of the Plaintiffs the younger natives of the district do not take the same interest in the said eel pas as did their predecessors and that a time would come when the said rights would practically be neglected" (Kent 1910).

The Court adjourned the case in August 1910, with the two parties meant to negotiate an agreement. However, this did not happen, and the Kawa Drainage Board proceeded with its planned drainage works. Māori landowners unsuccessfully sought a legal injunction against the works. By April 1914, when the Supreme Court heard the case again, the board had finished the majority of engineering works, with the Mangawhero Stream that flowed through the block of land diverted, and eel weirs removed (Unknown 1914a). The Supreme Court's decision once again reflective European understandings of the environment, wetlands, and resource usage. The "Kawa Swamp", Judge Cooper stated, "was noted for the very large number of edible eels which it contained ... flourished exceedingly" and was used "for the common benefit of all Natives living in the district, and were a very material part of their general food supply". The drainage scheme was, however, "now in a very advanced state", with the Mangawhero Stream dredged and straightened, and the wetlands partially drained. As a consequence of the drainage works, the number of eels was "materially diminished", with "the facility of catching eels by means of weirs greatly restricted". On completion of the drainage scheme, the entire area, the Judge stated, would cease to be wetlands that were only useful as a "fattening place for eels" and instead be "the most valuable dairy-farm land" (Te Anga 1914).

As the legal case continued through the court system, the solicitors and their Māori clients (Hone Te Anga and his kin) employed Western/Pākehā modernist legal framings of ownership and private property rights in support of their arguments about their freshwater, land, and resource rights. The adoption of modernist framing was an essential part of Māori efforts to receive financial compensation for the loss of wetlands and the decreased (or destroyed) capacities to harvest tuna and other freshwater biota used for food, medicine, art, and cultural activities. However, the legal case was reported in national and local newspapers as a story:

> of the clash of the modern with the ancient; of the dislocation that must almost inevitably occur when the advance of civilisation overtakes the lagging customs of the aboriginal. A local authority had drained a swamp; the success of the drainage works necessarily interfered with the fattening grounds of the eels; the Maoris whose food supply was then interfered with sought compensation. (Unknown 1914b)

The Supreme Court determined that the Māori owners were entitled to compensation, but it was restricted to the damage to their pā tuna but not the reduction in the number of tuna in the stream. The Drainage Board, the Supreme Court declared, possessed exclusive authority to alter waterways and drain wetlands. Thus the rights of Māori (and financial claims) were highly constrained. The Supreme Court referred the issue of compensation to the Compensation Court, which ultimately awarded the landowners £150 in compensation (far less than the £1500 sought by plaintiffs and most of the money likely went to paying the legal costs) (Unknown 1914a; Unknown Author 1914).

Rather than a clash of civilisations, the case of Te Kawa wetlands demonstrates that Māori sought to retain their rangatiratanga and practice kaitiakitanga through political and legal forums. In their different engagements with drainage board operations and the courts, Māori discussed, debated and challenged commonly-held attitudes, values and relationships; sometimes articulating Te Ao Māori holism and communalism, and other times embracing Te Ao Pākehā individualism and forward-thinking time (M. Jackson 1993; O'Regan 1984; Ruru 2009; Salmond 2017; Tipa et al. 2016).

The Te Kawa wetlands case study highlights those different ways in which Māori (as individuals and members of whānau, hapū, and iwi) sought to challenge the dominant settler-colonial assumptions about wetlands as unhealthy, unproductive and unused spaces. In doing so, Māori articulated relational connections with their whenua, awa, repo, and other dimensions of their waterscapes that extended beyond Western liberal ontologies centred on binary divisions between nature/culture and land/water. They as members of whānau, hapū, iwi, linked through whakapapa (genealogy) to each other as well as their rohe and all those beings (including plants, animals, and supernatural) that lived within it, continued to advocate for the importance of activities that were not part of the market economy or those of Te Ao Pākehā. From this perspective, while land, rivers, wetlands and biota were fundamental to Māori subsistence activities, they were never capable of being reduced to merely exploitable resources. Instead, there were always social, ethical and spiritual dimensions that bound Māori as tangata whenua (which translates directly as people of the land) to their landscapes and waterscapes. Yet, not all Māori articulated these views and instead advocated for Māori to adopt Te Ao Pākehā (at least where it concerned economic development activities). Thus, individuals like John Ormsby and many of his descendants operated as intermediaries between Te Ao Māori and Te Ao Pākehā and took efforts to retain tribal lands as the Crown endeavoured to acquire more and more Māori land. Yet, at the same time, the rules of the game were those defined and implemented by the settler-colonial state and which privileged Western liberal worldviews and land use at the extent of modes of life, ways of knowing and being.

The example of Te Kawa wetlands illustrates that Western conceptualisations of what constitutes environmental justice (and injustice) do not adequately address Māori, other Indigenous and non-Western, relational ways of thinking about human-environment relationships. In 1908 and 1909, the plaintiffs (Hone Te Anga et al.), like many other Māori at the time as well as prior and subsequent, sought to articulate to non-Māori their ways of conceiving and experiencing their worlds, which rests in a system of reciprocal relationships with their whenua, awa, biota and the metaphysical beings that dwelled within them. Through their legal case, Māori emphasised how the loss of access to important mahinga kai was not something that could be simply replaced with other types of food, as those foods held social, cultural, and spiritual dimensions, which connected tangata whenua to their ancestors (human-ancestors, god-ancestors, and other more-than-human-ancestors). These different ways of seeing and living (underpinned by different ontologies and epistemologies) were illustrated in how Māori and Pākehā understood the Te Tiriti o Waitangi/the Treaty of Waitangi, as well as later conflicts over governance and resource management. These differences between Māori and Pākehā ways of seeing and interacting with their environments (their modes of life), we suggest. The failures of the Crown (the settler colonial state) and the settler dominated legal system to recognise Māori values, knowledge, and modes of life (including their ways of governing and managing whenua and awa) as valid ways of thinking and being in the world were one critical form of environmental injustice experienced by Māori.

When we take into account how Māori and Pākehā represented and interacted with wetlands, it raises important questions about how scholars define environmental justice and injustices. At the same time, most environmental justice theorising and activism focus on present-day examples of environmental (in)justice as opposed to historical studies, environment injustices rarely (if ever) pop into existence overnight. Indeed, societies, communities, and human-environmental relationships do not exist in a historic vacuum. Instead, the social and environmental injustices and a plethora of environmental crises and challenges we face at the start of the twenty-first century are a product of past decisions, policies, and practices over the years, decades, and centuries across multiple scales. In Aotearoa, we argue that Māori experiences of environmental injustices are inextricably bound up with the historical and continuing processes and practices of settler colonialism. Yet, traditional accounts of environmental justice do not adequately take into account Māori values, interests, knowledge, and their experiences of environmental injustices.

If we examine wetlands through a distributive environmental justice lens (which seeks to identify the distribution of environmental "goods" and "harms" across society and/or space) then environmental injustice was not evident (Arcury and Quandt 2009; Brook 1998; Bullard 1993; Hockman and Morris 1998; Hofrichter 2002; Lee 2002; Pastor et al. 2001). From this perspective, a wetland drainage scheme that provided (supposedly) equal distribution of "benefits" (as was argued by government officials about Te Kawa) would not be an example of environmental injustice (as equal benefits and harms were distributed across groups). However, as we outlined in Chap. 2, such a narrow framing of environmental justice ignores the social, cultural, and institutional contexts in which environmental injustices take place and the systematic acts of discrimination against marginalised populations that all play substantial roles in creating and sustaining environmental injustices. A distributive environmental justice lens, therefore, misses an important opportunity to critique the roles of capitalism and colonialism across multiple intersecting temporal and spatial scales (see Swyngedouw and Heynen 2003). All of which is particularly pertinent when thinking about the specific environmental injustices faced by Indigenous peoples, and how settler-colonialism manifested itself as social and eco-violence against Indigenous bodies and spaces (Hendlin 2019; S. Jackson 2018; Whyte 2014, 2016).

Recent work by decolonial scholars rejects environmental justice as distributive equity for being underpinned by Western ontologies that position 'nature' as something capable of being classified, objectified, exploited (Álvarez and Coolsaet 2018). Nature as something quantifiable and therefore distributable—be it as environmental goods (such as clean water, land, fisheries) or harms (such as air pollution or hazardous waste)—however, stands at odds with Māori and many other Indigenous peoples' ways of seeing the world. From such a reading of environmental justice as distributive, Māori within the Waipā catchment did not necessarily experience any environmental injustice as a consequence of their wetlands being drained. Any losses or damage that Māori communities suffered as a consequence (including diminished access to traditional food sources) could, as both Native Minister Balance and Judge Kent informed members of Ngāti Maniapoto iwi in 1885 and 1909 respectively, be simply replaced by other goods. For instance, tuna supplanted by cows. Indeed, anything of value, from the worldview of Pākehā, could be quantified and calculated into a monetary figure, and (if damaged) then financial compensation could be paid (even if it was only £150 in the case of pā tuna). However, even when Māori adopted the lexicon of settler colonialism (of modernity, capitalism, and individual property rights) in their complaints and legal cases about the damage of drainage works, their arguments were challenged and rejected by Pākehā on the basis that the things Māori valued (wetlands, tuna, pā tuna) were of minimal or no financial worth. The refusal of Pākehā (be it settlers, local government and central government officials, and representatives of the judiciary) to acknowledge Māori interests and their rights to maintain their ways of life, natural resource management regimes, and food and water cultures, was later rearticulated in debates about flood controls (discussed in the next chapter) and accusations that Māori received economic benefits from using "white man's utilities" (roads, bridges, urban water supplies, drainage works, and flood controls) without paying their local taxes (Álvarez and Coolsaet 2018). Indeed, as Walker (2009) has argued scholars need to look beyond where environmental harms and goods were and are located to consider the diversity of environmental risks and the multiple types of environmental injustices experienced by different communities in different ways, which work at different scales (Walker 2009, p. 615). He defines three conceptualisations of environmental justice in pluralistic terms; distributive (distribution of goods and bads); procedural (policies and processes including the equitable capacities to participate in decision-making); and recognition (of different knowledges, values, and peoples). In the Waipā freshwater system, the cumulative impacts of inequitable government policies and practices that marginalised Māori voices and modes of life, and the failure to acknowledge (misrecognition) of Māori values, laws and knowledge, and entire waterscapes were fundamental to the settler state-sponsored wetland drainage project which caused multiple environmental injustices for Māori. One major injustice related to the loss of pā tuna and tuna. Since tuna served (and still does serve) as socio-culturally important for riverine Māori for multiple reasons; tuna was a commonly harvested food source (that provided whānau/hapū with a healthy source of protein), given as a gift to demonstrate one's hospitality to guests (tuna was served at feasts and given to visitors as part of manaakitanga (showing hospitality, generosity and support for others), and was considered as kin to local hāpu (through whakapapa connections). Another injustice related to the separation of people from their whenua, awa, and repo with the establish-ment of drainage canals, construction of flood levees, and pollution of the waterways (which we discuss in the next chapter); these acts of eco-logical dispossession meant that tangata whenua were not able to access the products (food, water, and medicinal) that established socio-cultural, spiritual and economic connections.

The work of Indigenous scholar Coulthard (2014) similarly highlights how the historical and contemporary struggles of the Dene people, who comprise First Nations groups living in the Western Subarctic area of Canada, were not just against settler colonialism, but also capitalism. Dene communities, which includes Yellowknives, Sathu, Salvey, Tlicho and Chipewyan, not only challenge the distribution of environmental risks and impacts of dispossession but also demand the rights to live "in relation to one another and the natural world in non-dominating ways and nonexploitative terms" (Coulthard 2014, p. 13). Likewise, Escobar argues that present-day Colombian Indigenous and Afro-Colombian social movements are waged in the name of different ways of life (Escobar 2015, 2016). Such movements (both historical and contemporary) amongst Indigenous and marginalised non-Indigenous non-European communities in colonial and post-colonial societies conceptualise human-environmental relationships in non-binary terms and situate develop-ment in terms of reciprocal relations between ecological and human communities. The environmental injustices experienced by Māori were (and still are) grounded not only in economic, ecological systems, but also socio-cultural structures that marginalise Māori ways of knowing, knowledge, values, and ways of interacting with different worlds. Thus, acts of dispossession (which included drainage works) deprived Māori communities not only of their material modes of subsistence but also negatively impacted their physical and spiritual health (hauora) and well-being (encapsulated in the principles of mauri and wairua). Dispossession and colonial structures of power and control (which were diverse and permeated all society) involved the (mis)recognition of Indigenous val-ues, identities and modes of life, and if ever acknowledged were simply ascribed Western ideas of development, economic values, and land own-ership rights.

Conclusion

Settler colonialism was (and still is) a system of socio-cultural, political, and economic domination that involves violent disruptions to Indigenous peoples' relationships with their waterscapes and landscapes. In the context of Aotearoa, settler colonialism manifested itself through a series of social, political, and ecological interventions that sought to dominate and radically remake the rohe (traditional territories) of Māori iwi (tribe) and hapū (sub-tribe). Like in other settler societies, settlers in Aotearoa (sometimes incidentally, other times deliberately) sought to establish their own homes, farms, factories, communities, institutions, and the entire settler-colonial state through the marginalisation and erasure of Indigenous places, which involved ecological remodelling. In this chapter, then, we explored just one element of settler-colonial domination: the transformations of wetlands into grasslands. In doing so, we are forced to exclude a wealth of other interrelated histories but direct readers attention to the work of scholars who explore colonial challenges to Māori sovereignty, health and wellbeing, language and education (Boast 2008; Anderson et al. 2014; Keenan 2014; Lange 1999; Mahuika 2019; Salesa 2001; Wanhalla 2006). However, we recognise that settler-colonial domination consists of insidious loops and sedimentation that seriously disrupted (but did not destroy) all elements of Indigenous cultural continuance, which encompasses social, cultural, ecological, spiritual, political and economic domains (Whyte 2018; Whyte et al. 2019).

We must think about how wetlands (loss, health, restoration) figure into how we think about healthy rivers and how we (Indigenous and non-Indigenous) respond to the challenges of the Anthropocene (Hatvany 2008; Parsons and Nalau 2016; Romero Lankao 2010; Vileisis 1999). The past and continuing efforts to remake the wetlands of Aotearoa is the manifestation of settler colonialism, which is enacted elsewhere using a variety of mechanisms but all aim to achieve the same end. Settler-colonial projects are directed at the appropriation of lands, waters, minerals, and other resources as well as the jurisdiction of Indigenous peoples not only to exploit natural resources and for establishing settlements, but also to facilitate the territorial foundation of dominant (hegemonic) neo-European societies (Bacon 2019; Belich 2009; Hiller 2017; Parsons and Nalau 2016; Meg Parsons 2019).

References

AJHR. (1875). *H-22 Boards of Health in the Various Provinces. Appendices of the Journal of the House of Representatives*. Wellington: Government Printer.

AJHR. (1885a). *G-02a Reports from Native Medical Officers. Appendix to the Journals of the House of Representatives, 1 January 1885. Appendices to the Journals of the House of Representatives*. Wellington: Government Printer.

AJHR. (1885b). *G-2. Notes of a meeting between Hon. Mr. Ballance and the Natives at the Public Hall at Kihikihi, on 4th February, 1885. Appendices of the Journal of the House of Representatives*. Wellington: New Zealand Parliament.

AJHR. (1911). *C-11 Drainage Operations in the Rangitaiki Plains Report for the Year Ending 31 March 1911 (AJHR No. C–11). Appendices of the Journal of the House of Representatives*. Wellington: Government Printer. Retrieved May 5, 2019, from http://paperspast.natlib.govt.nz/parliamentary/AJHR1911-I.2.2.3.26.

AJHR. (1913). C-9 Drainage Operations in Hauraki Plains: Report for the Year Ended 31st March 1913. Appendices of the Journal of the House of Representatives. Wellington: New Zealand Parliament. Retrieved May 5, 2019, from http://paperspast.natlib.govt.nz/parliamentary/AJHR1913-I.2.3.2.18.

AJHR. (1927). *I-03 NATIVE AFFAIRS COMMITTEE (REPORTS OF THE). NGA RIPOATA A TE KOMITI MO NGA MEA MAORI. (Hon. Sir APIRANA NGATA, Chairman.), Untitled, 1 January 1927* (Appendices to the House of Reprsentatives No. I–03). Wellington: Government Printer. Retrieved May 5, 2019, from http://paperspast.natlib.govt.nz/parliamentary/AJHR1927-I.2.3.3.3.

Álvarez, L., & Coolsaet, B. (2018). Decolonizing Environmental Justice Studies: A Latin American Perspective. *Capitalism Nature Socialism*, 1–20.

Anderson, A., Binney, J., & Harris, A. (2014). *Tangata Whenua: An Illustrated History*. Wellington: Bridget Williams Books.

Arcury, T. A., & Quandt, S. A. (2009). *Latinx Farmworkers in the Eastern United States: Health, Safety and Justice*. Cham: Springer Science & Business Media.

Ausseil, A.-G. E., Jamali, H., Clarkson, B. R., & Golubiewski, N. E. (2015). Soil Carbon Stocks in Wetlands of New Zealand and Impact of Land Conversion since European Settlement. *Wetlands Ecology and Management, 23*(5), 947–961.

Azarnivand, A., Chitsaz, N., & Malekian, A. (2017). Designing a Risk-based Multi Criteria Framework for River Health Assessment: A Case Study of

Taleghan Basin, Iran. *International Journal of Hydrology Science and Technology, 7*(1), 63–76.

Bacon, J. M. (2019). Settler Colonialism as Eco-Social Structure and the Production of Colonial Ecological Violence. *Environmental Sociology, 5*(1), 59–69.

Bamford, H. D. (1910). *Sworn Statement of Harry Dean Bamford, 13 May 1910, Para 5. BCDG A1492 Box 1, A16.* Archives New Zealand: Auckland.

Bamford & Brown. (1909). *Letter: Bamford & Brown to Kawa Drainage Board Clerk, 4 December 1909, BCDG A1492 Box 1, A16.* Auckland: Archives New Zealand.

Barugil, J. J. (1883, August 2). Mr J. J. Barugil on the Waikato. *Waikato Times*, p. 3.

Beattie, J. J. (2005). *Environmental Anxiety in New Zealand, 1850–1920: Settlers, Climate, Conservation, Health, Environment.* (Thesis). University of Otago. Retrieved May 31, 2020, from http://ourarchive.otago.ac.nz/handle/10523/347.

Beattie, J. (2008). Colonial Geographies of Settlement: Vegetation, Towns, Disease and Well-being in Aotearoa/New Zealand, 1830s–1930s. *Environmental History, 14*(4), 583–610.

Belgrave, M. (2017). *Dancing with the King: The Rise and Fall of the King Country, 1864–1885.* Auckland: Auckland University Press.

Belich, J. (2009). *Replenishing the Earth: The Settler Revolution and the Rise of the Angloworld.* Oxford: Oxford University Press.

Boast, R. (2008). *Buying the Land, Selling the Land: Governments and Māori Land in the North Island 1865–1921.* Wellington: Victoria University Press. Retrieved April 21, 2017, from https://books.google.co.nz/books?hl=en&lr=&id=VpUSvrc7snkC&oi=fnd&pg=PR11&dq=related:rMTvJd8z9nQJ:scholar.google.com/&ots=66dD0cl-A3&sig=Vl0zRTAyBx2qm98QtTUL_Ak7ZLY.

Bradford, L. E., Bharadwaj, L. A., Okpalauwaekwe, U., & Waldner, C. L. (2016). Drinking Water Quality in Indigenous Communities in Canada and Health Outcomes: A Scoping Review. *International Journal of Circumpolar Health, 75*(1), 32336.

Brealey, K. G. (1995). Mapping Them 'Out': Euro-Canadian Cartography and the Appropriation of the Nuxalk and Ts'ilhqot'in First Nations' Territories, 1793–1916. *The Canadian Geographer / Le Géographe canadien, 39*(2), 140–156.

Brook, D. (1998). Environmental Genocide. *American Journal of Economics and Sociology, 57*(1), 105–113. https://doi.org/10.1111/j.1536-7150.1998.tb03260.x.

Bullard, R. D. (1993). Race and Environmental Justice in the United States. *Yale Journal of International Law, 18*, 319.

Byrnes, G. (2001). *Boundary Markers: Land Surveying and the Colonisation of New Zealand.* Wellington: Bridget Williams Books. Retrieved April 21, 2017, from https://books.google.co.nz/books?hl=en&lr=&id=be6IHXxBhZEC&o i=fnd&pg=PP1&dq=related:rMTvJd8z9nQJ:scholar.google. com/&ots=VWgbTl7cCv&sig=-BgBay-40Kg59vp3MaPRMRX06pg.

Cameron, E. (2011). Copper Stories: Imaginative Geographies and Material Orderings of the Central Canadian Arctic. *Rethinking the Great White North: Race, Nature and the Historical Geographies of Whiteness in Canada*, 169–190.

Carlson, A. (2010). *Drain the Swamps for Health and Home: Wetlands Drainage, Land Conservation, and National Water Policy, 1850–1917* (PhD Thesis). University of Oklahoma, Oklahoma.

Chetty, S., & Pillay, L. (2019). Assessing the Influence of Human Activities on River Health: A Case for Two South African Rivers with Differing Pollutant Sources. *Environmental Monitoring and Assessment, 191*(3), 168.

Chiang, C. Y. (2008). The Nose Knows: The Sense of Smell in American History. *The Journal of American History, 95*(2), 405–416.

Clarkson, B. R., Ausseil, A.-G. E., Gerbeaux, P., et al. (2013). Wetland Ecosystem Services. In *Ecosystem Services in New Zealand: Conditions and Trends* (pp. 192–202). Lincoln: Manaaki Whenua Press.

Coulthard, G. S. (2014). *Red Skin, White Masks: Rejecting the Colonial Politics of Recognition*. Minneapolis: University of Minnesota Press. Retrieved May 19, 2019, from https://muse.jhu.edu/book/35470.

Cowan, J. (1928). The Romance of the Rail: A Descriptive and Historical Story of the North Island Main Trunk Railway. *The New Zealand Railways Magazine, 3*(3). Retrieved May 26, 2018, from http://nzetc.victoria.ac.nz/ tm/scholarly/tei-Gov03_03Rail-t1-body-d16-d6.html.

Cumberland, K. B. (1941). A Century's Change: Natural to Cultural Vegetation in New Zealand. *Geographical Review, 31*(4), 529–554.

Davidson, N. C. (2014). How Much Wetland has the World Lost? Long-Term and Recent Trends in Global Wetland Area. *Marine and Freshwater Research, 65*(10), 934–941.

Dench, S. (2011). Invading the Waikato: A Postcolonial Re-view. *New Zealand Journal of History, 45*(1), 33–49.

Dench, S. J. (2018). *Imaging and Imagining the Waikato: A Spatial History c.1800–c.1914* (Thesis). The University of Waikato, Hamilton. Retrieved May 4, 2019, from https://researchcommons.waikato.ac.nz/handle/10289/ 11856.

Denyer, K., & Robertson, H. (2016). Wetlands of New Zealand. In C. M. Finlayson, G. R. Milton, R. C. Prentice, & N. C. Davidson (Eds.), *The Wetland Book (pp. 1–15)*. Springer.

Durette, M., Nesus, C., Nesus, G., & Barcham, M. (2009). *Māori Perspectives on Water Allocation*. Wellington: Ministry for the Environment.

Edmundson, A. (2019). 'Preserving the Papuan': JHP Murray and Doomed Race Theory in Papua New Guinea. *History and Anthropology, 0*(0), 1–20.

Ellinghaus, K. (2006). Indigenous Assimilation and Absorption in the United States and Australia. *Pacific Historical Review, 75*(4), 563–585.

Escobar, A. (2015). Degrowth, Postdevelopment, and Transitions: A Preliminary Conversation. *Sustainability Science, 10*(3), 451–462.

Escobar, A. (2016). Thinking-Feeling with the Earth: Territorial Struggles and the Ontological Dimension of the Epistemologies of the South. *AIBR, Revista de Antropología Iberoamericana, 11*(1), 11–32.

Fisher, M. (1909). *Handwritten Note Fisher to Grace, 14 January 1909, Te Kawa Swamp, Protest against Drainage (Effects of Eel Weirs), MA1 973*. Wellington: National Archives.

Flikke, R. (2016). South African Eucalypts: Health, Trees, and Atmospheres in the Colonial Contact Zone. *Geoforum, 76*, 20–27.

Flint, N., Rolfe, J., Jones, C. E., Sellens, C., Johnston, N. D., & Ukkola, L. (2017). An Ecosystem Health Index for a Large and Variable River Basin: Methodology, Challenges and Continuous Improvement in Queensland's Fitzroy Basin. *Ecological Indicators, 73*, 626–636.

Grey, G. (1849). *Letter: Sir George Grey to Daniel Bolton. 3 March 1849. GLNZ G1.B6a. Grey New Zealand Letters*. Auckland: Auckland Libraries.

Guthrie-Smith, H. (1969). *Tutira: The Story of a New Zealand Sheep Station* (4th ed.). Wellington: AH& AWReed.

Halliday, S. (2001). Death and Miasma in Victorian London: An Obstinate Belief. *BMJ [British Medical Journal], 323*(7327), 1469–1471.

Hamilton, R. B. (1968). *Military Vision and Economic Reality: The Failure of the Military Settlement Scheme in the Waikato, 1863–1880*.

Hammer, G. E. J. (1991). *A Pioneer Missionary: Raglan to Mokau 1844–1880: Cort Henry Schnackenberg*. Auckland: Wesley Historical Society (New Zealand).

Hanrahan, M. (2003). Water Rights and Wrongs: Safe Drinking Water Remains a Distant Hope for Residents of Black Tickle and Many Other Indigenous People in Canada. *Alternatives Journal, 29*(1), 31–35.

Hargreaves, R. P. (1959). The Maori Agriculture of the Auckland Province in the Mid-nineteenth Century. *The Journal of the Polynesian Society, 68*(2), 61–79.

Hargreaves, R. P. (1961). Maori Flour Mills of the Auckland Province: 1846–1860. *The Journal of the Polynesian Society, 227–232*.

Hatvany, M. (2008). Environmental Failure, Success and Sustainable Development: The Hauraki Plains Wetlands through Four Generations of New Zealanders. *Environmental History, 14*(4), 469–495.

Hemming, S., Rigney, D., Muller, S., Rigney, G., & Campbell, I. (2017). A New Direction for Water Management? Indigenous Nation Building as a Strategy for River Health. *Ecology and Society, 22*(2).

Henderson, J., Germann, P., Macky, M., Finch, J., & Fentress, J. (1994). *Paterangi Remembers*. Hamilton: Federated Farmers of New Zealand. Retrieved January 23, 2018, from https://natlib.govt.nz/records/31194692.

Hendlin, Y. H. (2019). Environmental Justice as a (Potentially) Hegemonic Concept: A Historical Look at Competing Interests between the MST and Indigenous People in Brazil. *Local Environment, 24*(2), 113–128.

Hiller, C. (2017). Tracing the Spirals of Unsettlement: Euro-Canadian Narratives of Coming to Grips with Indigenous Sovereignty, Title, and Rights. *Settler Colonial Studies, 7*(4), 415–440.

Hockman, E. M., & Morris, C. M. (1998). Progress towards Environmental Justice: A Five-year Perspective of Toxicity, Race and Poverty in Michigan, 1990–1995 1. *Journal of Environmental Planning and Management, 41*(2), 157–176.

Hofrichter, R. (2002). *Toxic Struggles: The Theory and Practice of Environmental Justice*. University of Utah Press.

Howe, K. R. (1970). *Missionaries, Maoris, and "Civilization" in the Upper-Waikato, 1833–1863: A Study in Culture Contact: With Special Reference to the Attitudes and Activities of the Reverend John Morgan of Otawhao* (Thesis). University of Auckland. Retrieved May 31, 2020, from https://catalogue. library.auckland.ac.nz/permalink/f/gkavgk/uoa_alma21159775720002091.

Hursthouse, C. F. (1861). *New Zealand: The "Britain of the South"*. London: E. Stanford.

Jackson, M. (1993). Land Loss and the Treaty of Waitangi. *Te Ao Märama: Regaining Aotearoa. Mäori Writers Speak Out, 2*.

Jackson, S. (2018). Indigenous Peoples and Water Justice in a Globalizing World. *The Oxford Handbook of Water Politics and Policy, 120*.

Johnstone, B. M., & Roberts, P. R. (2004). *Not a Pioneer!: A Memoir of Waipa and Raglan, 1871–1960: Memories of Bernice Monrath Johnstone of Three Oaks, Whatawhata, New Zealand*. Ottawa: P.R. Roberts.

Jones, V. (1996, May 9). Shirley Finlayson. Retrieved September 20, 2018.

Kansal, M. L. (2018). Issues and Challenges of River Health Assessment in India. In *Water Resources Management* (pp. 105–119). Springer.

Kawa Drainage Board Clerk. (1909). *Kawa Drainage Board Clerk to the Owners of Kakepuku 8C, 13 November 1909, BCDG A1492 Box 1, A16.* Auckland: Archives New Zealand.

Keenan, E. (2014). Māori Urban Migrations and Identities, 'Ko Ngā Iwi Nuku Whenua': A Study of Urbanisation in the Wellington Region during the Twentieth Century. Retrieved March 19, 2020, from http://researcharchive.vuw.ac.nz/handle/10063/3576.

Kelm, M. E. (2005). Diagnosing the Discursive Indian: Medicine, Gender, and the "Dying Race". *Ethnohistory, 52*(2), 371–406.

Kent, G. S. (1910). *Sworn Statement of George Sedgwick Kent, 24 June 1910, para 6, BCDG A1492 Box 1, A16.* Auckland: Archives New Zealand.

Kerry-Nicholls, J. H. (1884). *The King Country: Or, Explorations in New Zealand. A Narrative of 600 Miles of Travel through Maoriland.* London: S. Low, Marston, Searle & Rivington.

Kiechle, M. A. (2017). *Smell Detectives: An Olfactory History of Nineteenth-Century Urban America.* Seattle: University of Washington Press.

Lange, R. (1999). *May the People Live: A History of Maori Health Development 1900–1920.* Auckland University Press.

Lee, C. (2002). Environmental Justice: Building a Unified Vision of Health and the Environment. *Environmental Health Perspectives, 110*(Suppl 2), 141–144.

Levy, E. B. (1970). *Grasslands of New Zealand.* Wellington: Government Printer.

Mahuika, N. (2019). *Rethinking Oral History and Tradition: An Indigenous Perspective.* Oxford: Oxford University Press.

McGovern, F. E. (1986, February 18). A Hard Life: Reminisces 1902–1919. Interview with Mrs Flo (Florence) McGovern of Peachgrove Road Hamilton by Jeff Downs. Retrieved October 1, 2018.

McGregor, R. (1997). *Imagined Destinies: Aboriginal Australians and the Doomed Race Theory, 1880–1939.* Carlton, VIC: Melbourne University Press.

McLellan, J. M. (2017). *Soldiers and Colonists* (Master of Arts). Victoria University of Wellington, Wellington.

Meyer, M. (1996). Interview with Mary Meyer by Vicki Jones in 1996. Retrieved October 1, 2018.

Ministry of Culture and Heritage. (2018). Ormsby, John. Dictionary of New Zealand Biography—Te Ara. Retrieved January 24, 2018 https://teara.govt.nz/en/biographies/2o8/ormsby-john/print.

Mitsch, W. J., & Gosselink, J. G. (2000). The Value of Wetlands: Importance of Scale and Landscape Setting. *Ecological Economics, 35*(1), 25–33.

Morgan, J. (1849). Letter: Rev. John Morgan to Sir Geogre Grey. 31 December 1849. GLNZ M44.2. Grey New Zealand Letters. Auckland Libraries, Auckland.

Morgan, J. (1862). Letter: Rev. John Morgan to Sir George Grey. 11 August 1862. GLNZ M44.23. Grey New Zealand Letters. Auckland Libraries, Auckland.

Morgan, J. (1864). Reverend John Morgan to Browne, 29 December 1864, Gore Browne 1/2d, Archives New Zealand, Wellington.

Morris, J. D. K., & Ruru, J. (2010). Giving Voice to Rivers: Legal Personality as a Vehicle for Recognising Indigenous Peoples' Relationships to Water? *Australian Indigenous Law Review, 14*(2), 49–62.

New Zealand Parliament. (1858). Waste Lands Act 1858 (21 and 22 Victoriae 1858 No 75). Retrieved March 12, 2020, from http://www.nzlii.org/nz/legis/hist_act/wla185821a22v1858n75265/.

New Zealand Parliament. (1875). Napier Swamp Nuisance Act. Retrieved February 12, 2018, from http://www.nzlii.org/nz/legis/hist_act/nsna187539v1875n4343/.

New Zealand Parliament. (1881). Drainage Act. Retrieved April 9, 2020, from http://nzlii.org/nz/legis/hist_act/da188145v1881n27149/.

New Zealand Parliament. (1893). Land Drainage Act. Retrieved September 9, 2018, from http://www.nzlii.org/nz/legis/hist_act/lda189357v1893n46176/.

New Zealand Parliament. (1904). Land Drainage Act. http://www.nzlii.org/nz/legis/hist_act/lda19044ev1904n13192/.

New Zealand Parliament. (1908a). Hauraki Plains Act. Retrieved April 9, 2020, from http://www.nzlii.org/nz/legis/hist_act/hpa19088ev1908n21279/.

New Zealand Parliament. (1908b). Land Drainage Act. http://www.legislation.govt.nz/act/public/1908/0096/latest/whole.html.

New Zealand Parliament. (1910). Rangitaiki Lands Drainage Act.

New Zealand Parliament. (1924). Reserves and other Lands Disposal and Public Bodies Empowering Act. Retrieved January 24, 2018, from http://legislation.govt.nz/act/public/1924/0055/latest/DLM198760.html.

O'Donnell, E., & Macpherson, E. (2019). Voice, Power and Legitimacy: The Role of the Legal Person in River Management in New Zealand, Chile and Australia. *Australasian Journal of Water Resources, 23*(1), 35–44.

O'Regan, S. (1984). *Māori Perceptions of Water in the Environment: An Overview.* Waiora, Waimaori, Waikino, Waimate, Waitai: Maori Perceptions of Water and Environment.

Ormsby, A. (1907). Government Dealings with Maori Lands: From a Maori Point of View. 1907. *King Country Chronicle.*

Ormsby, A. S. (1911). Natives and Rates. *King Country Chronicle*, 9 September. Te Kuiti.

Ormsby, A. S. (1920). Taxing Native Lands. *Waipa Post*, 7 September, p. 6. Te Awamutu.

Park, G. (2001). *Effective exclusion. An Exploratory Overview of the Crown's Actions and Maori Responses Concerning Flora and Fauna, 1912–1983. Report Prepared for the Waitangi Tribunal.* Wellington: Legislation Direct.

Park, G. (2002). *Swamps which might Doubtless Easily be Drained: Swamp Drainage and Its Impact on the Indigenous* (Environmental Histories of New Zealand) (pp. 176–185). Melbourne: Oxford University Press.

Parsons, M. (2019). Environmental Uncertainty and Muddy Blue Spaces: Health, History and Wetland Geographies of Aotearoa New Zealand. In R. Foley, R. Kearns, T. Kistemann, & B. Wheeler (Eds.), *Blue Space, Health and Wellbeing: Hydrophilia Unbounded* (pp. 205–227). London: Routledge.

Parsons, M., & Nalau, J. (2016). Historical Analogies as Tools in Understanding Transformation. *Global Environmental Change, 38*, 82–96.

Pastor, M., Sadd, J., & Hipp, J. (2001). Which Came First? Toxic Facilities, Minority Move-In, and Environmental Justice. *Journal of Urban Affairs, 23*(1), 1–21.

Patrick, R. J. (2011). Uneven Access to Safe Drinking Water for First Nations in Canada: Connecting Health and Place through Source Water Protection. *Health & Place, 17*(1), 386–389.

Phillips, C., Johns, D., & Allen, H. (2002). Why did Maori Bury Artefacts in the Wetlands of Pre-contact Aotearoa/New Zealand? *Journal of Wetland Archaeology, 2*(1), 39–60.

Poelina, A., Taylor, K. S., & Perdrisat, I. (2019). Martuwarra Fitzroy River Council: an Indigenous Cultural Approach to Collaborative Water Governance. *Australasian Journal of Environmental Management, 26*(3), 236–254.

President of the Maniapoto-Tuwharetoa District Maori Land Board. (1909). President of the Maniapoto-Tuwharetoa District Maori Land Board to the Under Secretary of the Native Affairs Department, 11 January 1909, Te Kawa Swamp, Protest against Drainage (effects of eel weirs), MA1 973, Archives New Zealand, Wellington. Unpublished.

Romero Lankao, P. (2010). Water in Mexico City: What Will Climate Change Bring to Its History of Water-Related Hazards and Vulnerabilities? *Environment and Urbanization, 22*(1), 157–178.

Ruru, J. (2009). *Property Rights and Maori: A Right to Own a River?* In New Zealand Centre for Environmental Law Conference.

Said, E. W. (1978). *Orientalism* (1st ed.). London: Pantheon Books.

Salesa, T. D. (2001). "The Power of the Physician": Doctors and the 'Dying Maori' in Early Colonial New Zealand. *Health and History, 3*(1), 13–40.

Salmond, A. (2017). *Tears of Rangi: Experiments Across Worlds*. Auckland University Press.

Savage, A. G. F. (1847). Chapter 1: Journey into the Interior of New Zealand— The Waikato. In *Life and Scenes in Australia and New Zealand*. London: Smith, Elder, and Co. Retrieved July 18, 2018, from http://www.enzb.auckland.ac.nz/document/?wid=610&page=0&action=searchresult&target.

Simmonds, J. C. (1938). Letter from: J.C. Simmonds, Engineer, Mangapu Drainage Board, to T. Wightman Otorohanga. 15 December 1938. Folder: Rivers and Drainage—Waipa River. 1935–1953. Container Code: C 58 395. Archives Reference Number: BAAS A269 5113 Box 62. Item Reference: C. Record Number: 96/434220. Archives New Zealand, Auckland. Unpublished.

St. John, J. H. A. (1873). *Pakeha Rambles through Maori Lands*. Wellington: Robert Burrett. Retrieved November 27, 2018, from http://www.enzb.auckland.ac.nz/document/?wid=1471&page=0&action=null.

Swyngedouw, E., & Heynen, N. C. (2003). Urban Political Ecology, Justice and the Politics of Scale. *Antipode, 35*(5), 898–918. https://doi.org/10.1111/j.1467-8330.2003.00364.x.

Te Anga, H. Hone Te Anga v Kawa Drainage Board—(1914) 33 NZLR 1139 (High Court 1914). Retrieved January 7, 2018, from http://www.lawreports.nz/hone-te-anga-v-kawa-drainage-board-1914-33-nzlr-1139/.

Te Koro, N. (1908). Letter from Ngawere Te Koro and others to Thomas Fisher, 23 October 1908, Te Kawa Swamp, Protest against Drainage (Effects of Eel Weirs), MA1 973, Archives New Zealand, Wellington. Unpublished.

Thomson, J. T. (1867). *Rambles with a Philosopher or, Views at the Antipodes by an Otagonian*. Dunedin: Mills, Dick and Company.

Tipa, G., Nelson, K., Home, M., & Tipa, M. (2016). Policy Responses to the Identification by Maori of Flows Necessary to Maintain their Cultural Values. In *37th Hydrology & Water Resources Symposium 2016: Water, Infrastructure and the Environment* (p. 552). Engineers Australia.

Tua Tua Moana Swamp. (1915a, May 25). *Waikato Times*, p. 6.

Tua Tua Moana Swamp. (1915b, May 26). *New Zealand Herald*, p. 8.

Unknown. (1876, April 1). The Health of the District. *Waikato Times*, p. 2.

Unknown. (1914a, June 6). Eel Swamp or Dairy Farm. *Waikato Times*, p. 2.

Unknown. (1914b, April 24). A Migration of Eels. *Waikato Argus*, p. 2.

Unknown Author. (1846, March 14). Progress in Native Civilization. New Zealander. *New Zealander*, p. 3.

Unknown Author. (1854, February 25). Native Flour Mills. *Maori Messenger: Te Karere Maori*, p. 1.

Unknown Author. (1859, December 24). Notes of a Visit to the Thames and Waikato Districts. *New Zealander*, p. 5.

Unknown Author. (1862, July 7). Upper Waipa. From Our Own Correspondent. July 2nd, 1862. *Daily Southern Cross*, p. 3.

Unknown Author. (1863, September 5). Travels in Walkato.—Continued. *Taranaki Herald*, p. 2.

Unknown Author. (1864a, January 27). First Military Settlement in Waikato. *Lake Wakatipu Mail*, p. 3.

Unknown Author. (1864b, March 1). Military Settlers. *New Zealander*, p. 5.

Unknown Author. (1864c, May 12). Military Settlement of the North Island. *Nelson Examiner and New Zealand Chronicle*, p. 7.

Unknown Author. (1865a, February 28). The State and Prospects of the Waikato Settlements. *New Zealander*, p. 3.

Unknown Author. (1865b, March 24). Waikato Military Settlement. 24 March 1865. *Daily Southern Cross*, p. 4.

Unknown Author. (1865c, June 7). Lecture on the Waikato by Mr. Whyman. 7 June 1865. *Daily Southern Cross*, p. 4.

Unknown Author. (1867, May 29). Recollections of a Waikato Missionary. *Daily Southern Cross*, p. 5.

Unknown Author. (1873, February 22). Health in Perfumes. *Press*, p. 5.

Unknown Author. (1875, July 1). Health from Flowers. *Daily Southern Cross*, p. 1.

Unknown Author. (1876a, April 18). Tree Planting in Our Townships. *Waikato Times*, p. 3.

Unknown Author. (1876b, April 29). Planting Trees in the Waikato. *Waikato Times*, p. 2.

Unknown Author. (1878a). The Swamp Nuisance. *Hawke's Bay Herald*, p. 2.

Unknown Author. (1878b, May 9). The Journey to Alexandra. *Auckland Star*, p. 2.

Unknown Author. (1878c, May 9). The Soldier's Grave's. *Auckland Star*, p. 3.

Unknown Author. (1879, March 1). Tourist Notes on Hawke's Bay. *New Zealand Country Journal, 3*(2), 87.

Unknown Author. (1884). Cussin's Survey, Survey of the King Country. *Evening Post.*

Unknown Author. (1885, March 28). Ladies' column. Hints for Housewives. *Waikato Times*, p. 2.

Unknown Author. (1887, August 19). Waikato Swamps and Swampers. *Press*, p. 6.

Unknown Author. (1907a, December 28). Kawa Drainage Scheme. *King Country Chronicle*, p. 2.

Unknown Author. (1907b, October 22). Mcfarlane's Swamp. *Waikato Independent*, p. 4.

Unknown Author. (1907c, December 13). Untitled. *Kawhia Settler*. Kawhia.

Unknown Author. (1908, 4 September). District Pars. *King Country Chronicle*, p. 2.

Unknown Author. (1909, September 30). Dismal Swamp. *Waikato Independent*, p. 4.

Unknown Author. (1910, May 11). Kawa Drainage Board. *King Country Chronicle*, p. 3.

Unknown Author. (1914, June 12). The Eel Pa Case. *Waikato Argus*, p. 2.

Unknown Author. (1927). Obituary: John Ormsby. *Waipa Post*, p. 5.

Unknown Author. (1935, August 23). Waipa Drainage. *New Zealand Herald*, p. 13.

Unknown Author. (1970). Folder: Rivers and Drainage—Lower Waikato Waipa Control Scheme Raglan County—Rakaumanga Houses. 1970–1975. Archives Reference Number: BAPP 5113 Box 1674. Container Code: C 13 388. Record Number: 96/434000/0/105. Archives New Zealand, Auckland. Unpublished.

Vileisis, A. (1999). *Discovering the Unknown Landscape: A History of America's Wetlands*. Washington, DC: Island Press. Retrieved May 12, 2019, from https://books.google.co.nz/books?hl=en&lr=&id=gHRO32i6et0C&oi=fnd&pg=PR2&dq=east+anglia+wetlands+history+drainage&ots=X_A2_Hzqk N&sig=ZMtH7COAyILXiEQbOkftPilOWpo#v=onepage&q&f=false.

Von Hochstetter, F. (1867). *New Zealand and its Physical Geography, Geology and Natural History with Special Reference to the Results of Government Expeditions in the Provinces of Auckland and Nelson*. Stuttgart: J. G. Cotta. Retrieved July 18, 2018, from http://www.enzb.auckland.ac.nz/document/?wid=434&action=null.

Waitangi Tribunal. (2018). *Te Mana Whatu Ahuru: Report on Te Rohe Pōtae Claims Pre-Publication Version Parts I and II*. Wellington: Unpublished.

Walker, G. (2009). Beyond Distribution and Proximity: Exploring the Multiple Spatialities of Environmental Justice. *Antipode, 41*(4), 614–636.

Wanhalla, A. (2006). Housing Un/healthy Bodies: Native Housing Surveys and Maori Health in New Zealand 1930–45. *Health and History*, 100–120.

Westmacott, S. (1977). *The After-breakfast Cigar*. AH & AW Reed.

Whyte, K. P. (2014). Indigenous Women, Climate Change Impacts, and Collective Action. *Hypatia, 29*(3), 599–616.

Whyte, K. P. (2016). Is it Colonial Déjà Vu? Indigenous Peoples and Climate Injustice. In *Humanities for the Environment* (pp. 102–119). London: Routledge.

Whyte, K. P. (2018). Settler Colonialism, Ecology, and Environmental Injustice. *Environment and Society, 9*(1), 125–144.

Whyte, K. P., Talley, L., & Gibson, J. D. (2019). Indigenous mobility traditions, colonialism, and the anthropocene. *Mobilities, 14*(3), 319–335.

Wilkinson, G. T. (1892). G.T. Wilkinson, Otorohanga. 23 November 1892. Subject: Forwarding letter from Wekerua Ngamuka re closing of track through Kakepuku No. [Number] 12 Block. R22400219, ACHI, 16036, 1892, MA 1 863 1892–2099, National Archives, Wellington.

Wolfe, P. (2006). Settler Colonialism and the Elimination of the Native. *Journal of Genocide Research, 8*(4), 387–409.

Worboys, M. (1994). From Miasmas to Germs: Malaria 1850–1879. *Parassitologia, 36*(1–2), 61–68.

5

A History of the Settler-Colonial Freshwater Impure-Ment: Water Pollution and the Creation of Multiple Environmental Injustices Along the Waipā River

Water pollution offers significant openings for interrogating what environmental justice (EJ) and injustice means from an Indigenous perspective, and the implications of water pollution on Indigenous peoples' ways of life, values, and sovereign rights. Activist and scholarly literature from Aotearoa suggests there is widespread opposition amongst Māori to the disposal of human waste into water bodies (Broughton et al. 2015; Greensill 2010; Pauling and Ataria 2010). The discharge of wastewater (both treated and untreated) into waterways can be seen as unhealthy, unethical, and a physical and metaphysical attack on Indigenous bodies and their sovereignty. These critiques are embedded in Indigenous sovereignty discourses globally and nationally (concerning Te Tiriti o Waitangi—The Treaty of Waitangi and rangatiratanga—sovereignty, political authority), as well as Māori ways of seeing the world (premised on understandings of tapu—sacredness and noa—normal, ordinary, safe, not subject to restrictions). The history of the pollution of the Waipā River offers a particularly compelling case of the significance of recognising Māori iwi rangatiratanga, worldview, and values as part of environmental governance and management decision-making processes.

© The Author(s) 2021
M. Parsons et al., *Decolonising Blue Spaces in the Anthropocene*, Palgrave Studies in Natural Resource Management, https://doi.org/10.1007/978-3-030-61071-5_5

Māori experiences of environmental (in)justice, therefore, cannot be disconnected from the historic and contemporary injustices of settler-colonialism. Moreover, injustices are often not one single thing (distributive, recognition or procedural) and, therefore, singular interventions that seek to tackle only one driver of injustice are insufficient. Māori experiences of environmental injustices are not simply evidence of distributive inequities, or failure to recognise Māori values, or poorly designed (or deliberately exclusionary) decision-making procedures but are more often a combination of all three. Through an examination of water pollution of the Waipā River, in this chapter, we demonstrate that histories of environmental (in)justices are complex and intertwined with national and local histories, politics, and identities. Injustices, thus, are a product of direct and indirect policies and practices that build up over time (what First Nations scholar Kyle Whyte terms "sedimentation" and "insidious loops") (Whyte 2018, p. 130).

Water Pollution: An Unacknowledged Problem

Unlike Te Ao Māori, where steps are taken to avoid polluting water with human waste (as discussed in Chap. 3), water pollution due to the discharge of waste products barely registered as an issue within settler society throughout the nineteenth and early-to-mid-twentieth centuries. Instead attention was firmly fixated on the persistent problems of unruly rivers and unproductive swamps (Knight 2016), with only sporadic reporting of cases (in the early twentieth century) by parliamentarians and communities noting the contamination of waterways due to human activities (most notably mining operations). Throughout this period, towns pumped untreated human waste into waterways and seas, as did factories and farms (effluent mixed with chemicals). Until 1953, there was no specific national legislation nor government policies specifically focused on monitoring and mitigating water pollution (AJHR 1900, 1910, 1925). In 1953, the Water Pollution Act was introduced (Waters Pollution Act 1953) which, as its name suggests, was directed at governing water pollution (New Zealand Parliament 1953). The new act introduced by-laws for industrial waste products and provided for the creation of a new institution—the Pollution Advisory Council (PAC)—to prevent and mitigate water pollution throughout the country. The PAC was responsible for designing a national classification system for water quality and the development of by-laws about the disposal of wastewater. Initially, PAC lacked any power to monitor or control water pollution; however, from 1963 the Water Pollution Regulations allowed it to undertake investigations into the drivers and extent of water pollution. In 1956, the PAC released its provisions for acceptable inland and coastal water standards (related to effluent, pH balances, and minimum water treatment) (Cunningham 2014; Knight 2016).

In 1956, the Department of Health, the Ministry of Works, and the Department of Scientific and Industrial Research prepared a report for PAC on water pollution in the Waikato River catchment. The report was the first substantive environmental study into water pollution in the Waikato River and its tributaries and asserted that no "serious thought was given to [the] pollution problem ... [the rivers and streams] were grossly polluted" (Pollution Advisory Council 1956, p. 1). In regard to the Waipā River and its tributaries, the report found evidence of "serious pollution" in the form of extremely high faecal coliform counts downstream of the townships of Te Kuiti and Otorohanga and poor visual condition of the waterways (see Fig. 5.1) (Pollution Advisory Council 1956, p. 40). The report noted obvious examples of point source pollution from both townships' sewage systems and industry (specifically the Otorohanga dairy factory). In 1956, for instance, 70 per cent of residents living in Otorohanga district were connected to Otorohanga township's sewage system, which consisted of a single septic tank (capable of handling 110 000 gallons daily) that discharged directly into the Waipā River. "Conditions at the outfall" were reported to be "very bad" with large "quantities of paper, rag and fecal matter litter[ing] the river for a long distance downstream" (Pollution Advisory Council 1956, p. 20). Similarly, the two septic tanks, that serviced residents in the township of Te Kuiti (see Fig. 5.2), were observed to be operating ineffectively with the "soapy coloured effluent" staining the banks of the Mangaokewa stream (Pollution Advisory Council 1956, p. 20). Non-point source pollution, such as agricultural run-off, was identified as likely to be the most "widespread source of pollution" in the Waipā catchment, but it was not

Fig. 5.1 View of the Mangaokewa Stream, near Te Kuiti. Photograph taken by William Archer Price, 1866–1948. Collection of post card negatives. Ref: 1/2-000698-G. Alexander Turnbull Library, Wellington, New Zealand

THE RISING TOWNSHIP OF TE KUITI, KING COUNTRY, AUCKLAND.

Fig. 5.2 Te Kuiti township 1909. (Source: AWNS 19091202 6 3, Auckland City Libraries, Auckland, New Zealand)

THE PROGRESS OF THE DAIRYING INDUSTRY IN THE KING COUNTRY: THE OPENING OF A BIG NEW
FACTORY AT TE KUITI RECENTLY.   A. S. Hawley, Photo.

Fig. 5.3 Newly built factory near Te Kuiti (circa 1912). AWNS 19120208 10 4, Auckland City Libraries, Auckland, New Zealand

possible to "ascertain the degree of pollution" due to discharges from dairy farms (Fig. 5.3) (Pollution Advisory Council 1956, p. 48). The report did not acknowledge the significance of the waterways for Māori (as sources of water, food, medicinal, and art supplies as well as a treasure and an ancestor), though the authors noted that many Māori swam in the river.

Consequences of Pollution on Health

Despite being largely erased in official narratives about the river, Māori throughout the Waipā catchment continued to use and live beside the river; harvesting its freshwater biota such as tuna (freshwater eels), pūhā (stow thistles *Asteraceae*), and piharau (lamprey *Geotria australis*) for kai (food) for themselves, their whānau (family, extended family), their extended kin groups, and visitors. Sporadic mention in government reports and newspaper articles (typically accompanied by disparaging remarks about Māori 'uncivilised' ways of life) highlight that Pākehā (New Zealand European) were fully aware Māori continued to harvest resources from the Waipā freshwater system. These accounts also show Pākehā viewed Māori harvesting activities as being of lesser value than Pākehā-led economic development activities (extracting gravel from the riverbed, watering livestock, and discharging wastewater and dumping garbage into and beside rivers) and recreational activities (sailing and duck hunting in the peat lakes, rowing, boating, and sport fishing for trout) (Dixon 1937; Finlay 1923). Accordingly, the negative consequences of water pollution and other activities on Māori harvesting was of no concern to government officials (Department of Public Works 1928; Unknown Author 1900, 1935; Sullivan 1998).[1]

---

[1] It should be noted that from 1907 Pākehā (also referred to as New Zealand European) comprised the ethnic majority in the majority of the Rohe Pōtae (which encompassed the middle and upper Waipā River) and lower Waipā (within the wider Waikato region). In 1926 Māori were a minority in every part of Rohe Pōtae (except the small community of Kāwhia located on the West Coast of the district); and by 1936 Māori were the minority there as well (Robinson 2011). In many areas, particularly the towns of Otorohanga and Te Kuiti, they had become an even smaller minority. The New Zealand Census in 2013 reported that 71.6 per cent of the district's population identified as

During the first half of the twentieth century, as untreated sewage consistently flowed into the waterways of the Waipā and Māori continued to collect water and aquatic food supplies, government health officials noted that Māori were experiencing higher incidence of infectious diseases as well as higher infant and adult mortality rates than Pākehā. Of particular concern was outbreaks of typhoid fever (a bacterial infection linked to exposure to water and food contaminated with faecal products) (Unknown Author 1936). However, health officials did not necessarily ascribe typhoid to the discharge of untreated human waste in the waterways (despite international medical knowledge demonstrating contaminated water and food were the cause of typhoid outbreaks). Instead, officials blamed Māori refusal to abandon communalism (such as tangi or funeral practices), poor hygiene practices, and continued use of (supposedly) unhealthy wetlands as reasons for Māori becoming unwell (and sometimes dying) from typhoid (Anonymous 1884; Gott 1916; Unknown Author 1897, 1916, 1926, 1937). In reality, a significant portion of Māori households in Te Rohe Potāe, in the early-to-mid twentieth century, relied on the waterways for their drinking and cleaning waters (as they were not connected to town water supplies) and also harvested a significant portion of their foodstuffs from the rivers and wetlands. Accordingly, Māori were more likely to be exposed to any bacteria in the waterways than Pākehā (see Figs. 5.4 and 5.5) (Unknown Author 1916) (Wood 1950). Despite this, the government took no specific actions to reduce the health risks that polluted water supplies posed to Māori during the first three decades of the twentieth century. Eventually, in 1936, the Chief Medical Officer of the Waikato and Rohe Potāe districts (Turbott) sought and gained approval from the central government for the provision and installation of water tanks for Māori in the district as a way of improving Māori health (Unknown Author 1936). However, the Native Minister warned that the government support was conditional and would only be provided on the basis that "care should be taken to see that Maori [were] not relieved of the responsibility for providing for [themselves] those essential amenities that [were] well within [their]

Pākehā, 36.4 per cent as Māori (compared to the national average of 14.9), 2.9 per cent as Pacific peoples, and 2.3 per cent as Asian peoples (Taonga 2020).

Fig. 5.4 Unidentified Maori woman washing clothes on a rock, Te Kauri, Otorohanga. (Source: Ref/1/2-140360-G. Alexander Turnbull Library, Wellington, New Zealand)

capacity to provide". The Native Minister warned that there was a "danger that a benefit conferred on the people today might be considered their right tomorrow", one should not give "something for nothing" (Cunningham 2014, p. 214). Whereas, the state deemed that Pākehā residents were automatically entitled to portable water supplies, for Māori it was deemed a privilege (rather than a right) that they needed to prove themselves worthy of.

Early definitions of environmental racism, emerging from the early US EJ movement led by Black civil rights leaders, framed it as intentional, overt, and malicious acts of environmental injustice on "communities of colour" (Figueroa 2001; Pulido 2016). From this definition, Māori's higher exposure to polluted waters (than non-Māori) in the Waipā cannot be read as intentional or malicious acts. Although, the failure to provide Māori with potable water supplies does indicate a lack

Fig. 5.5 Unidentified Maori woman washing clothes by a small pool, Te Kauri, Otorohanga District. Ref/1/2-140364-G. Alexander Turnbull Library, Wellington, New Zealand

of care towards Māori by the state, which breached the third article of Te Tiriti o Waitangi, which promised Māori the same privileges and protections given to all British subjects, and treated Māori as second-class citizens.

More recently EJ scholarship, however, extends understandings of environmental racism beyond the narrow definition of malicious intent by government, industries, individual actors (Pulido 2017a, b; Pulido and Peña 1998). Instead, human geographer Laura Pulido and others, argue that environmental racism can be seen as a critical component of racial capitalism (discussed earlier in Chap. 4). Pulido work, in particular, documents how the US settler-colonial state consistently failed to address the environmental racism gap (between Black, Latinx and Indigenous peoples and the White US population) throughout the late twentieth and early twenty-first centuries, despite widespread awareness of it (recognised in federal and state policies), because of the financial and political costs. Meaningful actions to address these environmental injustices would not only be disruptive to industries, and the broader political system, and the settler-colonial state itself (Pulido 2017b). Instead, the US government (such as those run by the Environmental Protection Authority) developed a diversity of programs and policies in which it went through the motions (performances of regulatory activities), especially the participation of groups (Pulido et al. 2016), yet no meaningful changes occurred (Pulido 1998, 2017c). The issue was, therefore, not a lack of skill or knowledge about the nature of the problem, but rather a lack of political resolve that can be ascribed to racial capitalism (Pulido 2017b). Environmental racism, Pulido (2017b) argues, must be viewed in the context of a history of various arrangements of state-sanctioned violence which enables racial capitalism. Parallels can be seen in the Waipā context, despite the vast historical, socio-cultural, and political differences between Aotearoa and the US, in that Māori were expected to bear multiple burdens associated with state-led interventions. In addition to areas of land that were confiscated by the state (including the lower portion of the Waipā River catchment), Māori in the middle and upper catchment were expected to sell their lands at discounted prices to the government (as discussed in Chap. 4) for state-facilitated development efforts (including settlements and infrastructure) but received minimal benefits from such development projects (including limited access to freshwater supplies). Indeed, Māori (as individuals and members of whānau, hapū—sub-tribes, and iwi) suffered poverty, poor health, and inequitable access to basic social services as a consequence of state actions, yet the state consistently justified such actions on the basis that they were cost effective and necessary to ensure the development of a productive economy and prosperous (Pākehā) communities. Indeed, the state's actions (and inactions) that allowed for the ongoing contamination of the Waipā waterways with waste products (effluent from people, livestock, and industries) were essential components of the "ecology of capitalism" in Aotearoa (Moore 2015). Water pollution and its negative impacts of Māori were not simply incidental by-products of human habitation and development, but rather interwoven into the fabric of the racial capitalistic settler-state.

Disposal of Waste

According to Euro-Western understandings of waste management, industries and settlements required "sinks" to which they could remove and deposit polluted materials. These sinks were typically spaces (land, water, and air) accorded limited value. Just as specific places were devalued so too were bodies (human and more-than-human beings) which could function as "sinks" (Pulido 2015, 2017b). The Waipā River, its various tributaries, and its more-than-human entities were all sinks that sewage was discharge into alongside agricultural and industrial waste. Neither central nor local government authorities sought to implement the main recommendations of the 1956 report into the pollution of the Waikato River and the Waipā River (New Zealand Government 1956a, b; New Zealand Parliament 1953). The regional government authority (Waikato Valley Authority) argued that it was not responsible for water pollution (only flood risk), and instead devolved water pollution monitoring and regulation to local councils (Borough and County) (Unknown Author 1957). Accordingly, each borough and county council operated separate urban water infrastructure schemes with no consideration to what was happening upstream or downstream. For instance, the Otorohanga County Council extracted water for consumptive purposes from the Waipā River upstream of the township of Otorohanga, piped water to residents and businesses, and discharged the used water (including untreated sewage) back into the Waipā River downstream of the township. The same thing happened upstream at Te Kuiti and further downstream at Te Awamutu, and so on along the whole of the Waipā and Waikato rivers. Each council, therefore, was only concerned about water quality in so far as it pertained to the direct security of their township's water supplies; so long as there was no obvious health risk to local residents from town water supplies, then councils were unconcerned with ongoing pollution of the waterways.

Untreated human waste, therefore, continued to be pumped into the waterways with little consideration of consequences on people or biota. In the late 1960s Otorohanga Borough Council was still using the same septic tank system that was criticised in the 1956 report as was Te Kuiti

Borough Council; (later the institutions were renamed in the 1980s the Otorohanga District Council ODC and Waitomo District Council WDC) (Unknown Author 1966a, 1967, 1968). In 1970, the Inspector of Health noted that the Waipā River "holds no immediate potential for human consumption, untreated, or stock watering" due to human effluent (Unknown Author 1970).

In 1974, ODC filed yet another application with the Waikato Valley Authority to permit it to continue to discharge untreated wastewater until the oxidation ponds were completed (Unknown Author 1974a). The application was, however, opposed by the Environmental Defence Society (formed in 1971 by a group of concerned lawyers and students) who submitted that the "discharge of effluent would detrimentally affect the receiving waters" and would negatively impact the recreational use, "scenic and natural features and fisheries" of the river, which were contrary to Section 20 (6) of the Water and Soil Conservation Act (1967) (Unknown Author 1974a, b, c, d, 1975a). The submission's referenced after recreational use, fisheries, and natural features; no mention was made to Māori connections and usage of the river and its resources. Despite the Environmental Defence Society's submission, the Authority granted Otorohanga County Council the right to discharge wastewater, which was extended to allow (for up to 600,000 litres of wastewater per day) once its new oxidation ponds were finished in mid-1975 for a period of ten years (Unknown Author 1975b).

No specific recognition (by the state or environmental groups) was given to how water pollution was (or could) negatively impact Māori lives, livelihoods, and their unique relationships with their whenua (land), awa (river), and moana (sea). Indeed, government legislation and assessments of the pollution status of the Waipā River (and other rivers in the country) were produced in the continuing haze of settler privilege. Western scientific knowledge, and the values, concerns and priorities of Pākehā entirely dominated discussions of water management, and mātauranga Māori (Māori knowledge), ways of seeing the world, and management approaches were wholeheartedly excluded from scientific, political and public discussions about water management until the closely decade of the twentieth century. The views of Ngāti Maniapoto and other iwi about water pollution were rendered largely silent in the written records of government officials in the first seven decades of the twentieth century; it seems that Pākehā officials did not deem Māori complaints about water pollution as worthy of writing down in their departmental memos. However, other sources (oral histories, memoirs, legal cases, and protests) highlight how Māori around Aotearoa continued to hold fundamentally different understandings of what constituted clean water and sought to challenge the settler state (and of Pākehā more generally) once even a voice within planning forums.

In addition to the settler-colonial violence and discriminatory policies, the reality of being an ethnic minority in their rohe (aka to Australian Aboriginal *country* and First Nations homelands) meant that Māori were more and more excluded (deliberately or incidentally) from decision-making processes about freshwater management. Thus, distributive inequities (higher exposure to environmental risks) were compounded by procedural injustices (institutional arrangements that prevented or restricted Māori participation in environmental planning processes), which were further exuberated by the failure to recognise Māori (values, knowledge and authority). In terms of Māori understandings of the Treaty, Māori are Treaty partners with decision-making authority meant to be shared between iwi and Crown; in terms of tikanga Māori (customary laws), iwi authority rests in their status as tangata whenua (who possess decision-making authority over an extended area of tribal lands/waters that they shared with other iwi) and as mana whenua (who hold spiritual authority over a narrow area of tribal lands/waters that they do not share with other iwi). Misrecognition (of specific connections between iwi and their whenua and awa) then, for Māori, often overlapped with procedural injustices, wherein government policies and processes regarding river management (as were seen earlier in regard to wetland drainage) did not provide any space for Māori to be able to meaningfully participate in and be able to shape decisions.

The Resource Management Act and the Limits of Recognition

By the 1980s, there were growing concerns about the effectiveness of environmental administration within Aotearoa in light of ongoing environmental degradation. For Māori, these concerns also extended to the limited opportunity for Māori to participate formally in environmental decision-making and management (Burton and Cocklin 1996). In the 1990s, more than 150 years after the signing of Te Tiriti o Waitangi, changes to legislation (most notably, the Resource Management Act 1991 (RMA)) and to local government sought to improve environmental management and decision-making (Harmsworth et al. 2016; Jacobson et al. 2016; New Zealand Parliament 1991; Thompson-Fawcett et al. 2017; Tipa et al. 2016). The passing of the RMA, by the Fourth Labour Government (who also were the only government to announce Treaty principles), ushered in fundamental changes to environmental management in Aotearoa by replacing 59 statutes and amending more than 150 others (Knight 2016, 2019).

The purpose of the RMA is "to promote the sustainable management of natural and physical resources" (section 5) (Fig. 5.6). To achieve this, the act mandates an effects-based approach to sustainable management and provides an integrated regime for managing land, water, air and ecosystems. By shifting the focus from the causes of degradation (as was the case with previous legislation such as the Water and Soil Conservation Act 1967) to the effects of activities on the environment, the RMA provided a regulatory regime (supposedly) capable of addressing the degradation and pollution affecting freshwater systems (Crow et al. 2018; Knight 2016; New Zealand Parliament 1967). It also allowed for greater public participation than previous legislation (Burton and Cocklin 1996; Lowry and Simon-Kumar 2017). For more than a century, Maori groups had demanded central government allow them to participate formally in environmental management; the RMA, for the first time, provided formal mechanisms for Māori to participate in planning processes (Burton and Cocklin 1996).

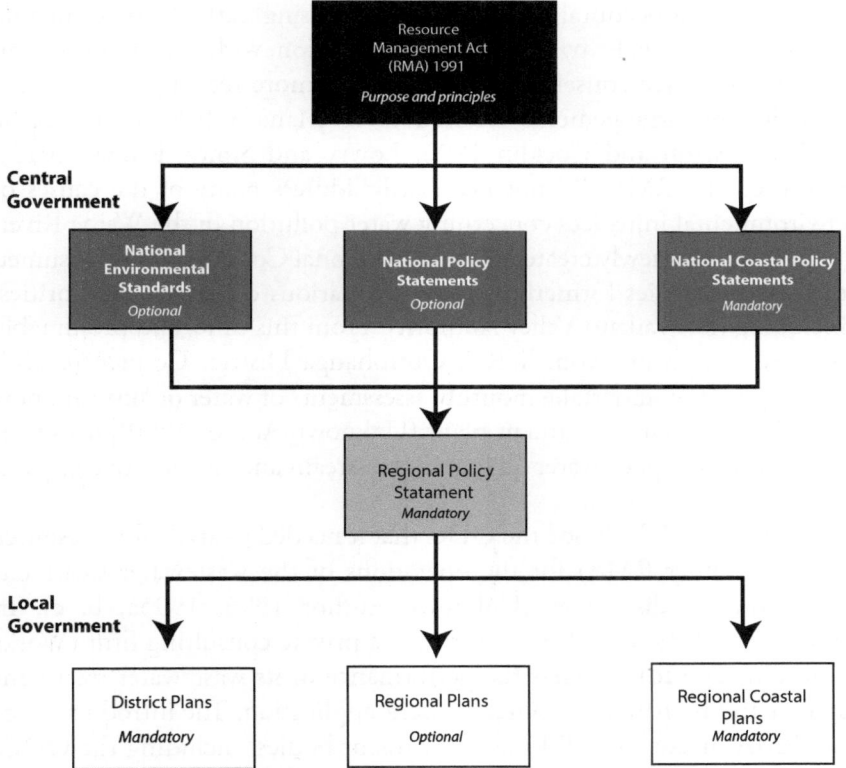

Fig. 5.6 Diagram outlining relationship between Resource Management Act and other local government mechanisms. (Source: Authors' Own)

The RMA includes specific provisions related to Māori in achieving the purpose of the act whereby all persons exercising functions and powers under it: shall recognise and provide for the relationship of Maori and their culture and traditions with their ancestral lands, water, sites, wāhi tapu (sacred sites), and other taonga (treasures, treasured possessions) (section 6e); shall have particular regard to kaitiakitanga (section 7a); and shall take into account the principles of the Treaty (section 8). Language of this kind was glaringly absent from earlier environmental (and other) legislation throughout the twentieth century (Burton and Cocklin 1996). Despite some limitations, the RMA did go some way to addressing problems of procedural inclusion characterising earlier environmental legislation especially by requiring consultation with Māori over plan changes, resource consent applications and, more recently, through iwi participation arrangements referred to as Mana Whakahono a Rohe (s58L) (Burton and Cocklin 1996; Lowry and Simon-Kumar 2017). However, the RMA did not necessarily address many of the causes of environmental injustices concerning water pollution in the Waipā River.

In 1989, the newly created Waikato Regional Council (WRC) assumed the responsibilities formerly held by the various catchment authorities, including the Waikato Valley Authority. From this time, and presumably because of pressure from WRC, Otorohanga District Council (ODC) was required to undertake monthly assessments of water quality and flow from its wastewater treatment plant (Unknown Author 1990); tests later highlighted the poor water quality in the stream and the negative impacts of aquatic biodiversity.

In 1995, WRC advised the ODC that it needed to apply for a resource consent (under RMA) for the operations of the wastewater treatment plant and its discharges (Unknown Author 1993, 1995a, b, c). In November 1995, the ODC contracted a private consulting firm (Works Consultancy Ltd) to assess the performance of its wastewater treatment system and prepare its resource consent application. The introduction of the RMA meant that all local government bodies, including the ODC, were required to "demonstrate active consultation with any party that may have an interest in the effects of your operation on the environment" (Unknown Author 1993, 1995a, b, c). The ODC decided to delegate its responsibilities for actively consulting with Māori and stakeholders to the consultants (led by engineer Peter Askey). The consultants formed an "Oxidation Ponds Working Party", made up of various representatives from different groups: community groups (Otorohanga Community Board, the Ratepayers Association); the farming industry (Federated Farmers); recreational advocacy bodies (Fish and Game); central government agencies (the Department of Conservation); environmental non-government organisations (the Royal Forest and Bird Society); and local Māori (Askey 1995a; Unknown Author 1995d).

The consultants' consultation process with iwi began with an initial meeting held at Te Kawa Marae on 11 October 1995 with Te Nehenehenui

Regional Management Committee (Nehenehenu RMC). At the meeting, attendees from local iwi and hapū (affiliated to Ngāti Maniapoto) spoke at length about their views about the treatment plant and how waste was being inappropriately managed by the council. They reported how unpleasant smells were emitted from the plant and the treat wastewater it released into the stream was of poor quality (still filled with solid materials). By far and away the most significant issue, iwi attendees informed the consultant (Askey), was the practice of discharging human waste directly in the waterways. Human waste (irrespective of how it was treated using scientific methods and technologies) should not be disposed of into waterways. The iwi did not challenge the council on the grounds of science or engineering approaches (such as the treatment plant's use of oxidation ponds and filters) but instead argued that the discharge of waste into waster was wrong (on cultural, spiritual, and ethical grounds). One iwi representative, Bronwen Hughes, argued that the oxidation ponds and discharge into the waterways were "culturally insensitive" to tikanga because "they expose human waste" to everything in the river and breached rules regarding tapu (Unknown Author 1995e; Works Consulting Ltd 1995).

Iwi attendees spoke of how the discharges of wastewater were damaging fish life within the Waipā River catchment. The practices were decreasing the health and wellbeing of both humans and biota. Any "humans who might consume the fish or plants harvested from the waterways downstream of the discharge" from the Otorohanga plant were risking their physical and spiritual wellbeing (Unknown Author 1995e; Works Consulting Ltd 1995). The polluted waters meant Māori capacities to harvest foods from their mahinga kai (food gathering sites) were diminished, and they reported feelings of loss; in particular, members of local hapū were meant to harvest and cook foods sourced from their mahinga kai (such as tuna) to guests when they visited their marae and not being able to was a source of deep sadness and could diminish the mana (social status, power) of the hapū.

The 1995 Otorohanga working party, formed in haste by the Works Consultancy Ltd but representing the ODC, conducted its first meeting with tangata whenua at Te Kawa Marae. Two people from the Nehenehenui RMC (Richard Rangitaawa and Victor Tapara) were asked to join the working party as representatives of tangata whenua (Unknown Author 1995e). Nehenehenui RMC represented a number of hapū affiliated to Ngāti Maniapoto, who (as detailed in Chap. 2) trace their whakapapa (genealogical connections) to the Waipā River and consider the awa to be their kin. The consultants, whether by design or by ignorance of Māori protocols, did not invite other hapū or iwi (who possessed connections to the Waipā River) to join the working group. Later in December 1995, one representative from Te Mauri o Maniapoto was invited to join the working party (presumably because of complaints that the working party was excluding some tangata whenua groups). After inspecting the Otorohanga treatment plant one of the iwi representatives on the working group expressed serious concerns about the water quality in Mangaorongo Stream (a tributary of the Waipā). The impacts of wastewater discharged by the plant into the stream was negatively impacting both the quality of the water and health of aquatic fauna (fish species and tuna) health. And in doing so "food sources such as fish and eels for human consumption" were being contaminated, eating food contaminated by human waste threatened the health and wellbeing of tangata whenua, posing material risks (in terms of infectious diseases) and cosmological risks (in terms of breaching tapu and damage to mauri and wairua) (Unknown Author 1995f).

The consultants informed iwi representatives of the working group that they were right to be concerned about the health of their awa, as the discharged wastewater was causing adverse effects on the fish life in the Mangaorongo Stream. However, the consultants' reported that the majority of water pollution within the catchment was a product of run-off from farms (non-point of sources pollution) and therefore the discharge of wastewater was only a small contributor (Askey 1995b).

As a consequence of their investigations and consultations, the consultants recommended to the ODC various remedial engineering solutions to address the issues at the ODC treatment plant, including increased sludge removal, pond aeration, and the construction of a wetland. Iwi preferences for treated waste to be discharged first into onto the land were briefly noted ("earth contact for final treatment prior to mixing with water in the stream"), however no specific engineering solutions were discussed as the idea was deemed too financially costs (Unknown Author

1966b). The ODC and WRC, on receiving the report in July 1996, was supportive of the consultants' recommendations; however, iwi were less euthasiastic (Environment Waikato 1996).

The consultants distributed their draft report and recommendations to iwi as part of consultation processes. Nehenehenui RMC, although supportive of the general aims of the proposed engineering works to improve water quality, raised questions about how the council was going to mitigate the odorous gases, monitor the discharges, address the impacts of discharges on the health of biota and people, and taken into account Māori values. Nehenehenui RMC once again raised the question of why it was necessary to discharge human effluent into the waterways and why an alternative approach could not be adopted. There were "many other cultural considerations" that were being missed by consultations, and it was "culturally … quite offensive" for Māori to drink water that was (even in part) contained human waste products (Unknown Author 1996a). Other iwi representatives, from Te Mauri o Maniapoto, meet with the consultants in Otorohanga in September 1996 and reiterated the concerns of Nehenehenui RMC discharge of treated wastewater into the stream.

The consultants rejected Māori requests for land-based waste disposal due to financial considerations (it would require large areas of land and would cost the council a lot of money) (Askey 1996). However, they did eventually propose the replacement the final 10 to 20 metres of the effluent line with an earth trench (before the waste flowed in the stream) and to establish a wetland to go some way to address Māori preferences for land-based disposal. Te Mauri o Maniapoto representatives were pleased with the proposed measure as it would "allow for the treated effluent to contact with [Papatūānuku—Earth Mother] before entering the stream, and the treated effluent would look like a natural spring flowing into the earth" (Unknown Author 1996b). And despite their ongoing concerns about water quality, ultimately, both Te Mauri o Maniapoto and Nehenehenui RMC agreed (in March 1997) to support the resource consent application of the ODC. Their support was conditional on the basis that: the proposed engineering works included the construction of both an earth trench and wetlands; the resource consent duration was reduced (from 15 years to 10 years); and ongoing water quality monitoring were conducted.

Even after both groups agreed to support the ODC, many iwi members expressed continuing opposition. Three individuals from the Nehenehenui RMC (Massey Ormsby, Rachael Ormsby, and Jacqui Amohanga) filed an additional submission to the WRC in November 1997 that outlined their continuing concerns about human waste entering their awa.

The submission declared the discharge of wastewater "into the Mangaorongo Stream must cease completely!" and called for the council to embark on a long-term shift in wastewater management focused on land-based disposal to address the environmental degradation of their awa (Amohanga et al. 1997, p. 9).

Ngāti Maniapoto opposition to the disposal of waste products remained fundamentally interwoven with Māori understandings of waste as tapu and the need to ensure that tapu was kept apart (which we outlined in Chap. 3). With the risks posed by breaching tapu by discharge human waste into waters extending beyond simply scientific assessments of water quality, but this did not mean that Māori only drew on mātauranga to justify their concerns but rather multiple knowledges were used to try to communicate to non-Māori audiences (predominately Pākehā) and get them (be it government officials, consultants, scientists, and other decision-makers) to take their perspectives seriously. However, Māori understandings (of water, waste, health, and wellbeing) were frequently lost (or misrecognised) by non-Māori (who were primarily Pākehā) in discussions over not only the operations of the Otorohanga treatment plant but waste management schemes that were taking place throughout the Waikato region and the entire country.

Throughout the 1990s, Māori individuals and groups filed petitions, staged protests, submitted claims to the Waitangi Tribunal, and mounted legal cases that challenged government environmental management approaches. The Waitangi Tribunal was established as a permanent commission of inquiry to investigate Māori claims that the Crown was not honouring the terms of the Treaty of Waitangi. Tribunal responsibilities include researching and holding public inquiries into historic and contemporary claims filed by any Māori individual or group regarding

Crown breaches of the Treaty, reporting back to claimants and the Crown about inquiry findings, and making recommendations to the Crown regarding how it can address Treaty breaches (reconciliation and restorative justice) (Jones 2016; Mutu 2018, 2019; Wheen and Hayward 2012).

Many Māori complaints drew attention to how freshwater and saltwater spaces were continually being polluted by human waste, and how the discharge into water was a fundamental breach of rules around tapu. Māori vocally campaigned for their values to be respected and that direct actions be taken to stop the discharge of waste products into waterways. Waste products, once discharged into waterways, result in both the receiving waters as well as all those beings that are connected to those waters (through whakapapa) become unhealthy; their mauri (life force) and wairua (spiritual integrity) diminished by the tapu of human waste. Accordingly, from a tikanga perspective, human waste products (even if treated using the best scientific and technological methods) should always be kept away from bodies of water (be it a river, harbor, or sea). As Māori scholar, an expert on tikanga (customary laws), Sidney (Hirini) Moko Mead (from iwi Ngāti Awa) articulates:

> The rules of tapu advise Maori to separate the clothes one wears from cloths associated with food such as table clothes and tea-towels. Babies' napkins and cloths associated with menstruation are kept away from food utensils. By extension these rules apply to the separation of sewage which include some human body parts .... This very tapu mixtures needs to be separated from the food we eat not only because of its spiritual attributes but also for health reasons. The institution of tapu operates for the well being of people ... Break the rules and immediately people are unsettled in the minds, are fearful of their well being because [their] basic beliefs are being transgressed. Blood is tapu ... A body part of a living person is tapu. Excreta is tapu ... There is no problem [in terms of Māori customary laws] with the return of excreta or body parts to Papatūānuku [the Earth Mother] ... What is abhorrent is the idea of associating biosolids with the food chain. (Mead 2016)

Mead's quote is in line with those articulated by Māori scholars and leaders. For instance, Aila Taylor, a key spokesperson for Te Atiawa during the Waitangi Tribunal's inquiry into the Crown's failure to honour its Treaty obligations, expressed similar views about waste disposal:

> What comes from the earth goes back to the earth. We believe that human waste should go back to the earth. We believe that anything to do with human waste should have nothing to do with food … seafood should not be gathered from reefs polluted by an [sewage disposal] outfall. This belief is not just related to 'scientifically detectable' pollution; even if scientists 'proved' that an outfall was not polluting, we would be unhappy gathering seafoods from a reef near such an outfall. (Taylor and Te Taha Māori 1986, p. 2; Waitangi Tribunal 1993, pp. 12–14)

Both Taylor's and Mead's quotes (and those earlier from representatives of Ngāti Maniapoto) highlight Māori understandings of human waste continued to be found up in the principle of tapu. The ways in which the tapu of waste pollutes water and food (both of which carried their own tapu status) is not able to be quantifiable or measured by Western scientific knowledge but instead represents a fundamentally different way of knowing the world; such ontological and epistemological differences, as we and other scholars previously argued, should not be positioned in opposition, integrated together, or privileged one over the other, but instead allowed to exist as equally separate and important systems of knowing, thinking and doing) (Hopkins et al. 2019; Howitt and Suchet-Pearson 2006; Parsons et al. 2017) (see Fig. 5.7). Accordingly, throughout Aotearoa Māori individuals and groups (since the passage of the RMA allowed them processes to voice their concerns) articulated over and over again to decision-makers their opposition to waste being discharged into water.

Through an autoethnography study, for instance, Māori scholar-activist Angeline Greensill (iwi Waikato-Tainui, hapū Tainui) documents her own personal (and those of her whānau and hapū) involvement in efforts to prevent discharge of waste into Whaingaroa Harbour (now more known by its settler-name of Raglan). Located on the West Coast of the Waikato region, the township of Raglan's treatment plant parallels those in operation along the Waipā River. The local council constructed a new sewage system, which consisted of two oxidation ponds on top of

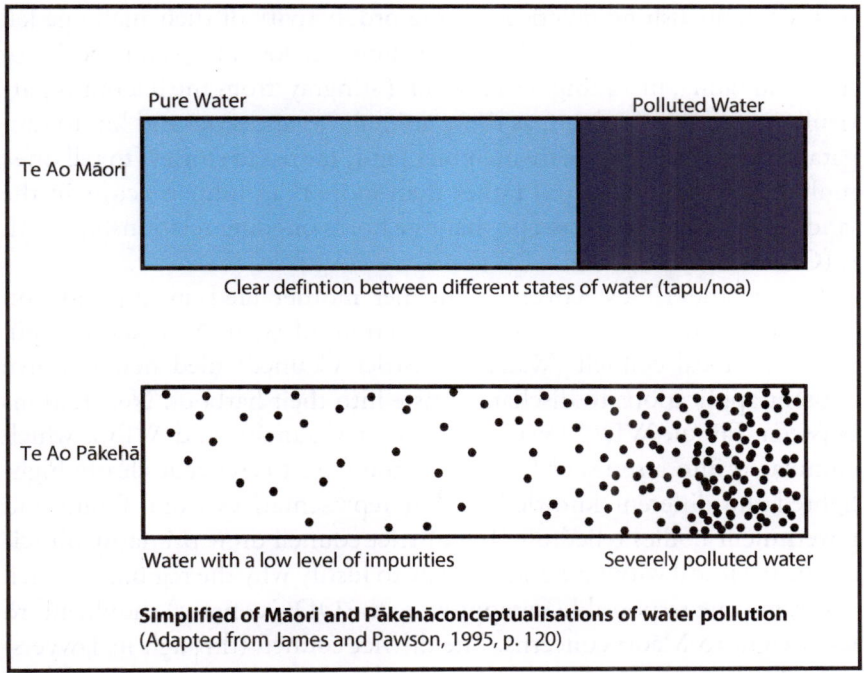

Fig. 5.7 Māori and Pākehā representations of Water, adapted diagram from James & Pawson (1995, p. 120)

an important wāhi tapu (Te Rua o Te Ata) and built the discharge pipe over land belonging to the local hapū (Tainui). Hapū members, including Greensill and her mother (the Māori activist and leader Evia Rickard), protested against the council's actions, but were unable to stop the new plant being built. The pipeline discharged treated wastewater (which included human waste) into the mouth of the harbour. Greensill reports that, as a consequence of the discharge of sewage and other acts of environmental violence committed against their waters, hapū no longer could swim in their waterways (due to health concerns). Likewise, hapū could no longer harvest culturally important foodstuffs as a consequence of the plethora of acts of ecological violence (including effluent discharge, agricultural run-off, sedimentation, removal of native plants and animals, unsustainable fishing practices) as the productivity of their mahinga kai noticeably declined. Greensill records how the loss of certain species of flora and fauna, including seahorses and stingray, from the habour is particular distressing for hapū as those animals (as ancestors and kin to tangata whenua) become "mere memoirs [and] more sad stories" to tell one's mokopuna (grandchildren) rather than more-than-human actors in the landscape/waterscape/seascape that one holds ongoing relationships with it (Greensill 2010, p. 33).

During the 1990s, Greensill and her mother filed an objection on behalf of Tainui hapū, paralleling the actions of Ngāti Maniapoto hapū, when the local council (Waikato District Council) filed new resource consent applications to discharge waste into their harbour. Negotiations between Tainui, WDC (Waikato District Council), and WRC, which cumulated into a series of Environmental Court cases that clearly highlighted the different knowledges that representatives from Tainui and government bodies relied on. The district council drew primarily on scientific technical witnesses and case law to justify why the regional council (the same one responsible for approving the ODC consent) should afford less weight to Māori concerns. The district council (through its lawyers) argued that: "Māori cultural and spiritual beliefs as to protection of water from discharges should not be accorded an absolute entitlement". Moreover, the economic development needs of the entire district and its population, lawyers for the district council argued, needed to take precedence over the minor concerns of tangata whenua. Similarly, the district council's engineer downplayed Māori concerns about the wastewater disposal as mere "perceptions", with the "cultural issue" of waste disposal something that could not be provided through scientific assessments and therefore held no weighting (be it in terms of water management practices nor legal standards).

In 2004, the Environment Court ruled in favour of the Waikato District Council and against the Tainui hapū to permit the council to continue to discharge waste into the harbour. The court announced that there was scientific evidence that showed that the disposal of treated effluent would only cause a small amount of negative effects on the environment and most environmental degradation was caused by farming and deforestation practices (Greensill 2010, p. 61). The Environment

Court's decision stated that the evidence presented by Tainui was "merely assertions" (cultural and spiritual rather than scientific) "not so much of effects as such but of indirect or consequential results of the discharge". The judge, therefore, placed far greater weighting on the words of non-Māori 'experts' (who were overwhelming Pākehā men with engineering and planning degrees) and New Zealand (settler-colonial) case law rather than Māori 'experts' (rangatira/chiefs, kaumatua/elders) and tikanga Māori.

The decision of the judge in 2004 was unsurprisingly given the state's legal system's foundations in British colonial laws as well as the continued dominance of settler values that privileged scientific (Pākehā) knowledge and technological interventions over other ways of knowing. It reveals strong parallels to what was happening in the Waipā River. The Tainui and Otorohanga examples are just two of countless others from across Aotearoa where Māori continue to experience injustices in environmental planning regimes (despite sections of the RMA being designed to include Māori in decision-making processes and recognise Māori values). Māori values were (and still are) often dismissed and denigrated as merely (trivial, uncivilised) cultural or spiritual perspectives, which hold less weight than the ideas, values, and practices of the dominant settler society (originating from Western Enlightenment thought) (Greensill 2010, p. 33). Thus, while the RMA did signal a significant change in legislation that allowed for some degree of Māori participation in decision-making (procedural inclusion) and recognition of Māori as a culturally distinct group with specific interests in their local environment, it did not translate into any substantive changes to environmental management. Indeed, how the RMA was interpreted and applied by local government bodies across the country (including within the Waipā River Catchment) did not necessarily resolve many of the environmental injustices experienced by iwi (as settler-colonialism and capitalism remained hegemonic).

The recognition of Māori interests in water and their concerns about water pollution failed to challenge the status quo and did not necessitate that the state (be it local government or central government) reform in any significant way. We trace this failure (tied to the persistent recognitional injustices as well as lack of participatory parity) partly to the structure and operations of local governments. There is an ever expanding literature on the relationships between Māori and local government (Bargh 2016; Hayward 2011; Ryks et al. 2010), including those focused on partnerships (Lewis et al. 2009) and environmental co-management arrangements, highlights the problematics of recognition based politics in Aotearoa (Coombes and Hil 2005; Forster 2016; Lowry and Simon-Kumar 2017; Morgan and Te Aho 2013; Muru-Lanning 2012; Te Aho 2015). Avril Bell, for instance, highlights two significant interconnected problems in the structural underpinnings of iwi-local government relations: (1) the issue of the political representation of Māori (or the lack therefore) in elected local government bodies; (2) local government not being given the status of Treaty partners (Bell 2018). At the national level, there are seven Māori seats (out of a total of 120 members of parliament), with Māori voters given the choice every five years if they enrol on the general electoral roll or the Māori roll (with the number of Māori seats adjusted to reflect how many people are on the Māori roll as a proportion of total population). Local government bodies, since the Local Electoral Amendment Act (2002), have had the choice of creating dedicated Māori seats on local councils; however, only one has done so. The Bay of Plenty Regional Council established three Māori seats alongside four general seats, thus ensuring that there was fair representation for the 28 per cent of the Bay of Plenty population who were Māori. Māori individuals can stand for local body elections and some do so. However, Māori are "chronically underrepresented" in local councils as only 3.6 per cent of local councillors in 2007 were Māori (even though Māori made up 15 per cent of the population) (Hayward 2011, p. 187). Moreover, local councillors, who were elected to general seats, were not required to represent Māori but rather the entire community. Thus, "Māori issues and interests [were] even more seriously underrepresented within the local government sector than even these figures suggest" (Bell 2018, p. 82). In addition, while the central government by 2002 did recognise iwi (even if it was and still is inadequate), they did not place the same obligations on local government (as they were not classified as Treaty Partners). From the perspective of local government, Bell (2018) argues, hapū and iwi were simply accessible and convenient organisations to consult with to meet their duties (under the RMA and Local Government Act) vis-à-vis consulting with and involving Māori in decision-making processes. Thus, while most local authorities did possess some form of consultation processes (as demonstrated in the ODC wastewater treatment plant example) in place with iwi and hapū, these processes fell short of the Treaty governance partnerships that iwi sought.

Unsurprisingly (given the lack of Māori representation), local governments continued to place greater weighting on the submissions of non-Māori groups, even when groups employed non-scientific arguments to justify their concerns. For instance, the only submission the WRC received (aside from Māori submissions) that expressed opposition to the ODC's resource consent application about the Otorohanga treatment plant, in 1996–1997, came from the local chapter of Fish & Game. The not-for-profit organisation, Fish & Game (originally formed as acclimatisation societies that established to introduce exotic plants and animals to Aotearoa), sought to ensure that additional conditions were added to the resource consent to protect sport fish (trout) and game birds (ducks). Fish & Game requested that the ODC planted riparian vegetation downstream of the discharge point (which was incorporated into the resource consent conditions) (Auckland and Waikato Fish and Game 1997). It is worth noting that while Fish & Game's request for riparian planting was simply incorporated into the council's policy for the treatment plan, the numerous requests made by Māori (lodged in the days, months, years, and decades before and after Fish & Game's submission) to protect and/or restore vegetation along the riverbanks were rejected on the basis that vegetation impeded the flow of water and created flood hazards.

Procedural and Recognition Environmental (In) Justices: Continuity and Change

In the first decade of the twenty-first century, Ngāti Maniapoto Trust Board (as the representative body of the wider Ngāti Maniapoto iwi) articulated their concerns about the declining water quality of the Waipā River. Two iwi reports, the State of the Environment Report (Kowhai Consulting Ltd and Ministry for the Environment 2002) and the Ngāti Maniapoto Management Plan (Kowhai Consulting Ltd 2007), highlight changes within the Waipā river and the negative implications on their mahinga kai. The 2002 state of environment report summarises:

> Looking at the Waipa River today, it is hard to believe that our tupuna spoke of a time not so long ago when the waters of the Waipa were clear, deep and blue. Within the clear, clean waters of the Waipa were fat eels, large crayfish, and a variety of fish, a plentiful source of food. Children in the vicinity of the Waipa made regular trips to the river for swimming, eeling and fishing. For most of us today, this is a dim memory or a legend of times past. The waters of the Waipa now run muddy brown, polluted with farm run-off, industry discharges, sewerage spills and stormwater drainage. Many food species have disappeared from the river, and the remaining tuna/eels within the river may not be safe for eating. (Kowhai Consulting Ltd and Ministry for the Environment 2002, p. 7)

Ngāti Maniapoto highlights how the degradation of the Waipā River and its tributaries is an indicator of the state of not only the hauora (health) and mauri of the river, but also the interconnections of awa, whenua, biota, and people (specifically mana whenua):

> The streams and rivers are the lifeblood of our environment, and they tell us about the state of our environment, the forests, lakes, oceans and seashore. As with the human body, if the blood of the environment is poisoned, the rest of the body will also suffer. (Kowhai Consulting Ltd and Ministry for the Environment 2002, p. 11)

"While not … opposed to development", Ngāti Maniapoto stress, that the historic costs of development "to the environment is unacceptable. It is time to—restore some balance" (Kowhai Consulting Ltd and Ministry for the Environment 2002, p. 11). The above quotes indicate, once again, how Māori ways of thinking (in this instance those of Ngāti Maniapoto) are place-based and kin-centric and premised on the reciprocal relationships between human and more-than-human actors, which differed markedly from Western ways of thinking (particularly those held by decision-makers who were and still are the ones making decisions about how water and waste are managed in Aotearoa).

The Otorohanga District council, following on from the 1996 consultants' recommendations and its successful resource consent application, did acknowledge (to some degree) hapū and iwi concerns about river health and Māori opposition to water-based waste disposal methods. As a requirement of its resource consent to continue to discharge wastewater into the stream, which was granted by the WRC, the district council constructed an earth trench (consisting of a channel with rocks placed it and a wetland) to address Ngāti Maniapoto concerns (Unknown Author 1966b, 1995f, 2010a). The ODC was also meant to investigate land-based disposal options as per Māori requests (but it did not undertake these investigations before 2010). Indeed, the works that ODC did undertake, to respond to Māori concerns (about declining water quality, loss of biodiversity, noxious smells, and breaches of tapu), were all haphazard and made with minimal investment (in terms of time and money). The earth trench and wetlands appear to be constructed as an afterthought or a tokenistic act to appease Māori; simply a box ticking exercise that the district council went through to ensure it could get its resource consent to continue to operate its waste management plant with minimal changes to the status quo. The council instead spent the vast majority of its resources (financial and human) on upgrading the oxidation plant and limited resources on the fundamental problem (in terms of tikanga Māori) of the disposal of waste into waterways. Unsurprisingly, a few years later (in 2010) when a new consultancy firm (this time Cliff Boyt Consulting) was employed by ODC to prepare yet another resource consent application, the consultants found both the earth trench and wetlands were substandard (Stammers 2008; Unknown Author 2008, 2009a). Neither were functioning as intended and water quality remained low. The earth trench was, the new engineering consultant reported, unsalvageable due to the large amount of organic matter blocking the flow of water (Boyt 2010). The wetlands (planted less than five years prior) were declared a financially imprudent approach to water quality improvement and an unusable expense. Wetlands, in his view, did not purify wastewater as fast, as cheaply, or as effectively as the use of other hard infrastructure (engineering-based) options. A far more sensible option (from a cost/benefit analysis) was for the ODC to invest in new "in-pond or after-pond treatments" and entirely de-commissioned the wetlands within two years (Boyt 2010). The consultant did not specifically address how these proposed waste management strategies (to abandon the earth trench in favour of a drop line, to decommission the wetlands, and to construct new oxidation ponds) would mitigate Māori concerns about tapu. The consultant, however, did note that the council still needed to conduct an investigation of land-based disposal strategies to appease local Māori (as promised until its previous resource consent) to demonstrate it could be simply merged into the consultancy firm's "assessment of the environmental effects" of wastewater (Boyt 2010; Unknown Author 2010a). Herein, Māori values were merged by the consultant to be one and the same as environmental values.

Despite the RMA specifically recognising and providing for the values and concerns of Māori to be taken into account in planning decisions, this legislative recognition did not translate into substantive changes to water (and waste) governance and management in the Waipā. The ODC, like other councils in the Waikato region, continued to discharge wastewater into the stream despite vocal and persistent opposition from Ngāti Maniapoto and other iwi. Māori academic Sydney Mead outlined, in 1998, the potential dangers associated with human waste for people in his submission to the Wellington Council about waste discharges into the region (cited by Pauling and Ataria 2010, p. 9): "excreta was tapu and for health reasons this waste product of the human body needed to be kept as far away as possible from where the villagers cooked their food, ate, talked and slept", and it effluent was always kept away from water (be it in a river, lake, or sea). Indeed, the settler-state (local and central government) and the court system remained largely ignorant of Māori perspectives about water pollution. Both the settler-colonial legal order (discussed further in Chap. 6) and the governing bodies who were responsible for monitoring and (supposedly) maintaining the 'quality' of freshwater defined pollution through Western scientific knowledge and did not provide for Indigenous ways of knowing (centred in the Waipā context on mātauranga and tikanga). As Coulthard aptly summarises: "one does not expend much effort to elicit the countless ways in which the liberal discourse of recognition has been limited and constrained by the state, the courts, corporate interests, and policy makers so as to help preserve the colonial status quo" (Coulthard 2007, p. 451). Indeed, he argues, that the "colonial powers … only recognize the collective rights and identities of Indigenous peoples insofar as this recognition does not throw into question the background legal, political and economic framework of the colonial relationship itself" (Coulthard 2007, p. 451). Indeed, none of the consultative processes instituted as a consequence of the RMA nor the Local Government Act and the tokenistic actions taken to address Māori opposed to water-based modes of waste disposal challenged the political, legal, economic and social framework on which settler-colonialism rested (New Zealand Parliament 1991, 2002).

Although councils were obligated (as consent authorities) to recognise and provide for Māori values under the RMA, what this meant in practice was far from ideal (or just). Indeed, since acts of recognition (of Indigenous peoples as culturally distinct groups holding specific values and connections to particular lands/waters) remains largely vested in the apparatus of the settler nation, recognition remains only partial, inadequate, and amounted to misrecognition at times. The RMA provisions, however, as the ODR noted did not given Māori the right to veto developments but instead required the council to take into account the cultural preferences of Māori (clear evidence of where recognition-based justice fails to deliver EJ for Indigenous peoples as Coulthard asserts) (Coulthard 2014). In the case of the Otorohanga wastewater system, the WRC simply recommended that the ODC consider replacing the system when it was practicable. The ODC, as with other district councils in the Waikato Region, argued that were no land-based schemes that were not a viable option (due to financial expense) and also that the waste discharged into waterways already being of a very high standard.

Later reports by the ODC declared that land-based disposal approaches were not feasible for the Otorohanga district due to the high financial costs to the council and its ratepayers; the council noted it would need to purchase hectares of prime agricultural land for private landowners which would be exceedingly expensive. Instead, the council suggested that it could cut drainage channels across the re-established wetlands (constructed as a consequence of consultation in 1995) to allow the treated wastewater to be discharge into the stream; with the malfunctioning earth trench replaced by a drop pipe (Unknown Author 2011a). The consultant made no reference to how the new measures would specifically address Māori concerns about whenua-based waste disposal, their reports of continuing poor quality of water in the steam, and declines in biodiversity. Money rather than mauri was the main focus of council.

Just as the appropriation and control over land, labour arrangements, and social and economic policy are all significant components of sociopolitical and economic systems in settler-states, so too are the ways water resources are extracted, treated, and disposed of. A wealth of EJ scholarship conceptualises racism and the practices of waste disposal as externalities, rather than key parts of racial capitalism. However, since racism persistently produces differential value, it stands to reason, Pulido argues (Pulido 2017b, p. 529), that capital "would incorporate this uneven geography of value into its calculus". The ways in which government persistently chose to (mis)manage the waters of the Waipā River catchment (draining its wetlands, realigning and dredging its riverbeds, constructing flood levees, and discharging (un)treated sewage into its waters) highlight the low status afforded to both rivers and Māori bodies (and values) by the settler-state (embedded within racial capitalism). The decisions of councils and courts to oppose Māori proposals about how wastewater should be disposed of were consistently rejected on the basis that either their knowledge (mātauranga) was not scientific evidence (just spiritual or cultural beliefs) or (in situations where mātauranga and science were in agreement) the argument turned to the high financial costs of alternative approaches. Indeed, this is a familiar strategy within recognition-based approaches, noted by scholars including Bargh and Coulthard, whereby the state recognises Indigenous identities and articulates the importance of including Indigenous peoples' within frameworks (neoliberal, settler-state, co-governance) that are designed and sanctioned by the state and in doing so, other alternative social, economic, political, and ecological arrangements are excluded (Bell 2018; Coulthard 2014; McCormack 2018).

Elsewhere in the country, however, land-based disposal (such as Rotorua) of waste was positioned as a way to address Māori concerns about the disposal of human waste into waterways. Proposed alternatives to replace the trench included (once-again) disposal on the land, the construction of a larger rock-lined channel, or a special Papa-tū-ānuku channel (like that constructed in another waste treatment plant in Hastings).

In the township of Hasting (East Coast of the North Island), for instance, a Papa-tū-ānuku channel was specifically designed to include aspects of Western scientific and mātauranga Māori practices and address iwi concerns. The Hasting District Council describes the treatment processes as comprising:

> fine ... screening, screening washing and compaction; grit removal ... wastewater pumping ... Biological Trickling Filters ... a Papatuanuku (rock) passage to restore the mauri of the treated human waste (kuparu) before discharge ... [as well as a] bark bed biofilter [that] captured air ... to remove odour. (Boyt 2010)

Both iwi and council reported that they were satisfied with the channel. The Hasting example highlights that, despite substantive ontological and epistemological differences, common ground can be found between Te Ao Māori (the Māori world) and Te Ao Pākehā (the New Zealand European world) when it comes to water management. Yet, it requires an approach (procedural inclusion, recognition, and distributive equity) that embraces both legal and ontological pluralism, and extends beyond the confines of Western science and socio-cultural values. However, no in-depth investigations were conducted into land-based disposal options in Otorohanga (or in the neighbouring district of Te Kuiti, which also discharged into a stream that feeds into the Waipā River), and councils remained wedded to long-standing approaches.

In the years 2008 and 2009, three emergency discharges of sewage into Mangaorongo Stream occurred when the oxidation ponds become full and threatened to overflow due to high rainfall events. These and other incidences prompted the WRC to issue a formal warning letter to the ODC about its failure to comply with its resource consent condition. The WRC frequently sent warning letters to district councils (including ODC and even more frequently to Waitomo District Council that operated the Te Kuiti wastewater treatment plant) because of compliance issues (see Tables 5.1 and 5.2); including failures to: conduct required water testing; submit monitoring reports to WRC and iwi; consult with iwi; construct upgrades to treatment plant; and maintain or improve water quality in streams. Throughout 2011 and 2012, for instance, the

ODC received only warnings from the WRC about its failure to comply with its resource consent for the wastewater treatment plant; most notably, the continuing discharges of high volumes of wastewater that contained faecal coliform counts in excess of the allowed standards as well as

Table 5.1 Waikato Regional Council's assessment of the Otorohanga wastewater treatment level of compliance (as per its permit to discharge waste into environment) between the years 1981 and 2012

Date	Status given to plant	Enforcement action
1981	Wastewater treatment plant deemed to be well operated and maintained	No enforcement action recommended
November 1984	Plant well maintained and operated	No enforcement action recommended
July 1989	Wastewater discharge from oxidation ponds considered satisfactory	No enforcement action recommended
February 1993	Minor compliance issues reported	No enforcement action recommended
March 2002	Significant non-compliance reported	No enforcement action recommended
December 2002	High level of compliance	No enforcement action recommended
April 2004	Partial compliance	No enforcement action recommended
June 2004	High level of compliance	No enforcement action recommended
June 2005	High level of compliance	No enforcement action recommended
June 2006	High level of compliance	No enforcement action recommended
August 2007	High level of compliance	No enforcement action recommended
October 2008	Significant level of non-compliance	First formal warning letter issued by Waikato Regional Council
November 2009	Significant non-compliance	Second formal warning letter issued
September 2010	Significant non-compliance	Third formal warning issued
August 2011	Significant non-compliance	Fourth formal warning issued
August 2012	Significant non-compliance	No enforcement action recommended

Table 5.2 Waikato Regional Council's assessment of the Te Kuiti wastewater treatment plant level of compliance (as per its permit to discharge waste into environment) between the years 1989 and 2012

Date	Status awarded to treatment plant	Enforcement action
June 1989	Plant met its consent conditions, although the facilities were old and needed replacing	No enforcement action taken
March 1993	Plant well maintained and operated, however there was no facilities or technologies to measure the daily flow of water or bacterial counts	No enforcement action taken
February 2000	Plant awarded a non-compliant status	Matter referred to Waikato Regional Council's Regulatory Committee. Action plan developed and an abatement notice issued in October 2000 (due to failures of the Waitomo District Council to meet all the conditions of the action plan)
May 2001	Significant level of non-compliance	No enforcement action recommended
June 2003	Significant level of non-compliance	No enforcement action recommended
June 2004	Significant level of non-compliance	No enforcement action recommended
June 2005	Significant level of non-compliance	No enforcement action recommended
May 2006	Significant level of non-compliance	No enforcement action recommended
September 2007	Significant level of non-compliance	First formal warning letter issued
November 2008	Significant level of non-compliance	Second formal warning letter
July 2010	Significant non-compliance	Third formal warning letter issued; referred to Waikato Regional Council's Enforcement Decision Group who issued an abatement notice (pending the completion of wastewater treatment plant upgrades)

(continued)

Table 5.2 (continued)

Date	Status awarded to treatment plant	Enforcement action
July 2011	Significant level of non-compliance	No enforcement action recommended
July 2012	Significant level of non-compliance	No enforcement action recommended

the discharge of water directly into the stream (bypassing the wetlands) (Unknown Author 2012a, b). However, the WRC chose not to adopt any strong mechanisms of enforcement (such as issuing fines as it was empowered to do under the RMA) and still continued to resort to letter writing (Unknown Author 2009b, c, 2010b, c). Yet, words did not translate into actions and the water continued to be polluted (Unknown Author 1998, 1999a, b).

In July 2011, the ODC undertook another round of consultation with iwi, as part of its efforts to secure another set of resource consent application for the operations of the Otorohanga treatment plant (set to expire in 2012) (Unknown Author 2011b, c). The Nehenehenui RMC was appointed to conduct a 'cultural assessment' for the council and formed a working group; the group was made up of four members from Nehenehenui RMC, one representative from Maniapoto Māori Trust Board and one from Whariki Business Services. The working group conducted a site visit to the Otorohanga wastewater treatment plant in September 2011 and consulted with other tangata whenua in the area. The group reported that there was a general lack of maintenance evident throughout the entire treatment plant, including its oxidation pond, wetland, and surroundings. In addition, they observed that cattle were freely able to access the stream, and there was a complete lack of native vegetation along the banks of the stream. Ultimately, the group concluded that the current operations failed to address the long-term goal of Māori, which was "to restore the waterways ... to a level acceptable to the iwi and where there is an abundance of food which is safe to eat and water is suitable as drinking water" (Unknown Author 2011d).

Although the Nehenehenui RMC supported the ODC's application, they made a long list of recommendations to the council on how it should improve the water quality of its treated wastewater (Unknown Author 2011d). The first and second recommendations centred on Nehenehenui RMC being actively involved (encapsulating the Treaty principle of part-nership) in drafting the management plan and monitoring the operations of the Otorohanga waste treatment plant. The ODC response was that it would provide a copy of the operations and management plan within six months of the ODC being issued with a new resource consent, and invite Nehenehenui RMC to undertake a site visit to the treatment plant once a year during which time they could discuss the previous year's monitor-ing results. Other recommendations focused on the operations of the treatment plant including improvements to the maintenance of wetlands, drains, and oxidation pond. Nehenehenui RMC wanted suitable native vegetation planted along the drains (within the treatment plant) as well as along the stream (including at the outlet where the wastewater was discharged) as well as the area fenced off from cattle. Indeed, an earlier scientific assessment conducted by the National Institute of Water and Atmospheric Research (NIWA) in 2002 recommended that riparian planting take place along the banks of the Mangaorongo Stream (includ-ing the area where the wastewater was discharged into the stream). In response to Nehenehenui RMC requests, the ODC declared it was unwilling to plant the stream surrounds with vegetation as it was too financially expensive, difficult to achieve (being in public and private ownership and on steep banks), and ineffective (which would do make no difference to water quality and the number of aquatic fauna). Yet, earlier, when Fish & Game requested similar action be taken, the council agreed to plantings (to provide habitat for trout fishing and duck hunt-ing) (Cunningham 2014; Unknown Author 2011d). Another Nehenehenui RMC recommendation was that the district council adopt a report card approach to the management and monitoring of the waste treatment plant to ensure iwi were kept fully informed about the perfor-mance of the plant as well as the health of their awa. Once again, the ODC likewise sought to sidestep this recommendation and suggested instead that the council provide a copy of its annual monitoring report (which it was required to submit to the WRC) to the Nehenehenui RMC.

The issue of water pollution was further exacerbated by district coun-cils' reluctance to provide accurate information about water quality and treatment plant operations to local iwi and hapū. Access to information is a critical part of a groups' abilities to participate in decision-making about environmental issues, with procedural justice closely tied to people being able to access information about their environment, which includes any hazardous materials and practices that may impact communities. District councils frequently failed to: conduct the necessary tests of river/stream water quality; submit its monitoring reports to the WRC and iwi/hapū; investigate land-based waste disposal; and take into account tikanga Māori (that is to say Māori values and practices). All of which, in various ways, were legal requirements under the RMA and councils' resource consents. The conflict between what local governments were saying and what they were actually doing in practice, along with the general perceived secrecy of the local councils, rearticulated Ngāti Maniapoto's and other iwi's wider concerns about the settler nation's failure to honour Te Tiriti o Waitangi and its apparent inabilities to address freshwater degradation. If we return once again to EJ as procedure, there is clear evidence that the government (despite the introduction of new legislation aimed to address Māori lack of inclusion) did not provide open and inclusive processes of decision-making. Such processes involve enabling access to spaces and information that were previously restricted. In this way, the continued lack of procedural justice for Ngāti Maniapoto is intimately interwoven with the settler-colonial-state (and it's various agencies) "closed geography of information, access and power" (Walker 2009, p. 628). Procedural fairness, Walker and other EJ scholars observe, is built on fluidity of movement of ideas, perspectives, knowledges, and peoples across institutional boundaries as well as between different worlds (plural ontologies and epistemologies), allowing for "open rather than constrained networks and deliberation" (Walker 2009, p. 628). In the Waipā River case study, the RMA established processes that required governments to exchange information with members of the public (including iwi and hapū) but did not achieve procedural fairness because far too often information supplied was incomplete (with monitoring and reporting infrequent). The real-world geographies of flows (or lack) of information, iwi encounters within decision-makers, and power relations between groups are critical tests of procedural fairness, which highlight the continuation of injustices due to lack of participatory equity (between

Pākehā and Māori, and iwi and the Crown). The open provision of information about water, biodiversity, and land and the deliberative possibilities of participation in consultation processes (which we later highlight in our discussion of the new co-governance arrangements over the Waipā River in Chap. 7 are in practice highly socially, culturally, politically and spatially differentiated, with iwi, hapū and whānau still finding it difficult to access information about their local environment. In addition, as we explore in Chaps. 7 and 8, ways in which people can access to adequate resources (money, time, and social support) also present barriers for iwi being present in participatory spaces; from attending a hui (meeting) on a marae to discuss a resource consent application, to being able to give a presentation to a regional council's hearing about the application, or even taking part in an international summit on Indigenous water justice, substantive resources and time-space constraints make it harder for iwi to participate in council mandated procedures. Yet, as Schlosberg's work previously demonstrates, the experiences of injustice is rarely singular (Schlosberg 2003, 2013). In the case of Ngāti Maniapoto's experiences of EJ stemming from water pollution, inequitable distribution (of environmental harms and goods), limited participation, and a lack of recognition of Ngāti Maniapoto values and knowledge (and broader tikanga Māori) all worked to produce injustice.

Over the twentieth century and into the first decade of the twenty-first century the Crown and its agencies,[2] including local governments, consistently excluded or marginalised Māori knowledge and values (including those centred around the tapu of waste and water). In doing so, the settler state created a series of environmental injustices through misrecognition. Likewise, natural resource governance and management approaches (underpinned by settler-colonial laws, knowledge systems, and technologies) were created and operated in ways that did not allow for Māori voices to be heard for much of the twentieth century; even

---

[2] From the perspective of Māori iwi the Crown includes local government, whereas from the perspective of the Crown (as outlined in the Local Government Act, 2001) local governments are not part of the Crown (which only includes central government and central government departments) and are not Treaty partners with Māori. Due to the current distinct under legislation, local governments do not possess the same legal requirements as central government agencies to act in partnership with Māori. See Bell (2018).

when the RMA provided mechanisms for Māori to articulate their perspectives in planning regimes, government bodies often attempted to
bypass their legal obligations. Thus, another layer (this time procedural)
was added to the environmental injustices experienced by Māori. Lastly,
the pollution that continues to run off into waterways means that Ngāti
Maniapoto are unable to access the environmental goods (in the form of
the native freshwater food sources—such as tuna, inanga (whitebait), and
kākahi (freshwater eel)—as well as swimmable and drinkable rivers) that
they value the most. Since Māori ways of being (along the Waipā River
and its tributaries), despite the plethora of changes wrought by colonisation and capitalism, continue to centre on the capacities of whānau/
hapū/iwi to access and use their awa and its resources, we argue that
Māori experiences of water pollution in the Waipā were and are more
severe than experienced by non-Māori. Indeed, whereas Pākehā values
were taken into account by decision-makers and strategies adopted to
protect those biodiversity resources that Pākehā prioritised (most notably
trout and ducks), Māori values were not. Thus, another injustice (this
time a distributive injustice) was deposited onto Māori.

In response, the ODC requested that the WRC place their resource
consent application process on hold while it further consulted with mana
whenua. On the 28 April 2012 the ODC gave representatives from mana
whenua a tour of the Otorohonga wastewater treatment plant and
explained the proposed upgrades to the plant. The council reported that
mana whenua were comfortable with council's plans and Māori only continued to oppose the proposed resource consent condition that related to
the amount of E coli. concentrations that was allowed to be permitted in
the treated wastewater discharged into the stream. However, ODC
refused to modify this condition as it would too financially expensive for
the council to comply with the request of mana whenua (presumably
because it would involve further investments in treatment processes).
Moreover, local hapū whose rohe included the Mangaorongo Stream
(through the Nehenehenui RMC cultural assessment report) already
expressed their support for the ODC resource consent application (Boyt
2012; Unknown Author 2012c). Eventually, the WRC resource consent
panel concluded that the ODC had gone to: "significant lengths to
include the interests of tangata whenua in [its] application [for resource

council] and ensure that their concerns [were] met". The WRC con-cluded that the negative impacts on "tangata will be less than minor". Accordingly, the WRC ruled in favour of the ODC and granted it resource source for a period of 25 years (Unknown Author 2012d, e). The Waipā River catchment is defined as a vulnerable river accordingly to the Ministry of Environment (see Fig. 5.8) due to its high level of degrada-tion (which parallels those of many rivers in Aotearoa).

Conclusion

Regional councils were (and still are in 2020) responsible for issuing the resource consents in their territorial boundaries and ensuring that resource consent holders (in this instance the ODC and Waitomo District Council) complied with their resource consent provisions. WRC sought to ensure (to a limited degree) that the ODC acted in accordance to its resource consent for the Otorohanga wastewater treatment plant. However, the ODC continues to remain wielded to its past policy and waste management approaches, even when there were continuing prob-lems with poor water quality and Māori dissatisfaction with water-based disposal. The situation was paralleled almost exactly at the Te Kuiti treat-ment plant except worse. The plant, operated by the Waitomo District Council, discharged even worse quality water into the Mangaokewa Stream (filled with high levels of bacteria and nutrients). Submissions by representatives of Ngāti Maniapoto about the Te Kuiti treatment plant drew attention to issues of: poor water quality; depiction of native aquatic fauna; iwi preferences for land-based disposal; and the district council's failure to maintain the plant to the required standards (Hauauru Ki Uta Regional Management Committee 2011; Jensen 2011). Despite the ongoing non-compliance status at the Te Kuiti treatment plant, the WRC did not hold Waitomo District Council to account for its breaches of its resource consents and merely issued warnings (Hauauru Ki Uta Regional Management Committee 2011). Even though WRC could possessed the greater regulatory powers (such as imposing fines on district councils) to ensure the RMA was followed, it took no such action to protect the Waipā awa from further degradation. The archival resources of we

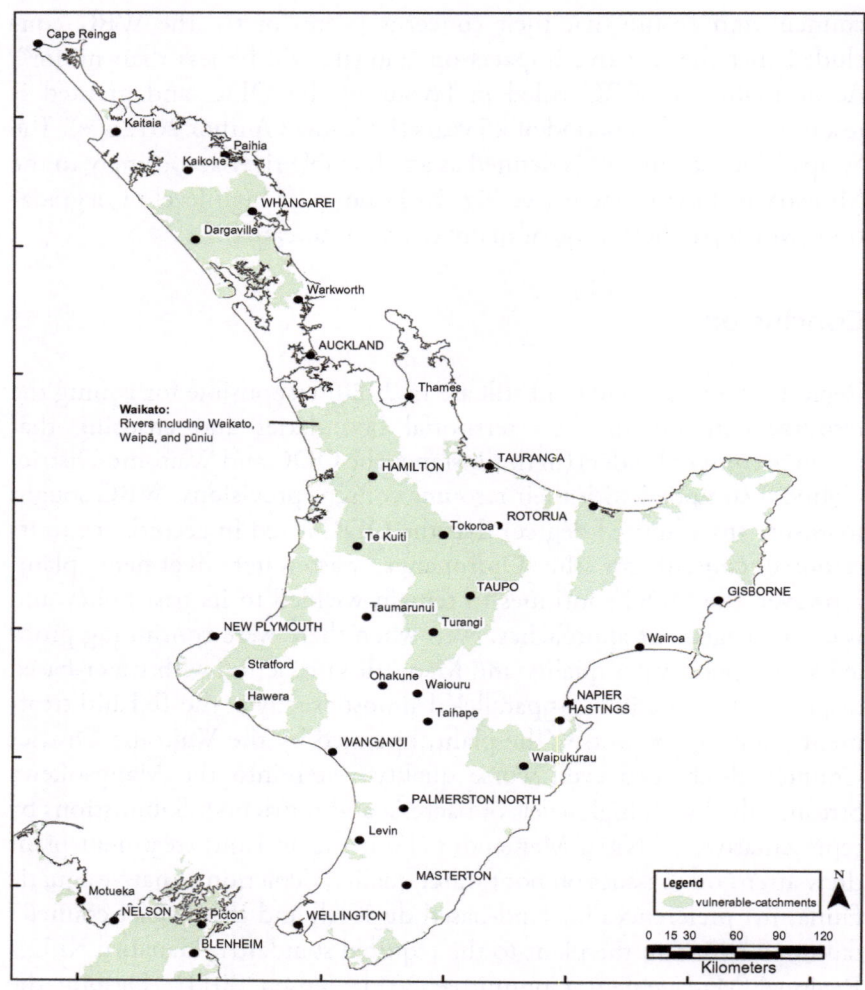

Fig. 5.8 Map showing locations of "vulnerable" river catchments in the North Island of Aotearoa New Zealand in 2017. Ministry of Environment defines vulnerable rivers as one that are significantly degraded. (Source: Authors own)

analysed ends in 2012, however, as of 2020, the water quality along the length of the Waipā River and its tributaries remains exceedingly poor. From its headwaters to when it joins with the Waikato River, the Waipā River ranks in the bottom 25 per cent of rivers in Aotearoa New Zealand in terms of water quality.

In the next chapter, we move into the post-Treaty settlement era in Aotearoa wherein certain iwi (including Ngāti Maniapoto and Waikato-Tainui) are forming co-governance and co-management arrangements (sharing authority between iwi Māori, Crown and local governments) about the management of rivers, forests, and other geo-entities. Such arrangements represent new hybridised governance structures that hold the potential to remake and address environmental injustices. However, the problems of local government-iwi relationships continue to come to the fore (as we outline further in Chaps. 9 and 10). In this chapter we highlight how Māori interests in and concerns about water pollution were consistently disregarded and downplayed by the settler-state. The attempts at procedural inclusion and recognition (under the RMA) consistently faltered (both at a local government level and within the Environment Court) because the rules of the game were set by those in power and thus reinforced the status quo. Council officials (elected and employed), consultants (engineers and scientists), and judges decided what knowledge was worthy of inclusion and what strategies should be approved. Despite consultation processes with iwi, wherein Māori tried to articulate their worldviews, tikanga, and mātauranga about their awa and their aspirations for improved water quality to largely non-Māori decision-makers, councils continued to privilege western scientific knowledge over mātauranga and Te Ao Pākehā over Te Ao Māori. The language of finance was used to back up council decision-making, Māori preferences for land-based disposal (be it on pastures or into wetlands) were declared too financially costly, too time consuming, and less ineffective that existing engineering approaches. Here we turn our focus back to the limits of both procedural and recognition theories of EJ and asked (paraphrasing Bargh 2018): How can relationships be restored (between iwi and the state, between people and the river, between humans and more-than-humans) when one side of the relationship, such as the settler-state, is defining the process and taking virtually no responsibility for changing their existing values, attitudes, and beliefs, let alone their behaviours? The act of recognition, we argue, needs to extend beyond the settler state recognising Māori iwi (tribes) as a culturally distinct groups who are significantly and/or uniquely impacted by an environmental issue (in this instance water pollution). It needs to be situated from 'below' rather than 'above': with Indigenous peoples themselves defining the terms of recognition and strategies. Central to this will be a broad scale recognition of Māori ontologies and epistemologies (including tikanga) about human-water relations, and the interconnectivity and dependency of all beings (human and more-than-human) which we discuss more in our next chapter.

References

AJHR. (1900). H-21 Rivers Commission: Interim Report of Commission to Inquire into Certain Matters Relating to Watercourses. In *Appendices of the Journal of the House of Representatives*. Wellington: New Zealand Parliament. Retrieved 20 May 2019, from http://paperspast.natlib.govt.nz/parliamentary/AJHR1900-I.2.3.2.48.

AJHR. (1910). C-14 Waihou and Ohinemuri Rivers (Report of Commission Appointed to Inquire into Silting of); Together with Minutes of Evidence and Exhibits. *Appendices of the Journal of the House of Representatives*. Wellington: New Zealand Parliament. Retrieved 20 May 2019, from http://paperspast.natlib.govt.nz/parliamentary/AJHR1906-II.2.3.3.2.

AJHR. (1925). *C-11a Rangitaiki Land Drainage (Report of the Commission Appointed to Inquire into and Report Upon)*. Wellington: Government Printer. Retrieved 16 July 2017, from http://www.atojs.natlib.govt.nz/cgi-bin/atojs?a=d&d=AJHR1925-I.2.2.2.15&e=%2D%2D%2D%2D%2D%2D-10%2D%2D1%2D%2D%2D%2D%2D%2D0--.

Amohanga, J., Ormsby, R., & Ormsby, M. (1997, November 24). Further Response on Behalf of Nehenehenui Regional Management Committee Submission by Jacqui Amohanga, Rachael Ormsby, and Massey Ormsby. 60 41 51A, Volume 2, Waikato Regional Council, Hamilton. Unpublished.

Anonymous. (1884, February 14). Typhoid Fever Amongs the Maoris a T Hokianqa. *New Zealand Herald*, Auckland, p. 6.

Askey, P. (1995a, November 23). Peter Askey, Principal Environmental Engineer for Works Consultancy Ltd to the Programme Manager, Environment Waikato. 60 41 51A, Volume 1, Waikato Regional Council, Hamilton. Unpublished.

Askey, P. (1995b, December 19). Peter Askey to Environment Waikato. 60 41 51A, Volume 1, Waikato Regional Council, Hamilton. Unpublished.

Askey, P. (1996, July 25). Works Consultancy Ltd to Nehenehenui Regional Management Committee. 60 41 51A, Volume 1, Waikato Regional Council, Hamilton. Unpublished.

Auckland and Waikato Fish and Game. (1997, March 20). Auckland/Waikato Fish & Game to Environment Waikato. 60 41 51A, Volume 1, Waikato Regional Council, Hamilton. Unpublished.

Bargh, M. (2016). Opportunities and Complexities for Māori and mana whenua Representation in Local Government. *Political Science, 68*(2), 143–160.

Bargh, M. (2018). Māori Political and Economic Recognition in a Diverse Economy. In D. Howard-Wagner, M. Bargh, & I. Altamirano-Jiménez (Eds.), *The Neoliberal State, Recognition and Indigenous Rights* (1st ed., pp. 293–308). Canberra: ANU Press. https://doi.org/10.22459/CAEPR40.07.2018.

Bell, A. (2018). A flawed Treaty Partner: The New Zealand State, Local Government and the Politics of Recognition. In D. Howard-Wagner, M. Bargh, & I. Altamirano-Jimenez (Eds.), *The Neoliberal State, Recognition and Indigenous Rights: New Paternalism to New Imaginings* (pp. 77–92). Canberra: ANU Press.

Boyt, C. (2010, November 19). Email from Cliff Boyt. 60 41 51A, Volume 4, Waikato Regional Council, Hamilton. Unpublished.

Boyt, C. (2012, May 16). Cliff Boyt to Waikato Regional Council. Document 2187740, Waikato Regional Council, Hamilton. Unpublished.

Broughton, D., (Te Aitanga-a-Hauiti, Taranaki, Ngā), McBreen, K., & (Waitaha, Kāti Māmoe, Ngāi Tahu). (2015). Mātauranga Māori, tino rangatiratanga and the future of New Zealand science. *Journal of the Royal Society of New Zealand, 45*(2), 83–88.

Burton, L., & Cocklin, C. (1996). Water Resource Management and Evironmental Policy Reform in New Zealand: Regionalism, Allocation, and Indigenous Relations. *Colorado Journal of International Environmental Law and Policy, 7*, 331.

Coombes, B., & Hil, S. (2005). "Na whenua, na Tuhoe. Ko D.o.C. te partner"—Prospects for Comanagement of Te Urewera National Park. *Society & Natural Resources, 18*(2), 135–152.

Coulthard, G. S. (2007). Subjects of Empire: Indigenous Peoples and the 'Politics of Recognition' in Canada. *Contemporary Political Theory, 6*(4), 437–460.

Coulthard, G. S. (2014). *Red Skin, White Masks: Rejecting the Colonial Politics of Recognition.* Minneapolis: University of Minnesota Press.

Crow, S. K., Tipa, G. T., Booker, D. J., & Nelson, K. D. (2018). Relationships between Maori Values and Streamflow: Tools for Incorporating Cultural Values Into Freshwater Management Decisions. *New Zealand Journal of Marine and Freshwater Research, 52*(4), 626–642.

Cunningham, M. (2014). *The Environmental Management of the Waipa River and Its Tributaries. Case-Study Commissioned by the Waitangi Tribunal for Te Rohe Potae District Inquiry (Wai 898)* (District Inquiry Research Report No. A150 (Wai 868_). Wellington: Waitangi Tribunal.

Department of Public Works. (1928). Removal of Gravel from Waipa River. ABKK W4 357 889 Box 132. National Archives, Wellington. Unpublished.

Dixon, J. (1937, February 27). Government Intervention. *King Country Chronicle*, p. 5.

Environment Waikato. (1996, July 31). Environment Waikato Memorandum. 60 41 51A, Volume 1, Waikato Regional Council, Hamilton. Unpublished.

Figueroa, R. M. (2001). Other Faces: Latinos and Environmental Justice. In L. Westra & B. E. Lawson (Eds.), *Faces of Environmental Racism: Confronting Issues of Global Justice* (pp. 167–184). Lanham: Rowman & Littlefield Publishers.

Finlay, G. P. (1923, May 4). King Country Tour. *Auckland Star*, Auckland. Retrieved from https://paperspast.natlib.govt.nz/newspapers/AS19230504.2.80?query=nativitis.

Forster, M. (2016). Indigenous-Environmental-Autonomy-in-Aotearoa-new-Zealand. *AlterNative: An International Journal of Indigenous Peoples, 12*(3), 316–330.

Gott, J. (1916, June 21). Devonport Refuse Tip. Letter to Editor. *NZ Herald*, p. 5.

Greensill, A. N. (2010). *Inside the Resource Management Act: A Tainui Case Study.* (Thesis). The University of Waikato. Retrieved 25 June 2019, from https://researchcommons.waikato.ac.nz/handle/10289/4922.

Harmsworth, G., Awatere, S., & Robb, M. (2016). Indigenous Māori Values and Perspectives to Inform Freshwater Management in Aotearoa-New Zealand. *Ecology and Society, 21*(4), 9.

Hauauru Ki Uta Regional Management Committee. (2011, July 26). Submission of Hauauru Ki Uta Regional Management Committee. Documber Number 2018335. Waikato Regional Council, Waikato Regional Council Archives, Hamilton. Unpublished.

Hayward, J. (2011). Mandatory MÄori Wards in Local Government: Active Crown Protection of MÄri Treaty Rights. *Political Science, 63*(2), 186–204.

Hopkins, D., Joly, T. L., Sykes, H., Waniandy, A., Grant, J., Gallagher, L., et al. (2019). "Learning Together": Braiding Indigenous and Western Knowledge Systems to Understand Freshwater Mussel Health in the Lower Athabasca Region of Alberta, Canada. *Journal of Ethnobiology, 39*(2), 315–336.

Howitt, R., & Suchet-Pearson, S. (2006). Rethinking the Building Blocks: Ontological Pluralism and the Idea of 'Management'. *Geografiska Annaler: Series B, Human Geography, 88*(3), 323–335.

Jacobson, C., Matunga, H., Ross, H., & Carter, R. W. (2016). Mainstreaming Indigenous Perspectives: 25 Years of New Zealand's Resource Management Act. *Australasian Journal of Environmental Management, 88*(4), 331–337.

James, P., & Pawson, E. (1995). Contested Places: The Significance of the Montunui-Waitara Claim to the Waitangi Tribunal. *Aboriginal History, 19*, 111.

Jensen, W. (2011, July 26). Submission of Wayne Jensen. Document Number 2019051. Waikato Regional Council, Waikato Regional Council Archives, Hamilton. Unpublished.

Jones, C. (2016). *New Treaty, New Tradition: Reconciling New Zealand and Maori Law*. Toronto: University of British Columbia.

Knight, C. (2016). *New Zealand's Rivers: An Environmental History*. Christchurch: Canterbury University Press.

Knight, C. (2019). A Potted History of Freshwater Management in New Zealand. *Policy Quarterly, 15*(3). https://doi.org/10.26686/pq.v15i3.5681.

Kowhai Consulting Ltd. (2007). *He Mahere Taiao: The Maniapoto Iwi Environmental Management Plan for Maniapoto Maori Trust Board*. Otorohanga: Kowhai Consulting Ltd..

Kowhai Consulting Ltd, & Ministry for the Environment. (2002). *Te Purongo: Ngati Maniapoto State of the Environment Report: A Tribal Perspective*. Otorohanga: Kowhai Consulting Ltd..

Lewis, N., Lewis, O., & Underhill-Sem, Y. (2009). Filling Hollowed Out Spaces with Localised Meanings, Practices and Hope: Progressive Neoliberal Spaces in Te Rarawa. *Asia Pacific Viewpoint, 50*(2), 166–184.

Lowry, A., & Simon-Kumar, R. (2017). The Paradoxes of Māori-State Inclusion: The Case Study of the Ōhiwa Harbour Strategy. *Political Science, 69*(3), 195–213.

McCormack, F. (2018). Indigenous Settlements and Market Environmentalism: An Untimely Coincidence? In D. Howard-Wagner, M. Bargh, & I. Altamirano-Jiménez (Eds.), *The Neoliberal State, Recognition and Indigenous Rights* (Vol. 40, pp. 273–292). Canberra: ANU Press.

Mead, H. M. (2016). *Tikanga Maori (Revised Edition): Living by Maori Values.* Wellington: Huia Publishers.

Moore, J. W. (2015). *Capitalism in the Web of Life: Ecology and the Accumulation of Capital.* Verso Books.

Morgan, T. K. K. B., & Te Aho, L. (2013). Waikato Taniwharau: Prioritising Competing Needs in the Management of the Waikato River. In J. Daniels (Ed.), *Advances in Environmental Research.* New York: Nova Science Publishers.

Muru-Lanning, M. (2012). The Key Actors of Waikato River Co-governance: Situational Analysis at Work. *AlterNative: An International Journal of Indigenous Peoples, 8*(2), 128–136.

Mutu, M. (2018). Behind the Smoke and Mirrors of the Treaty of Waitangi Claims Settlement Process in New Zealand: No Prospect for Justice and Reconciliation for Māori Without Constitutional Transformation. *Journal of Global Ethics, 14*(2), 208–221. https://doi.org/10.1080/17449626.2018.1507003.

Mutu, M. (2019). The Treaty Claims Settlement Process in New Zealand and Its Impact on Māori. *Land, 8*(10), 152. https://doi.org/10.3390/land8100152.

New Zealand Government. (1956a). Health Act. Retrieved 7 January 2020, from http://www.legislation.govt.nz/act/public/1956/0065/latest/whole.html.

New Zealand Government. (1956b). Waikato Valley Authority Act. Retrieved 22 January 2020, from http://www.nzlii.org/nz/legis/hist_act/wvaa19561956n104320/.

New Zealand Parliament. (1953). Waters Pollution Act. Retrieved 22 January 2020, from http://www.nzlii.org/nz/legis/hist_act/wpa19531953n104253/.

New Zealand Parliament. (1967). Water and Soil Conservation Act., Pub. L. No. 135. Retrieved 16 June 2020, from http://www.nzlii.org/nz/legis/hist_act/wasca19671967n135320/.

New Zealand Parliament. (1991). Resource Management Act. Retrieved 16 June 2020, from http://www.legislation.govt.nz/act/public/1991/0069/223.0/DLM230265.html.

New Zealand Parliament. (2002). Local Government Act. Retrieved 16 June 2020, from http://www.legislation.govt.nz/act/public/2002/0084/167.0/DLM170873.html.

Parsons, M., Nalau, J., & Fisher, K. (2017). Alternative Perspectives on Sustainability: Indigenous Knowledge and Methodologies. *Challenges in Sustainability, 5*(1), 7–14.

Pauling, C., & Ataria, J. (2010). *Tiaki Para: A Study of Ngāi Tahu Values and Issues Regarding Waste.* Lincoln: Manaaki Whenua Press.

Pollution Advisory Council. (1956). *Report on Pollution in the Waikato River Basin*. Wellington: Pollution Advisory Council.

Pulido, L. (1998). Environmentalism and Economic Justice: Two Chicano Struggles in the Southwest. *Senior Managing Editor, 5*(2), 70.

Pulido, L. (2015). Geographies of Race and Ethnicity 1: White Supremacy vs White Privilege in Environmental Racism Research. *Progress in Human Geography, 39*(6), 809–817.

Pulido, L. (2016). Environmental Racism. In *International Encyclopedia of Geography: People, the Earth, Environment and Technology: People, the Earth, Environment and Technology* (pp. 1–13). Hoboken: Wiley Online Library.

Pulido, L. (2017a). Evolving Racial Formations and the Environmental Justice Movement. In R. B. Holifield, J. Chakraborty, & G. P. Walker (Eds.), *The Routledge Handbook of Environmental Justice*. London and New York: Routledge and Taylor & Francis Group.

Pulido, L. (2017b). Geographies of Race and Ethnicity II: Environmental Racism, Racial Capitalism and State-Sanctioned Violence. *Progress in Human Geography, 41*(4), 524–533.

Pulido, L. (2017c). Geographies of Race and Ethnicity III: Settler Colonialism and Nonnative People of Color. *Progress in Human Geography, 42*(2), 309–318.

Pulido, L., & Peña, D. (1998). Environmentalism and Positionality: The Early Pesticide Campaign of the United Farm Workers' Organizing Committee, 1965–1971. *Race, Gender & Class, 6*, 33–50.

Pulido, L., Kohl, E., & Cotton, N.-M. (2016). State Regulation and Environmental Justice: The Need for Strategy Reassessment. *Capitalism Nature Socialism, 27*(2), 12–31. https://doi.org/10.1080/1045575 2.2016.1146782.

Robinson, H. (2011). *Te taha tin ana: maori health and the crown in te rohe potae inquiry district, 1840 to 1990* (A Report for Rohe Potae District Inquiry (Wai 898) No. A31). Wellington: Waitangi Tribunal.

Ryks, J., Wythe, J., Baldwin, S., & Kennedy, N. (2010). The Teeth of the Taniwha: Māori Representation and Participation in Local Government. *Planning Quarterly, 177*, 39–42.

Schlosberg, D. (2003). The Justice of Environmental Justice: Reconciling Equity, Recognition, and Participation in a Political Movement. *Moral and Political Reasoning in Environmental Practice, 77*, 106.

Schlosberg, D. (2013). Theorising Environmental Justice: The Expanding Sphere of a Discourse. *Environmental Politics, 22*(1), 37–55.

Stammers, J. (2008, April 15–16). Report on Discharge of Sewage to Water. 60 41 51A, Volume 3, Hamilton: Waikato Regional Council. Unpublished.

Sullivan, W. A. (1998). *Changing the Face of Eden: A History of Auckland Acclimatisation Societies, 1861–1990*. Hamilton: Auckland/Waikato Fish & Game Council.

Taonga, N. Z. M. for C. and H. T. M. (2020). Population and Society. *Te Ara Encyclopedia of New Zealand*. Web Page, Ministry for Culture and Heritage Te Manatu Taonga. Retrieved 2 June 2020, from https://teara.govt.nz/en/king-country-region/page-7.

Taylor, A., & Te Taha Māori. (1986, March). The Importance of the Sea and the Waterways to the Maori, and Why Outfalls are Unacceptable as Effluent Treatment. *Nature Conservation Council Newsletter*, p. 60.

Te Aho, L. (2015). *The Waikato River Settlement: Exploring a Model for Co-management and Protection of Natural and Cultural Resources*. Ka Hula Ao Center for Excellence in Native Hawaiian Law, Richardson School of Law. Retrieved 6 January 2019, from https://researchcommons.waikato.ac.nz/handle/10289/10414.

Thompson-Fawcett, M., Ruru, J., & Tipa, G. (2017). Indigenous Resource Management Plans: Transporting Non-Indigenous People into the Indigenous World. *Planning Practice & Research, 32*(3), 259–273.

Tipa, G., Nelson, K., Home, M., & Tipa, M. (2016). Policy Responses to the Identification by Maori of Flows Necessary to Maintain Their Cultural Values. *37th Hydrology & Water Resources Symposium 2016: Water, Infrastructure and the Environment*, p. 552.

Tribunal, W. (1993). *The Waitangi Tribunal and the Motunui-Waitara Claim*. Wellington: Waitangi Tribunal.

Unknown Author. (1897, January 27). Is Typhoid Preventible?. *Wanganui Chronicle*, p. 2.

Unknown Author. (1900). Duck Shooting in New Zealand. *New Zealand Illustrated Magazine, III*, p. 61.

Unknown Author. (1916, June 14). Health of the Borough. *King Country Chronicle*, p. 5.

Unknown Author. (1926, July 30). Death from Typhoid. *New Zealand Herald*, p. 10.

Unknown Author. (1935, August 24). New Trout Hatchery. *Waikato Independent*. Retrieved 5 June 2019, from https://paperspast.natlib.govt.nz/newspapers/WAIKIN19350824.2.4?query=flood%20waipa%20river&sort_by=byDA&items_per_page=100&page=6&snippet=true&title=KWE,NZ,

NZH,PWT,ROTWKG,TO,ACNZC,AKTIM,NZHAG,ALG,AS,DSC,KC
C,OG,PAKIOM,TAN,THA,THS,WAIGUS,WAIKIN,WHDT,WT,MAT
REC,TGMR&type=ARTICLE.

Unknown Author. (1936, September 11). High Death Rate. *New Zealand Herald*, p. 12.

Unknown Author. (1937, October 28). Maoris and Typhoid. *Waikato Independent*, p. 5.

Unknown Author. (1957). Minutes of a Meeting of the Waikato Valley Authority, 5 August, 1957, Series 1 13 60/1, Waikato Regional Council Archives. Hamilton: Waikato Regional Council. Unpublished.

Unknown Author. (1966a, January 11). Application for Registration of an Outfall and Permit Discharge Wastes Into Classified Waters. 60 41 51A, Volume 1. Hamilton: Waikato Regional Council. Unpublished.

Unknown Author. (1966b, May 24). Minutes of Otorohanga Oxidation Ponds Working Party Meeting. 60 41 51A, Volume 1. Hamilton: Waikato Regional Council. Unpublished.

Unknown Author. (1967, June 29). Temporary Permit No. 434/209/T. 60 41 51A, Volume 1. Hamilton: Waikato Regional Council. Unpublished.

Unknown Author. (1968, March 1). Temporary Permit No. 434/209/1T. 60 41 51A, Volume 1. Hamilton: Waikato Regional Council. Unpublished.

Unknown Author. (1970, May 2). Temporary Permit No. 434/726, 60 41 51A, Volume 1, Waikato Regional Council; Inspector of Health to the Medical Officer of Health. 60 41 51A, Volume 1. Hamilton: Waikato Regional Council. Unpublished.

Unknown Author. (1974a, May 6). Secretary of the Waikato Valley Authority to the Otorohanga County Clerk. 60 41 51A, Volume 1. Hamilton: Waikato Regional Council. Unpublished.

Unknown Author. (1974b, May 27). Otorohanga County Engineer to the Secretary of the Waikato Valley Authority. 60 41 51A, Volume 1. Hamilton: Waikato Regional Council. Unpublished.

Unknown Author. (1974c, July 23). Notice of Objection. 60 41 51A, Volume 1. Hamilton: Waikato Regional Council. Unpublished.

Unknown Author. (1974d, October 10). Water Right No. 2594. 60 41 51A, Volume 1. Hamilton: Waikato Regional Council. Unpublished.

Unknown Author. (1975a, July 24). Water Right Application. 60 41 51A, Volume 1. Hamilton: Waikato Regional Council. Unpublished.

Unknown Author. (1975b, October 29). Water Right No. 2787. 60 41 51A, Volume 1. Hamilton: Waikato Regional Council. Unpublished.

Unknown Author. (1990, September 28). Manager Resource Monitoring, Environment Waikato, to Otorohanga District Council. 60 41 51A, Volume 1. Hamilton: Waikato Regional Council. Unpublished.

Unknown Author. (1993, March). Internal Report—Resource Consent Monitoring. 60 41 51A, Volume 1. Hamilton: Waikato Regional Council. Unpublished.

Unknown Author. (1995a, February 13). Manager of Resource Use Monitoring, Environment Waikato, to Otorohanga District Council. 60 41 51A, Volume 1. Hamilton: Waikato Regional Council. Unpublished.

Unknown Author. (1995b, September 15). Inquiries Officer, Environment Waikato, to Otorohanga District Council. 60 41 51A, Volume 1. Hamilton: Waikato Regional Council. Unpublished.

Unknown Author. (1995c, August 10, 15). File Notes. 60 41 51A, Volume 1, Hamilton: Waikato Regional Council. Unpublished.

Unknown Author. (1995d, December 6). Programme Manager of Energy and Ultilies, Environment Waikato to the Chief Executive Officer of Otorohanga District Council. 60 41 51A, Volume 1. Hamilton: Waikato Regional Council. Unpublished.

Unknown Author. (1995e, October 11). Minutes of Meeting Held with the Nehenehenui Regional Management Committee. 60 41 51A, Volume 1. Hamilton: Waikato Regional Council. Unpublished.

Unknown Author. (1995f, October 12). Minutes of Otorohanga Oxidation Ponds Working Party Meeting. 60 41 51A, Volume 1. Hamilton: Waikato Regional Council. Unpublished.

Unknown Author. (1996a, July 16). Secretary of Te Nehenehenui Regional Management Committee to Works Consultancy Ltd. 60 41 51A, Volume 1. Hamilton: Waikato Regional Council. Unpublished.

Unknown Author. (1996b, September 10). Minutes of the meeting held with Te Mauri o Maniapoto. 60 41 51A, Volume 1. Hamilton: Waikato Regional Council. Unpublished.

Unknown Author. (1998). Attachment—Submission in Opposition. Series 60 61 24A, Volume 3. Hamilton: Waikato Regional Council. Unpublished.

Unknown Author. (1999a, March 23). Minutes of Meeting Held at Te Kuiti. Series 60 61 24A, Volume 3. Hamilton: Waikato Regional Council. Unknown Author.

Unknown Author. (1999b). Submissions Folder, Series 60 61 24A Volume 3. Hamilton: Waikato Regional Council. Unpublished.

Unknown Author. (2008, October 29). Consent Compliance Audit Report. 60 41 51A, Volume 3. Hamilton: Waikato Regional Council. Unpublished.

Unknown Author. (2009a, May 15). Resource Consent Monitoring Report. 60 41 51A, Volume 3. Hamilton: Waikato Regional Council. Unpublished.

Unknown Author. (2009b, October 1). Email from the Resource Officer, Environment Waikato, to Jon Stammers. 60 41 51A, Volume 4. Hamilton: Waikato Regional Council. Unpublished.

Unknown Author. (2009c, October 20). Application for s127 change to Consent—Evaluation Report. 60 41 51A, Volume 4. Hamilton: Waikato Regional Council. Unpublished.

Unknown Author. (2010a, October 19). Resource Officer, Environment Waikato, to Otorohanga District Council. 60 41 51A, Volume 4. Hamilton: Waikato Regional Council. Unpublished.

Unknown Author. (2010b, September 28). Consent Compliance Audit Report. 60 41 51A, Volume 4. Hamilton: Waikato Regional Council. Unpublished.

Unknown Author. (2010c, May). Resource Consent Monitoring Report. 60 41 51A, Volume 4. Hamilton: Waikato Regional Council. Unpublished.

Unknown Author. (2011a, October). Resource Consents Application and Assessment of Effects on the Environment: Otorohanga Wastewater Treatment Plant—Resource Consents Project, pp. 65–73. Document 2077260. Hamilton: Waikato Regional Council. Unpublished.

Unknown Author. (2011b, March 10). Otorohanga Wastewater Treatment Plant—Resource Consents Project, pp. 4–5, 10. 60 41 51A, Volume 5. Hamilton: Waikato Regional Council. Unpublished.

Unknown Author. (2011c, October). Resource Consents Application and Assessment of Effects on the Environment: Otorohanga Wastewater Treatment Plant—Resource Consents Project, Appendix K Consultation Plan, pp. 14–16. Document 2077260, 60 41 51A, Volume 5. Hamilton: Waikato Regional Council. Unpublished.

Unknown Author. (2011d, October). Resource Consents Application and Assessment of Effects on the Environment: Otorohanga Wastewater Treatment Plant—Resource Consents Project, Appendix J Cultural Impact Assessment Report, pp. 3, 11. Documber 2077260. Hamilton: Waikato Regional Council. Unpublished.

Unknown Author. (2012a, June). Resource Consent 953619 Monitoring Report. 60 41 51A, Volume 5. Hamilton: Waikato Regional Council. Unpublished.

Unknown Author. (2012b, August 17). Consent Compliance Audit Report. 60 41 51A, Volume 5. Hamilton: Waikato Regional Council. Unpublished.

Unknown Author. (2012c, August 23). Chief Executive of Otorohanga District Council to Waikato Regional Council. 60 41 51A, Volume 5. Hamilton: Waikato Regional Council. Unpublished.

Unknown Author. (2012d, August 21). Email from the Senior Resource Officer, Waikato Regional Council, to Cliff and Pat Boyt. 60 41 51A, Volume 5. Hamilton: Waikato Regional Council. Unpublished.

Unknown Author. (2012e, November 2). Report of the Waikato Regional Council. Documber 2292779. Hamilton: Waikato Regional Council. Unpublished.

Walker, G. (2009). Beyond Distribution and Proximity: Exploring the Multiple Spatialities of Environmental Justice. *Antipode, 41*(4), 614–636.

Wheen, N. R., & Hayward, J. (2012). *Treaty of Waitangi Settlements*. Wellington: Bridget Williams Books.

Whyte, K. P. (2018). Settler Colonialism, Ecology, and Environmental Injustice. *Environment and Society, 9*(1), 125–144.

Wood, G. E. (1950). *The New Zealand Official Year-Book, 1947–1949*. Wellington: Census and Statistics Department. Retrieved 2 June 2020, from https://www3.stats.govt.nz/New_Zealand_Official_Yearbooks/1947-49/NZOYB_1947-49.html#idsect1_1_57826.

Works Consulting Ltd. (1995, December 19). Works Consulting Ltd to Te Porerau Joseph, Te Mauri o Maniapoto. 60 41 51A, Volume 1. Hamilton: Waikato Regional Council. Unpublished.

6

Legal and Ontological Pluralism: Recognising Rivers as More-Than-Human Entities

Around the world, many peoples and societies are contending with the trials of creating and applying apparatuses recognise Indigenous interests and authority within freshwater governance and management (Berry et al. 2018; Castleden et al. 2017; Cosens and Chaffin 2016; Curran 2019; Jackson 2018; Muru-Lanning 2016a; Ruru 2018a; Wilson 2019). Many new policies and strategies specifically acknowledge the rights of Indigenous peoples, their interests in and values they attached to specific geo-regions (be it rivers, lakes, or forests) and environmental resources, including flora and fauna, and take the form of new legal agreements which are directed at reconciling diverse worldviews, values, and ways of life within particular environments (Daigle 2016; Johnston 2018; Nursey-Bray and Palmer 2018; Premauer and Berkes 2015). However, across the settler societies (including Aotearoa, Australia, USA and Canada) there are no consistent approaches to Indigenous freshwater governance and management being adopted to honour water resource agreements between Indigenous nations and settler-states or resolve Indigenous environmental injustices.

In this chapter we explore the ways in which the formal recognition (to some extent) of Indigenous knowledge systems within environmental

© The Author(s) 2021
M. Parsons et al., *Decolonising Blue Spaces in the Anthropocene*, Palgrave Studies in Natural Resource Management, https://doi.org/10.1007/978-3-030-61071-5_6

governance and the role of reconcilition in achieving environmental justice. We draw on McGregor's definition of reconcilation conceptualised as going beyond "the human dimension to include 'relationships with the Earth and all living beings'" (McGregor et al. 2020, p. 9).

In Aotearoa, tikanga (Māori legal order) are receiving greater focus amongst scholars, legal practitioners, and activists, with mātauranga and tikanga are increasingly recognised for holding practical methods for achieving justice for Māori. We examine whether recent agreements between the New Zealand Crown (Crown) and Māori tribal groups (iwi), known as Treaty 'settlements', to establish shared co-governance and management over rivers encapsulate and are capable of achieving environmental justice (as defined with Māori ontologies and epistemologies). In this chapter, more broadly, we explore how legal and ontological pluralism, amongst scholars as well as law- and policy-makers in Aotearoa and other (post)colonial contexts, can address environmental injustices. Rather than seek to provide a singular definition of Indigenous environmental justice (IEJ), we instead examine how Indigenous peoples are engaged in efforts to negotiate with and challenge the colonial legal orders, develop their laws, policies, and governance frameworks to achieve justice within the freshwater realm.

The structure of the chapter is as follows. First, we provide very brief overview of how respect for Indigenous ontologies and epsitemologies is a critical component of IEJ. Second, we discuss tikanga Māori (the laws of Māori) and the ways in which the settler-nation deliberately sought to exclude and supplant tikanga for more than a century. We emphasis that Aotearoa's exclusionary policies and legal processes were not unique but rather were a feature of settler colonialism around the globe. Third, we examine the emergence of legal pluralism in settler-colonial and former-colonial societies wherein legal traditions (based on different ontologies) are being incorporated laws and institutional arrangements. Fourth, we chart how increased to recognise Te Tiriti o Waitangi/the Treaty of Waitangi and the development of formal reconciliation processes between iwi (Māori tribes) and the Crown (Aotearoa New Zealand's Central Government) as fostering the development of new agreements and legislation founded on legal and ontological pluralism. Lastly, we analyse legislation that recognising tikanga (to some extent) and mātauranga (knowleddge) regarding how two rivers (Waipā and Whanganui) are governed and managed, and draw attention the stregnthens and weaknesses of the new laws for achieving IEJ.

Indigenous Knowledge, Laws, and Worldviews

Despite being focused on Indigenous peoples' lives, livelihoods, lands and waters, much of the existing scholarship on Indigenous injustice are not situated within Indigneous worldviews, epistemologies, and methodologically (Muir and Booth 2012; Shah and Rodina 2018; Vickery and Hunter 2016; Zwarteveen and Boelens 2014). Yet, as we articulate throughout this book, Indigenous theories, ontologies, epistemologies, and methodologies can inform discussions of what is IEJ and how it can be achieved. The argument builds on international scholars emerging within Indigenous research more broadly, in which Indigenous knowledge systems, philosophies, and legal systems are the building blocks for inquiries (McGregor 2018; Watene 2016; Whyte 2018; Winter 2019). Such approaches are intended to address the "lack of consideration and respect" shown for Indigenous intellectual traditions (Hunt 2014, p. 29).

One of the key features of Indigenous peoples' worldviews or ontologies (despite incredible diversity in cultural, linguistic, political, economic, historical and geographical contexts) is their conceptualisation of people being in relationships with 'more-than-humans' or 'other orders of beings' (Bergman 2006; McGregor 2018; Watene 2016; Whyte 2018; Winter 2019). The importance of these relationships are highlighted in how Indigenous peoples' conceptualisations of justice extend to include the "more-than-human world" (McGregor 2018; Salmond 2018). Indigenous systems of knowledge are premised on a set of assumptions (ontological, epistemological and metaphysical) about humanity's position in the world. Furthermore, these assumptions about how the world (or worlds) and all beings within it/them also conveyed key principles that underpin Indigenous laws and governance systems, including giving directions about how people should act in regard to others (including rivers, plants, and animals).

There is not a singular and universal Māori worldview, but rather a pluralities of ways of thinking, reflective partly of the localised and place-based nature of Māori iwi (tribal) groupings. Instead, scholars talk broadly of the ontologies and epistemologies encapsulated within the term Te Ao Māori (the world of Māori) as distinct from Te Ao Pākehā (the world of Pākehā aka the settler world). Te Ao Māori is characterised by non-linearity and relationships based on relationality and reciprocity between humans and non-humans (including metaphysical beings). Tikanga (laws)—which includes principles and values—are produced and sustained by rūnanga (tribal councils), iwi, hapū (sub-tribe), and whanau (extended family) (Salmond 2017; Thompson-Fawcett et al. 2017). Tikanga and principles are the foundations of iwi identities, duties, obligations and rights of individuals, whānau (family, extended family), hapū, iwi, and rūnanga. Māori identify themselves through their genealogical (whakapapa) connections and affiliated to whānau, hapū and iwi, but also to particular lands (whenua), mountains (maunga), rivers (awa) and seas (moana). These whakapapa relationships inextricably bind them (as kin) to their environment (taiao) encompassing all elements including rivers (awa) and land (whenua); as humans and more-than-humans (accordingly to Māori cosmology) alike are all descendants from Papatūānuku (Earth Mother) and Ranginui (Sky Father) (Harmsworth et al. 2016; Ruru 2013; Salmond 2017). Mātauranga Māori (Māori system of knowledge) is premised on this relational ontology, wherein awa are the living embodiment of whakapapa, the mana (prestige, power and sovereignty) of hapū and iwi, and possess their own distinct mauri (life force) and spiritual veracity (wairua), as well as their own agency (Whaanga et al. 2018). Explorations of such ideas highlight the ways in which Indigenous environmental injustices differ from those encountered by non-Indigenous peoples, and also demonstrate that efforts to achieve IEJ rests in actions that attend to the interwoven wellbeing of human and more-than-human beings.

Tikanga Māori: The First Legal Order of Aotearoa

Māori ways of thinking and being in their world, based on responsibilities of care for their more-than-human relatives, underpinned the laws (tikanga) and governance systems of Māori. Tikanga refers to traditions, protocols and laws that regulated behaviour within Māori iwi, hapū and whānau. These laws were (and are still) embedded in sources and practices including: (1) whanaungatanga (extended family, responsibilities, relationships, the centrality of kinship, whakapapa that binds the Māori world together); (2) mana (power, authority, control, prestige, power contributing to leadership); (3) tapu (respect, scared, forbidden) and its opposite noa (normal, ordinary) that pays different roles including social (keeping people safe), political (ceremony, leadership), spiritual (wairua/ spiritual integrity); (4) utu (retaliation and retribution); (5) kaitiakitanga (guardianship over environment, taking care of one's more-than-human kin). Tikanga, therefore, is premised on the need to maintain the balance between all things and thereby ensuring the protection and enhancement of mauri and wairua of human and more-than-human beings both now and in the future.

In terms of the contrast between British (and then settler colonial) legal order and tikanga Māori, Justice Eddie Durie underlined how Western law is rules-based (literate) whereas tikanga is governed by values which the community subscribed to (Durie 1994). While Euro-Western cultures generally ascribe to a clear distinction between law and mortality, tikanga Māori is rules, practices, values, and ethics based (Durie 1994, p. 3). Metge observed, however, that "Western laws are also values-based; the values concerned being interpreted by the law makers" (Metge 1997, p. 5). Mulgan added: "All law, Pākehā as well as Māori, arises out of social norms and the need to enforce these norms within society. The ultimate source of Pākehā law is not the courts or statutes but the social values reflected by Parliament by statutes and by judges in their decisions" (Mulgan 1997, p. 2). Metge concluded that the primary difference between tikanga Māori and settler legal order originates in the different sources and modes of communication (Metge 1997). Tikanga emerges "out of on-going community debate and practice and are communicated orally" and as a consequence "they are adapted to changing circumstances easily, quickly and without most people being consciously aware of the shift" (Metge 1997). In contrast, Western (settler) laws are "formulated and codified by a formal law-making body and are published in print; their amendment, while possible, is a complex and lengthy process" (Metge 1997). Yet, all societies possess laws that represent certain values and fulfil particular functions within society, most notably the preservation of social order and maintenance of collective security. Law is abided in diverse societies because individuals and communities obey the law (on the basis that they believe that the law is just, they seek protection from the law, or they fear sanctions as a consequence of non-observance) (Jones 2016; McGregor 2018).

Tikanga Māori was the legal order that operated in Aotearoa prior to Pākehā colonisation and continued to operate in various ways despite colonial efforts to denigrate and suppress it in the nineteenth and twentieth centuries (Dorsett 2017; Ruru 2009, 2012). For instance, the settler-state disregarded Māori laws about how waste products should be disposed of (as outlined in Chap. 3) and discharged human waste directly into waterways; as we describe in Chap. 6, this practice breached tikanga and caused negative impacts to the health of the wai (water) and its human and more-than-human beings whose mauri was intertwined with. The establishment of Aotearoa's legal order (which was heavily informed by that of Britain) and the ways in which tikanga Māori was disregarded and excluded parallels what happened in other colonial contexts.

Within the borders of the settler colonial states of Aotearoa, United States, Australia, and Canada—boundaries often newly defined with limited attention to existing Indigenous territories, governance regimes or practices—non-colonial state Indigenous law was assigned subordinate status to 'official' settler state law imposed by settler-colonial powers (Green and Hendry 2019; Hendry and Tatum 2018; Robinson and Graham 2018). Such a characterisation of Indigenous nations and legal orders as non-state were significant. The denial of Indigenous legal orders the standing of 'law' helped to facilitate the Indigenous laws being marginalised and suppressed on the basis of being 'mere' traditions or customs (Green and Hendry 2019), and also labelled such legal systems (such as tikanga) as unofficial and unworthy of the attention of legal scholars and practitioners. The work of recent non-positivist legal scholars is motivated at counteracting this negative framing by portraying Indigenous legal orders (*qua* law) as not only worthy of respect but more significantly as equal status of the legal orders of settler-states (Green and Hendry 2019, p. 10). For Margaret Davies, critical legal pluralism is about legal pluralities more broadly, not simply in the context of legal pluralism but a diversity of legal theory; a strategy focused on debunking the principle that "there is either an objective or true version of legal pluralism" (Hendry 2019, p. 171). For Davies, all law is generated through governance practices (Davies 2010; Hendry 2019), whereas for other non-positivist legal scholars, "law is *morally entailed by* practice" (Green and Hendry 2019, p. 10). Green and Hendry (2019) argue that while a non-positivist framing of law is not always necessary, such a conceptualisation does provide an important "explanatory power in relation to the settler-state legitimacy crises" (Green and Hendry 2019, pp. 10–11).

Scholars are increasingly focusing on how legal pluralism can offer new opportunities for transforming legal and governance regimes by challenging the dominant settler-state legal system, allowing for multiple legal systems to simultaneously operate (Indigenous and settler-colonial). For instance, Jones' (2016) exploration of legal pluralism in Aotearoa (consisting of legislation, case laws, and tikanga) provides renewed possibilities for Māori iwi to achieve some form of self-determination and autonomy within the overarching structure of the settler-colonial state (Jones 2016; O'Donnell and Macpherson 2019).

A key part of efforts, over the last three decades, to address Māori injustices involves reforms to Aotearoa NZ's contemporary (settler-colonial-based) legal order to recognise (to some extent) aspects of tikanga. Māori legal scholar Jacinta Ruru calls on Māori lawyers to continue this journey to reconciliation by considering the place of "our first laws—tikanga Māori—as law as part of our complete legal system" present-day (Ruru 2018b). In her research Ruru argues, following from countless other Māori (in their roles as academics, lawyers, leaders and politicians, activists, and members of particular iwi/hapū/whānau), that settler-nations like Aotearoa "need to look for new ways to meaningfully reconcile with Indigenous peoples to displace legal assumptions for

Crown ownership and the governance of land and water" (Ruru 2018b). Indeed, during the late twentieth and twenty-first centuries, tied to political recognition of the Treaty and the Crown's attempts to reconcile with Māori, there have been ongoing attempts to revive and reassert the applied usefulness of tikanga Māori as a legal order and process and, in doing so, articulate and define a place for that law within the settler-state of Aotearoa's legal system (Jackson 1995, 2007).

Legislation is now requiring that the Crown's legal order incorporate (to a limited degree) aspects of tikanga. The Resource Management Act (1991)) acknowledges that Māori exercise kaitiakitanga (environmental guardianship) and the significance of wāhi tapu (scared sites) and taonga (treasures) in waters and lands. Te Ture Whenua Māori Act 1993 (the Māori Land Act) recognises that a child adopted into a family in accordance with tikanga Māori practices of whāngai (customary adoption within the same hapū) can inherit land interests (from members of their adoptive family). Importantly, the Court of Appeal, in 2020, determined the Crown's allowance of offshore iron sand mining off the coast of Taranaki conflicted with iwi kaitiakitanga practices (Court of Appeal 2020). Likewise, the Supreme Court, in 2012, found that "Māori custom according to tikanga is therefore part of the values of the New Zealand common law" (New Zealand Supreme Court 2013, p. 94). At the forefront of these efforts to include tikanga into settler legal order are the "visions and aspirations of our Māori communities, iwi, whānau, and hapū" (Ruru 2018b). Treaty settlements, which we discuss later in this chapter, are perhaps the place where Māori voices (and their tikanga) are best and most powerfully encapsulated, which are providing the changes to the legal order of Aotearoa.

Limited Recognition: Indigenous Legal Traditions with Settler Legal Order

Globally a wealth of new legal pluralist research documents efforts to de-centre settler-colonial state law and concentrate on legal subjects and their capacities to produce new legal knowledge and implement frameworks that comprise their legal subjectivity (Bambridge 2016; Curran 2019; Hendry and Tatum 2018; Jones 2016). In British Columbia Canada, as the work of Curran (2019) and others demonstrate, a wide number of First Nations' are seeking to repoliticise water governance regimes by situating their legal traditions and laws and their expectations about what constitutes free prior and informed consent in the joint water arrangements they hold with the provincial government (Bakker et al. 2018; Curran 2019). Similarly, in the United States, numerous different Indigenous nations are continuing to challenge the settler-state legal order and expand on how "Indians reserved water rights" are defined within state and federal laws (Curran 2019, p. 19). For instance, the judge's decision in the case of Agua Calienta Band of Cahuilla Indians v Coachella Valley District found that the Tribe holds the rights to federal reserved groundwater and also that the Tribe's right to use the water took precedence over the state government of California's water allocation regime (Curran 2019). Likewise, the Standing Rock protest movement, started by Standing Rock Sioux to resist the Dakota Access Pipeline, is a declaration that Indigenous peoples' and their legal orders remain despite the ongoing colonial intrusions and dispossessions, and demand for IEJ (Baum 2019; Gilio-Whitaker 2019; LeQuesne 2019; Whyte 2017).

Researchers observe that destabilising modern politics and the reassertion of Indigenous laws, governance structures, and practices that rupture dominant political configurations are evidence of the wider disruption of hegemonic Western knowledge systems (Blaser et al. 2013, p. 20; Oslender 2019; Wilson 2019; Yates et al. 2017). Recent research investigates approaches, diversely referred to as collaborative and/or integrative models, including joint or co-governance agreements between Indigenous peoples and governments. These approaches seek to recognise (to a greater or lesser extent) Indigenous rights, knowledges and interests in water (as well as lands and seas), and to create processes of sharing responsibilities for decision-making, as well as ways that different parties can co-learn and co-produce new knowledge to improve freshwater management and/or health (Bischoff-Mattson et al. 2018; Bischoff-Mattson and Lynch 2017; Harmsworth et al. 2016; Memon and Kirk 2012; Wilson 2019). Based on their research in Australia, Howitt and Suchet-Pearson call for "ontological pluralism" whereby the dichotomy discourse and interlinked issues are defined and addressed. They argue that this naming and confronting can facilitate frameworks of environmental management scholarship and practical actions founded on mutual respect and plural value systems and enacted in ways that "acknowledges and respects Indigenous ontologies, or ways of being, and at the same time is attentive to the historical and current dominance of Eurocentric thinking within natural resource management" (Howitt and Suchet-Pearson 2006). The support for ontological pluralism (termed by some scholars as the pluriverse) enables possibilities and potentialities to bring about a transformation in freshwater governance and management by supporting Indigenous and hybrid governance structures and practices entrenched within settler-colonial systems of power and control (Blaser 2014; Wilson and Inkster 2018; Yates et al. 2017).

Research from Central and South America similarly demonstrates how different societies (all of which are dealing with ongoing legacies of colonialism) are grappling with recognition of more-than-human sentient entities and Indigenous peoples' ontologies and interests in their ancestral lands and waters through legislation and policies; which attests to the diverse possibilities of ontological and legal pluralism. Under the Ecuadorian Constitution of 2008, for instance, the rights of Nature (Pacha Mama) are recognised. Article 71 refers to the "nature or the Pacha Mama" possessing the right to have its "existence, maintenance and regeneration of vital cycles, structures, functions and evolutionary processes" respected. Pacha Mama also possess the right to legal restoration if any damage to its natural processes occurs. As a consequent of Article 71, any legal person (human and more-than-human) as well as any community (in Ecuador or from elsewhere) can insist that the Ecuadorian government honours and respects such rights. A well-known legal case (the "Vilcabmba River case") saw Nature being named as the plaintiff. The court ruled that Nature did possess rights and ordered the government to restore the riparian ecosystems of the degraded Vilcabmba River (Clark et al. 2018, pp. 796–797). The legal recognition of Pacha Mama resonates with the Andes concept of Buen Vivir vision (living well with the Earth), drawing on Indigenous intellectual traditions and knowledge systems, to demand the ontological and epistemological extension of living well within human communities to be extended to encompass the natural world (Cochrane 2014; Samuel 2019). Buen vivir "displaces the centrality of humans as the sole subject endowed with political representation and as the source of all valuation" (Chuji et al. 2019).

Such works' demonstrate the different ways in which such approaches can disrupt taken-for-granted views (the colonial status quo) about what or who has agency and how the world(s) are made and remade (Blaser 2014; Blaser et al. 2013; Chuji et al. 2019; Oslender 2019; Sieder and Barrera 2017). Accordingly, ontological politics are increasingly at the heart of analyses of the connections between multiple ways of thinking (and doing) (Chandler and Reid 2018). Likewise, research from Aotearoa and Australia examines the potential for Indigenous understandings of and engagements with rivers to foster transformations of the ways in which rivers are governed and managed (Bark et al. 2015; Bischoff-Mattson et al. 2018; Weir 2009, p. 119). In Aotearoa, as our next section explores further, this includes research directed at identifying ways of conceptualising rivers (and nature more broadly) that is ontologically and epistemologically inclusive, (as well as pragmatic and equitable) (Charpleix 2018; Salmond et al. 2014; Salmond 2017).

Decolonising Freshwater Governance: (Mis) Recognition of the Treaty and Tikanga

The significance of the Treaty has been the subject of intense legal and academic debate since it was signed in 1840 by representatives of the British Crown and more than 500 Māori rangatira (chiefs) (Jackson 1993; Orange 2015). Most Treaty scholars now concur that rangatira never intended to cede their sovereignty (absolute authority) over Māori to the Crown, nor did they intend to give up their tikanga (customary laws) and instead entered into a partnership agreement on which ongoing relationships with the British Crown were to be built (Healy et al. 2012; Jackson 1992, 1993; Mutu 2011; Orange 2015). The Treaty was a partnership agreement between the two different cultures and worlds (Te Ao Māori and Te Ao Pākehā), which implied (even if it did not explicitly state) that the Crown acknowledged tikanga Māori as the existing legal order and that some form of legal pluralism would operate in Aotearoa following the Treaty. However, for most of the nineteenth and twentieth centuries, the Treaty was denigrated (Anderson et al. 2015; Belich 1996, 2013; Ruru et al. 2017).

Soon after its signing, the Treaty, as we discuss in depth in Chaps. 3 and 4, the settler-colonial courts and successive settler-colonial governments did not recognise the Treaty nor acknowledge its legal, constitutional or political significance (Anderson et al. 2015; Belich 1996, 2013; Ruru et al. 2017). The statements made by Chief Justice Sir James Prendergast in 1877, when he issued his judgement in the case of *Wi Parata v The Bishop of Wellington* (finding in favour of the Bishop of Wellington's claims over a section of Māori land), highlighted broader Pākehā legal and public attitudes towards the Treaty as well as tikanga more generally. Prendergast declared the Treaty "worthless" on the basis it was "between a civilised nation and a group of savages" who were not sufficiently advanced enough to sign a treaty, furthermore since the Treaty was not enshrined into domestic law it was now a "simple nullity". Prendergast's ruling and statements (informed by earlier Court of Appeal decisions) helped shape decision-making on Treaty issues for decades to come and were used to justify the alienation of more and more Māori land (Prendergast 1877).

Since the mid-1970s, however, there has been a significant increase in references to the principles of the Treaty or to specific rights and interests within legislation, which represents an important shift in recognising the legitimacy and authority of Maori in a range of contexts. Since 1975, many laws in Aotearoa make reference to Treaty principles (Jones 2016; Waitangi Tribunal 1999, 2018). The first legislation to do so was the Treaty of Waitangi Act (1975). Since that legislation, many other government policies, laws, Waitangi Tribunal reports, and court cases make reference to the Treaty principles; however, there is no final or complete list of what those principles are and the principles are not codified in any laws. Instead, official government documents refer to the Treaty principles in vague terms, without any reference to the actual treaty text (be it the English or Māori version of the treaty). In 1989, the Fourth Labour government became the first central government to outline Treaty principles to guide its actions with regards to its relationships with Māori: (1)

The Crown (central government) possesses the right to govern and make laws; (2) Māori iwi possess the right to organise as iwi, and are legally able (through laws) to control their resources; (3) Legal equality (that all New Zealanders are equal under the laws); (4) The Crown and iwi are obliged to interact with each other with a reasonable level of cooperation on major issues that are of collective concern; (5) The Crown is responsible for providing effective institutional processes for the resolution of Māori grievances in the expectation that reconciliation can occur (Jones 2016; Palmer 1989). However, no later central government in Aotearoa defined any new Treaty principles, and the principles are at best vague ideas that governments are meant to follow rather than laws.

Within Aotearoa's (settler-colonial) legal order, Māori hold no general constitutional rights that give them special legal recognition as Indigenous people or as Treaty partners (under Tiriti o Waitangi/Treaty of Waitangi hereafter the Treaty) and allows them to be heard in a court setting (Jones 2016; Ruru 2012). Partly, because the country does not possess a specific written constitution that explicitly acknowledges Māori interests. The Treaty is still not part of the country's domestic law. The Treaty is now commonly referred to by legal scholars as the "informal constitution along with the New Zealand Bill of Rights Act 1990 and the Constitution Act 1986" (Ruru 2012, p. 112). For members of the Aotearoa judiciary and those acting under the law, the Treaty itself is only relevant when explicitly included within statutes. Thus, while Article Two of the Treaty guarantee to Māori that they would retain their tino rangatiratanga (authority) over to their whenua (land) and other taonga (treasures) including rivers (while agreeing to give the Crown kawantantanga/governorship over Aotearoa), the lack of legal recognition of the Treaty or attempts to explicitly articulate the Treaty principles (discussed in the next section) into laws means that Māori still lack constitutional rights to water. Nevertheless, there is a small degree of domestic legal acknowledgement of the relationships of Māori with water. For instance, the Resource Management Act (RMA 1991) requires local authorities to recognise the relationships of Māori with their ancestral waterbodies (rivers, lakes, seas) and take into account kaitiakitanga (environmental guardianship exercised by Māori) when exercising their functions and powers to managing the development, use, and protection of environments (Bargh

2020; Bell 2018; Ruru 2012). However, the RMA, as we detailed in our previous chapter, only provides a limited degree of recognition to Māori interests in water, and gives them the right to be included in local government decision-making regarding management and use of water, which is to say participatory inclusion. Yet, Māori lack the authority to shape and make decisions about their waterways (which is at the heart of Māori demands dating back to their signing of the Treaty which guarantee to them that their rangatiratanga would be preserved and protected by the Crown).

While Māori (and any other person in Aotearoa) can appeal decisions relating to resource consents (issued under the RMA) to the Environment Court, these appeals are restricted to matters of law (Ruru 2012). There are numerous instances where Māori objectors (such as Greensill and members of her hapū Tainui discussed in Chap. 5) appeals of regional and district council decisions about resource consents to discharge wastewater, take water, or dam water. In the majority of these legal cases Māori emphasise how water, specifically their ancestral rivers (their awa), underpin their cultural identity (through their whakapapa), their belief that all water possesses a mauri and the significance of waterbodies as food harvesting sites. Yet, in most instances Māori do not come out of the courts as victors, and many lose their cases outright. The courts, while aware of Māori relationships to awa, wai, and whenua, argue that section 6(e) of the RMA does not give Māori the right to veto resource consents or other decisions of local government, but merely the right to participate in decision-making processes (Greensill 2010; Ruru 2012, 2018a). The judiciary interpretations, however, clashes with those of Māori who for generations have been protesting for their rangatiratanga to be recognised, respected and honoured by the Crown; the views of Māori are recently endorsed by the findings of the Waitangi Tribunal.

In 1975, the Waitangi Tribunal was established as a permanent commission of inquiry to investigate Māori claims that the Crown was not honouring the terms of the Treaty of Waitangi. The responsibilities of the

Tribunal include researching and holding public inquiries into historic and contemporary claims filed by any Māori individual or group that the Crown breached the Treaty, reporting back to claimants and the Crown as to its inquiry findings, and making recommendations to the Crown as to how it can address Treaty breaches (reconciliation and restorative justice) (Jones 2016; Mutu 2018, 2019; Wheen and Hayward 2012). Each Waitangi Tribunal inquiry into Māori claims is required to determine whether a Crown action (or omission) was or is inconsistent with the Treaty principles (as which recently occurred with its Wai 898 inquiry into Te Rohe Pōtae). Each Tribunal panel, which always comprises Tribunal members including a Māori Land Court judge, a historian, a kaumatua, are required to determine not only if the Crown breached the Treaty principles, but also which principles apply for each claims being investigated. For this reason, the Tribunal does not keep a singular set of unchanging Treaty principles that it applies for each claim before it (highlighting the different experiences of iwi). Indeed, in 1983, the Waitangi Tribunal stated "The spirit of the Treaty transcends the sum total of its component written words and puts literal and narrow interpretations out of place" (Waitangi Tribunal 1983, p. 47). Over the decades since 1975, however, some key principles emerged from Tribunal reports that are often applied in various claims. These principles are derived not only from the terms of the Treaty's two texts (Māori and English language versions), but also from the socio-cultural and political circumstances in which the Treaty was created and signed by (some) Māori and representatives of the British Crown in 1840. To illustrate the Waitangi Tribunal approach to the Treaty principles we refer attention to the Waitangi Tribunal's inquiry into Te Rohe Pōtae (King Country) (Waitangi Tribunal 2018). We stress that the Treaty principles are those that Te Rohe Pōtae Tribunal viewed were relevant to that inquiry and differ from those applied to other inquiries.

Waitangi Tribunal's approach to Treaty principles: Te Rohe Pōtae **Inquiry**

Treaty Principle	Interpretation of the principle
Tino rangatiratanga, self-government and autonomy	Māori communities retained their tino rangatiratanga (under Article Two of the Treaty), which included their right to self-government and autonomy, "and their right to manage the full range of their affairs in accordance with their own tikanga" (Waitangi Tribunal 2018, p. 189). As part of the Treaty exchange, which included mutual recognition of kāwanatanga and tino rangatiratanga, the Crown guarantees to protect and provide for Māori autonomy and authority. Autonomy, was defined previously by the Turanga Tribunal as 'the ability of tribal communities to govern themselves as they had for centuries, to determine their own internal political, economic, and social rights and objectives, and to act collectively in accordance with those determinants'.
Kāwanatanga and good governance	The Crown possess the right to govern and make laws, which was first (in the decades post-1840) for the purpose of controlling settlers and settlement and regulating relationships with foreign powers. The power of kāwanatanga (governance), however, is qualified by the rights that continued to be reserved to Māori (under Article Two of the Treaty). "To the extent that it affects Māori communities, the right of kāwanatanga must be used to protect Māori interests" (Waitangi Tribunal 2018, p. 189). Related to kāwanatanga, the Crown is required to ensure it acts in accordance with its own laws, be held to account for its actions to Māori, and be subjected to independent scrutiny where appropriate.
Partnership	The Treaty created a relationship that was dependent on ongoing dialogue and negotiation, under which Māori and the Crown would work together to agree to the practical details of how tino rangatiratanga and kāwanatanga would co-exist. Both Treaty partners were duty bound to act honourably and in good faith with each other. The obligations of this partnership meant that neither partner can act in a way that "affects the other's sphere of influence without their consent" (Waitangi Tribunal 2018, p. 189), it also created a duty that the Crown consult with Māori and obtain free and informed consent from iwi before land and water management.

(continued)

(continued)

Treaty Principle	Interpretation of the principle
Reciprocity and mutual benefit:	Above all, the Treaty provided the basis on which two peoples (Māori and Pākehā) could share one country. It centred on reciprocal partnership relations that were (and still are) one that involved the exchanges for mutual benefits and advantages between the Crown and Māori. Māori granted the Crown the new power of kāwanatanga (governance) in return for a guarantee that the protection of their tino rangatiratanga over their land, people, and taonga would be safeguarded. Through this recognition of different powers, the Treaty was intended to provide for the mutual protection of both Te Ao Māori and Te Ao Pākehā. Accordingly, it was meant to ensure that relationships between Māori and Pākehā peoples would provide mutual advantages for both cultures.
Active protection	The Crown are obligated to employ its power of kāwanatanga to actively protect the interests and rights of Māori rights (as guaranteed under Articles Two and Three of the Treaty) which included Māori authority and autonomy (tino rangatiratanga).
Options	The Treaty envisaged a new country wherein two peoples (Māori and Pākehā) would live together with their own laws and customs. The interface between Te Ao Pākehā and Te Ao Māori was to be governed on the basis of mutual respect and partnership. Inherent in the Treaty relationship was that Māori, whose laws and autonomy were guaranteed and protected, would "have the right to continue to govern themselves along customary lines, or to engage with the developing settler and modern society, or a combination of both" (Waitangi Tribunal 2018, p. 189). Māori were meant to be able to choose to continue to live according to their tikanga (laws) and ways of life (within Te Ao Māori), to engage with Te Ao Pākehā society and economy, or to combine aspects of both worlds and walk in both. Their choices were meant to be free and unrestricted.

Treaty Principle	Interpretation of the principle
Equity and equal treatment	The principles of reciprocity, autonomy, active protection, and partnership required the Crown to act fairly in its treatment of Māori (and Pākehā). The Crown cannot use its powers of governance to provide unfair advantages to Pākehā at the expense of Māori interests. Likewise, the Crown must not provide equal treatment to Māori groups nor foster divisions between them.
Redress	In situations where the Crown acted in excess of its powers of kāwanatanga and/or breached the Treaty terms, and Māori suffered prejudice as a consequence, then the Crown possesses a clear duty to set matters right. The Crown must provide redress in the form of a remedy to compensate Māori and to resolve the grievance.

The Tribunal's findings, however, are not laws, and therefore it is left up to the Crown and Māori groups to directly negotiate as a means to seek to address Māori claims about Treaty breaches and injustices committed as a consequence which is a process undertaken by a separate institution. In 1994, the Office of Treaty Settlements (OTS) was established (located within the Ministry of Justice and entirely distinct from the Waitangi Tribunal) to negotiate with individual iwi (and sometimes larger pan-iwi groupings) about legal-financial reparation packages that acknowledge and sought to address the Crown's failures to honour the Treaty and as a means to reconcile with Māori (discussed further in Chap. 7) (Jones 2016, pp. 21–22).

As a consequence of the negotiations between OTS and iwi, a range of 'Treaty settlements' started to emerge (from the mid-1990s with the Waikato-Tainui Raupatu Settlement and continuing into the 2020s). These Treaty settlements include a formal apology from the Crown for historic and contemporary injustices against a particular iwi, financial reparations to the iwi (monetary payments and return of Crown land-holdings), and the introduction of new legislation (Jones 2016; Williams et al. 2018). The Treaty settlement statutes provide an additional legislative means by which Māori are seeking to protect and maintain their connections with their awa and whenua, which in many instances extends those provided under the RMA. Many statutes explicitly acknowledge the significance of lakes and rivers to specific iwi as well as incorporate elements of tikanga Māori. The Deed of Settlement that contributed to the Ngāi Tahu Claims Settlement Act (1998) includes aspects to tikanga. Embedded in the statute are pūrākau (traditions and stories) of the whenua, such as the origin story of Aoraki/Mount Cook and the naming of the South Island, which Ruru argues provides a catalyst for transforming legal education and public understandings of law. Likewise, the Ngāi Tahu Settlement Act includes statutory recognition of Ngāi Tahu social, cultural, spiritual, political and economic connections with the Mata-Au (Clutha) River. The Act records that the river is in possession of its own life force (mauri) and is a descendant of the atua (gods) of Māori. More recent Treaty settlements, including those with Waikato-Tainui and Ngāti Maniapoto, include specific provisions for Māori iwi to co-govern and co-manage culturally significant sites, including rivers, lakes and national parks. And, most notably, the recognition of the legal personhood of Indigenous ancestors (the forest of Te Urewera and the river of Whanganui) (New Zealand Parliament 2014, 2017).

Indeed, the emerging backbone of legal pluralism in Aotearoa is tied to recognition of the Māori interests under Treaty settlements. A range of new institutions were instituted from the mid-2000s to co-govern and co-manage a range of natural resources and geo-regions as a way of addressing injustices as well as meeting the Crown's obligations under the Treaty. These include a plethora of formal agreements that now position Māori as partners within formal decision-making processes relating to freshwater systems, which represents a radical departure from past practices of governing and managing rivers in Aotearoa (as demonstrated previously in Chaps. 4, 5, and 6). Although each agreement differs in its contents, including the institutional structures and functions it establishes, a common thread amongst all of these agreements is that Te Ao Māori is positioned at the heart rather than being excluded or marginalised. Emphasis within the new agreements is placed on mātauranga Māori (Māori knowledge) and tikanga (customary laws and correct protocols), which includes the inclusion of the specific values of different iwi, their knowledge, histories, and aspirations for the future within river co-governance and co-management. We will now turn our attention to recent Treaty settlements and how these settlements (accompanied by resulting legislation) are exemplars of both legal pluralism and ontological pluralism.

Treaty Settlement: Ngā wai o Maniapoto (Waipā River) Act and the Waiwaia Accord

In September 2010, a Deed of Settlement between the Crown and Ngāti Maniapoto (by the mandated negotiation party Maniapoto Māori Trust Board) was signed and (as with other Treaty settlements) contained the historical account and reasons for the claim, acknowledgements, and apology from the Crown (Jones 2016). The deed also extended the co-governance and co-management arrangements that operated in respect of the Waikato River and Lower Waipā River (established under the Crown's other deeds of settlement with neighbouring iwi Waikato-Tainui, and Ngāti Tuwharetoa, Raukawa and Te Arawa) to include Ngāti Maniapoto and the Upper Waipā River (discussed in further depth in Chap. 7). At the same time as the deed was signed, Ngāti Maniapoto and the Crown also signed the Waiwaia Accord, which further affirmed both parties' commitment to partnership through the co-governance and co-management of the Waipā River. Later legislation, introduced in 2012, established the institutional arrangements for co-governance and co-management (through the Waikato River Authority, which is discussed in further detail in Chap. 7).

The Deed of Settlement and Waiwaia Accord both incorporate tikanga Māori and demonstrate Maniapoto ways of thinking wherein their whakapapa (genealogical connections) is interwoven with the ebb and flow of wai (water) within their awa, its mauri (life force) and mana (power and authority). The Waiwaia Accord includes sections in Te Reo Māori (the Māori Language) that highlights this understanding:

> Ko te mauri, ko te waiora o te Waipa ko Waiwaia
> Ko Waipa te toto o te tangata! Ko Waipa te toto o te whenua,
> koia hoki he wai Manawa whenua!
> Ko Waipa tetehi o nga taonga o Maniapoto whanui.

Ancestral authority handed down from generation to generation
in respect of Waiwaia,
Guardian of the Waipa River. (Ngāti Maniapoto et al. 2010)

Waiwaia is a taniwha (supernatural creature) and kaitiaki (guardian) of
the Waipā River and the Ngāti Maniapoto people, and is identified as the
essence and wellbeing of the Waipā River, and the personification of the
waters of the Waipā River. The phrase 'mana tuku iho o Waiwaia', which
is included in the Deed of Settlement and Waiwaia Accord, means the
ancestral authority and prestige handed down from generation to generation in respect of Waiwaia. In the Deed of Settlement, Waiwaia Accord,
and subsequent legislation (introduced to parliament in 2012) the status
of Te Awa o Waipā as a taonga (treasure) to Maniapoto and tūpuna
(ancestor) is recognised by the Crown; similarly, recognition is given to
Maniapoto obligations as kaitiaki to restore, maintain and protect the
mana, mauri, and wairua of all the waters within the rohe of Manaipoto
(Ngā Wai o Maniapoto). In doing so, the ontological and epistemological
underpinnings of Ngāti Maniapoto (their values, worldviews, and
tikanga) are explicitly acknowledged and incorporated within the legal
agreements, which includes the Ngā Wai o Maniapoto (Waipā River) Act
2012. Part of the legislation is quoted below as it highlights the ways in
which (for the first time) the Crown formally recognised the values, and
tikanga of Ngāti Maniapoto with respect to the Waipā River:

(10) To Maniapoto, the Waipā River is a single indivisible entity that flows
from Pekepeke to its confluence with the Waikato River and includes its
waters, banks, bed (and all minerals under it) and its streams, waterways,
tributaries, lakes, fisheries, vegetation, floodplains, wetlands, islands,
springs, geothermal springs, water column, airspace and substratum as well
as its metaphysical elements with its own mauri. (New Zealand
Parliament 2012)

Ngāti Maniapoto ontological underpinnings are demonstrated in the
above words, articulating the concepts of reciprocity, caring, and belonging. Both Ngāti Maniapoto and Te Awa Waipā need each other, they are
indivisible, a relationship without a start or an end, within which those

Māori who are mana whenua possess responsibilities as kaitiaki (guardians) to work to ensure the health and wellbeing of all (human and non-humans alike). Human and more-than-human actors are in close and ongoing relationships with one another; birds, mountains, trees, fish, rivers, and taniwha all possess the same genealogical lines of descent as human beings. Iwi members articulate how damage to their ancestral river diminishes the mauri of the river and causes them (as mana whenua) and their more-than-human kin harm. As one Ngāti Maniapoto member states: [Ko] te wai te toto o te whenua, water is the blood of the land. The land is the mauri of the people, keeps the people alive. If the water goes bad, the land goes …bad, the people die" (Iwi Rep 6 2020). Accordingly, environmental injustices occur not only because of material manifestations of environmental degradation (distributive injustice) and the marginalisation of Māori from decision-making processes (procedural injustice), but also because of the misrecognition of mātauranga and tikanga which is premised on the non-divisible reciprocal relationships between humans and more-than-humans.

For Ngāti Maniapoto, the Waipā River Act goes some way to redress injustice by misrecognition by including Ngāti Maniapoto values and principles. The legislation includes sections in Te Reo Māori (the Māori language) as well as including reference to the taniwha. It also establishes co-governance and co-management arrangements between iwi and the Crown over the Upper Waipā River; however, the design and implementation of co-governance agreements are now being critiqued by iwi for disregarding Māori legal and political governance systems and providing inadequate means to achieve iwi environmental justice (as we outlined in Chap. 7). Yet, the legislation does show evidence of legal pluralism and is a significant marker of the shift in relationships between Ngāti Maniapoto and the Crown and the potential to expand the narrow confines of the settler-colonial legal order to include tikanga. It also hints at the possibilities of recognition and acts to empower the coexistence and flourishing of many worlds (Dunford 2020). The inclusion of Indigenous ontologies, as we later demonstrate in Chaps. 8 and 9, is a critical way of destabilising conventional scientific and technocratic approaches to river management, and provides new ways to address complex social-environmental issues within freshwater systems in a relational, and holistic manner

(Crow et al. 2018; Parsons et al. 2019). In the case of the Waipā River, achieving the overarching purpose of the Waipā River Act requires tolerance for ontological inconsistency rather than treating Māori and modernist ontologies as mutually exclusive and in opposition (Salmond et al. 2014).

Treaty Settlement: Te Awa Tupua (Whanganui River)

Another Treaty settlement, and resulting co-governance arrangement, is that of the Whanganui River is recognised as a legal personality. As part of their negogiations with the Crown to reach Treaty settlement the various iwi who whakapapa to the Whanganui River, requested that the river be officially given the status as a legal person. The 2014 Treaty settlement (Ruruka Whakatupa Te Mana o te Iwi o Whanganui) recognised iwi and hapū deep-seated and ongoing relationships to their river, provided an apology to iwi and hapū for Treaty breaches, as well as a financial settlement ($80 million NZD). Iwi requested that the river be given legal personhood as a means to reconcile Te Ao Māori conceptualisation of rivers as more-than-human actors with Te Ao Pākehā and Western legal traditions. It was also a deliebrate attempt to find a way to protect and restore the mauri of their awa, which (like the Waipā River) had become severely degraded as a consequence of settler-led land-use changes, governance regimes, and management systems focused on agricultural productivism at the expense of freshwater ecosystem functioning (Charpleix 2018; Forster 2016; Morris and Ruru 2010; Ruru 2012; Salmond 2017).

The common whakatāuki (proverb) "Ko au te awa, to te awa ko au" (I am the river and the river is me) summarises the relationships between Whanganui iwi and their river, as well as Whanganui iwi role as kaitiaki (Brierley et al. 2019; Bryan 2017; Wilson 2019; Youatt 2017). The Whanganui River approach is a legal hybrid that incorporates components of Māori tikanga (customary law) that perceive rivers to be ancestors and/or kin (connected through genealogical connections to specific hapū and iwi) and settler legal traditions in Aotearoa which incorporates the Treaty principle of partnership (Forster 2016; Ruru 2013; Winter 2018). It is a new legal framework that attempts to, Forster maintains, "secure the autonomy of both Māori and the Crown [the New Zealand Government] in relation to governance and management of natural resources associated with the river" (Forster 2016, p. 325).

In an approach that resonates with the framing of the Waipā River as a tūpuna (ancestor) and kin of Ngāti Manaipoto (under Waipā River Act), the Whanganui River, within Te Awa Tupuna legislation, conceptualises the Whanganui River as a more-than-human actor who has and still is suffering ongoing damage as a consequence of human activities. The 2017 Te Awa Tupua (Whanganui River Claims Settlement Act) declares that the Whanganui River is "an indivisible and living whole" and encompasses the river from its headwaters in the mountains to the Tasman Sea and incorporates all material and spiritual dimensions and is afforded legal personhood with all the powers, obligations, and rights as a person. The status of a legal person means that the river can (in theory) enforce its rights over other legal persons. There is the potential that legal cases could be launched where the river is a plaintiff (such as those taken in Ecuador on behalf of Nature or Pacha Mama in which the courts ruled in favour of upholding the rights of Nature and required government to take action to restore a degraded river) (Clark et al. 2018).

In addition to being made a legal person, the 2017 act also gave the river an independent voice within decision-making. Te Awa Tupua is to be represented by a two-person committee (Te Kōpuka nā Te Awa Tupua) made up of one person who represents local iwi and the other a person nominated by the Crown. The committee is meant to act as "the human face of Te Awa Tupua". These human actors then must speak on behalf of the voiceless Te Awa Tupua (Charpleix 2018; New Zealand Parliament 2017). Under Te Pou Tupua rests, (in descending order of influence and authority), an advisory group (Te Karewao) as well as a strategy group (Te Kōpuka) both of which are made up of iwi and Crown representatives. In addition, broader community representation is given space in a collaborative community group (Te Heke Ngahuru) those membership structure and overarching purpose is looser and includes any person with interests in the river. The institutional arrangements for the Whanganui River (designed to "support the health and well-being of Te Awa Tupua" the legal person and the right) include the committees listed earlier, which presents a new approach to co-governance and co-management in Aotearoa (which is funded through a separate grant, with an initial funding of $30 million NZD provided by the Crown) (Clark et al. 2018).

The frameworks for governing and managing Te Awa Tupua recognise, afford value to, and provide funding for Māori co-governance and co-management, and in doing so recognise and open up ontological and epistemological spaces within the settler-state for Māori ways of knowing, being, and interacting with more-than-human entities. The principles of legal personhood as well as co-governance and co-management arrangements all reinforce the indivisibility of Whanganui iwi and the river, including their rangatiratanga and wairua, and the interconnectedness of their sovereignty with that of the river. In 2020, Te Awa Tupua has yet to be a plaintiff in a legal case, and it remains to be seen how the legal personhood of Te Awa Tupua will play out within Aotearoa's courts (and if the decisions will parallel or challenge those made in Ecuador in regard to Pacha Mama) (Clark et al. 2018; Muller et al. 2019).

Three years earlier, in 2014, Te Urewera (mountain range covered by forest in the North Island) also received legal personhood through legislation as part of the Treaty settlement between Ngāi Tūhoe and the Crown (New Zealand Parliament 2014). The legislation means that no one owns Te Urewera (which was unlawfully taken from Ngāi Tūhoe and converted into a national park by the Crown) and it effectively own's itself. Te Urewera is similarly represented by a committee comprised of iwi and government agency representatives. While some legal scholars argue that Te Urewera did not receive legal personhood as a method to ensure environmental protection (as there were already laws in place to prevent or mitigate environmental degradation as it was a national park), we note that generations of Ngāi Tūhoe protested about the negative consequences of settler-colonial rule on their rohe; which included both material and metaphysical losses and damages linked to Crown actions to suppress the sovereignty and authority of Ngāi Tūhoe (Morris and Ruru 2010; New Zealand Parliament 2014; Ruru 2018b; Waitangi Tribunal 1999, 2009). For Ngāi Tūhoe, like other Indigenous peoples, decision-making authority is inextricably tied to their environmental justice.

With the enactment of such new legislation, the legal framework of Aotearoa is being stretched and incrementally or more radically reconfigured from singular to plural in viewpoint. Scholars Christine Winter (2018) and Anne Salmond (2019) argue that while this singular (Te Ao Pākehā) to plural (Te Ao Māori and Te Ao Pākehā) expansion is being deployed through the existing colonial legal order, it is still facilitating a far greater recognition of Māori knowledge and tikanga than previous legislation allowed for. Indeed, the acknowledgement of mātauranga and tikanga surrounding rivers (and other more-than-human actors) possessing both mauri and mana within legal agreements, legislation and co-governance arrangements is a significant shift from previous statutes (such as the original RMA introduced in 1991) that contained mentions to Māori cultural values and wāhi tapu (sacred sites). Such legal pluralism, Ruru (2017) and Hickford (2018) suggest, is an important and necessary step to decolonise environmental governance in Aotearoa by explicitly acknowledging Māori worldviews, cultural identities and continuance, mātauranga and tikanga. A key part of this involves recognising that, from a Te Ao Māori perspective, landscapes and waterscapes are inhabited by living generations of people as well as their ancestors (human and more-than-human kin). The duties and obligations to show reciprocity, hospitality, and care for one's kin extended are therefore intergenerational and are based on the need to ensure relationships between all beings (human and more-than-human) are balanced and mutually beneficial. These deeds of settlements, legislation and co-governance arrangements, which recognise (to some degree) the interests, agency, and rights of the more-than-human realm disrupt the anthropocentricism inherent in Western liberal conceptualisations of EJ. Relationships based on whakapapa that extend across generations highlights the ways in which justice is always (from Te Ao Māori perspective) encompass both the needs and responsibilities of humans and more-than-human with "generations to come [holding] as much interest in the land" and waters "as the individuals living at any point in time" (Stephenson 2001, p. 166).

All these statutes passed through New Zealand Parliament accompanied by a formal apology from the Crown (the New Zealand Government) for the long-term damage that rivers (and its Māori kin groups) suffered as a consequence of settler-colonialism (specifically government actions and inactions that breached the Treaty). In all instances, the resulting co-governance arrangements ensure that Māori roles as kaitiaki are formally recognised and incorporated into the co-governing models for these geo-features; Māori comprise one of the two representatives that were appointed to represent Te Awa Tupua, similarly they make up fifty per cent of Te Urewera Board (for first term and thereafter making up two-thirds), and fifty per cent of the Waikato River Authority (Collins and Esterling 2019; New Zealand Parliament 2014; Rangitāiki River Forum 2015; Waikato River Authority 2016).

Through these legal mechanisms, the reciprocal and ongoing connections between rivers, forests, lands and their Māori kin groups (whanau/family, hapū/sub-tribe, iwi/tribe) are recognised. These relationships are "an indivisible and living whole from the mountains to the sea and incorporating all its physical and metaphysical elements" (Ruru 2018b).

A pivotal part of the decolonising processes is the disruption and desta-bilisation of the privileging of Western ontologies and epistemologies and allowing space for different ways of thinking and being. Muller et al. (2019) argues that the Whanganui and Te Urewera examples demon-strate a profound shift in power to Māori iwi by enabling Māori world-views to be given status in environmental governance and management decisions whilst still being situated within the legal frameworks of the settler-state. Muller et al. (2019) interprets the agreements as evidence of 'nation-building' approaches to environmental governance and manage-ment wherein the settler-state of Aotearoa recognises Māori sovereignty (which was first acknowledged under Te Tiriti o Waitangi in 1840 but ignored until 1975). They argue that the new legal agreements are testa-ments to the importance of the value of "ontological pluralism through the assertation of Indigenous sovereignties" (Muller et al. 2019, p. 9). Indeed, as Whyte, Wildcat and other Indigenous scholars argue, achiev-ing environmental justice for Indigenous peoples requires "the recogni-tion and restoration of reciprocal relationships between people and places" which includes recognition of more-than-human beings and mul-tiple worlds (Wildcat 2013, p. 514).

Complexities of Enacting Legal Pluralism

It is critical to note that the different wording in Treaty settlements, once agreed on by iwi and the Crown and formularised within legislation, does not ensure consistency in understanding or application. The complexities of co-existence (between Indigenous and non-Indigenous worlds, world-views, and legal orders) remains (despite Treaty settlements) and invariably result in legacies of assorted legal rights, interests and uses arising from setter-colonialism. Thus, while the Whanganui River is defined under the Te Tupua Awa statue as a legal person that is an indivisible entity (waters, subsoil, riverbed, plants, airspace above its waters), other legislation still compartmentalises the river. As legal scholar Hickford notes, the coastal marine area (from the Whanganui River to the Cobham Street Bridge within the township of Whanganui) is subject to the Marine and Coastal Area (Takutai Moana) Act 2011, which states that neither the Crown nor any persons can own the common coastal and marine area (Hickford 2018, p. 168). Accordingly, Whanganui iwi aspirations for invisibility, which are embedded within their Deed of Settlement and the Tupua Awa Act, are still forced to contend with several statutory regimes (products of Western ontologies and epistemologies) that continue to compartmentalise river systems.

The realities of translating legislation (this came about from Treaty settlements) into meaningful actions that address environmental injustices against Māori and their more-than-human relatives remains a politically fraught and power-laden process. Hickford refers to the potential for "interpretive risk" which results when:

> strangers to the processes of [Treaty settlement] negotiations end up interpreting what was agreed at earlier moments in time and constructing different ways of understanding those concepts captured in the legislation and deeds of settlement. Possibilities of mutual incomprehension persist ... [in this] 'middle ground'. (Hickford 2018, p. 171)

While parties may be able to work together towards common goals, this did not mean that there is shared understandings of concepts and practices. However, since the Crown defines parameters of Treaty settlement processes (including negotiations, awarding of financial compensation packages and passing legislation) it is fair to say that the Crown is in a stronger bargaining position when it comes to later determinations of the meaning of concepts. Indeed, despite the progress made toward greater legal and ontological pluralism within Aotearoa, the settler-state continues to dictate the terms by which iwi can participate in environmental governance and management decision-making processes. Accordingly, it is the settler-state who determines how Māori tikanga, knowledge, and relationships with their rohe are defined and recognised (through legislation and governance arrangements), which may leave iwi open to further injustices (Whyte 2011, pp. 199–200).

Indigenous Canadian scholar Zoe Todd warns of the dangers of Indigneous knowledges and ideas being appropriated in Euro-Western contexts "*without Indigenous interlocutors present to hold the use of Indigenous stories and laws to account* flattens, distorts, and erases the embodied, legal-governance and spiritual aspects of Indigneous thinking" (Todd 2016, p. 9). Todd's warning was made in the context of non-Indigenous scholars employing Indigenous knowledges through Eurocentric theories and methods; such a critique was made earlier by Māori scholar Linda Smith in her seminal work *Decolonising Methodologies* first published in 1999 (Smith 2013; Todd 2016; Watts 2013). Yet in the context of the interpretation of deeds of settlement, legislation, and policies, we extend Todd's warning to include non-Indigenous decision-makers interpeting and employing mātauranga, tikanga and Māori principles (such as mauri and kaitiakitanga) without consideration of the embodied expressions of Indigenous laws, stories, songs, and practices as inter-threaded together in "Indigenous-Place Thought" and Indigenous self-determination (Todd 2016, p. 9). There is an "interpretive risk" (whereby strangers to the reconcilitation process interpret the meanings of terms, settlements, and statutes differently from those people who originally agreed to them) as a consequence of three key factors. First, high staff turnover (including replacement of elected officials) mean that few government officials remain in positions long enough to be involved in both the creation and implementation of agreements (Treaty settlements, legislation, co-governance arrangements). New elected officials and government employees are often unfamilar with local specifics (socio-political, cultural and historical contexts) in which the agreements between iwi and the settler-state were formed as well as the intended meaning of key terms and mechanisms within deeds of settlement and statutes. The second (inter-related) factor (associated with interpretive risk) is the potential that non-Māori decision-makers (who still make up the majority of the New Zealand Parliament, central government departments and local government bodies) misunderstand Māori concepts and ways of thinking and in doing so misrecognise Māori interests. Indeed, there is a threat that tikanga, mātauranga, and iwi requirements that are vital for Indigenous Environmental Justice (IEJ) are not acknowledged at all. As we note earlier in Chap. 5, lack of recognition can occur when decision-making powers rests in the hands on one culture who by design or accident marginalise other cultures' knowledge, laws, worldviews and modes of living. Within Aoteraoa the power to interpret and decide what a legislation means and how it should be applied still largely rests in the hands of non-Māori individuals (government officials and members of judicary) situated in the Te Ao Pākehā. Accordingly, there are multiple interpretative risks associated with the new agreements tied to the complete failure to or partial acknowledgement of Indigenous ontologies and epistemologies; the problematics of recognition and interpretation extend to include Indigenous legal orders, governance structures, as well as Indigenous demands for greater economic and political autonomy (Ahmad 2019; Grosfoguel 2015; Maldonado-Torres 2016).

Within the context of academia, Watts and Todd suggest that the non-Indigenous scholars' current interest in studying Indigenous ways of thinking (the so-called ontological turn) and representing more-than-human ontologies as the solution to the global planetary crises of the Anthropocene, more often than not takes place without any recognition given to Indigenous peoples' lived realities (of socio-economic deprivation, multiple forms of violence, political marginalisation, lack of access to basic services, and environmental degradation of their ancestral lands and waters) (Bécares et al. 2013; Harris et al. 2006; Leonard et al. 2020; Mascarenhas 2007; Tobias and Richmond 2014; Todd 2014; Watts 2013). Likewise, attempts by the settler-state to recognise those elements of Māori knowledge and tikanga that are easily consumable (less discomforting) for the dominant political and social group (Pākehā) holds the potential to rearticulate existing injustices and is yet another example of what American anthropologist Deborah Bird Rose calls "deep colonising" (Rose 2004). Our notes of caution, however, are not a critique of current efforts to expand Aotearoa's legislation and governance frameworks to embrace Indigenous ways of thinking and being, but rather that greater attention needs to be devoted to how pluralism can operate in situations where inequitable power relations between Indigenous and non-Indigenous peoples remain. Indeed, all of those parties (scholars, politicians, government officials, Indigenous leaders) involved in advocating for, creating, and implementing these legally and ontologically pluralistic agreements, aimed at reconciliation and addressing Indigenous injustices, need to continue to be attune to the multiple manifestations of colonialism.

Although many Māori describe themselves as ambicultural (who walk in the worlds of Māori and Pākehā), the legal and political structures of Aotearoa are still not ambicultural (Winter 2018, p. 207). It is critical that we recognise that despite the passage of new legislation:

> The colonial moment has not passed. The conditions that fostered it have not suddenly disappeared. ... The reality is that we are just an invasion or economic policy away from re-colonising at any moment. (Todd 2016, p. 16)

Therefore, it is important to think about how the turn towards ontological pluralism within legislation, policies and governance structures may reinforce inequitable power arrangements (Todd 2016, p. 9). Māori legal scholar Ani Mikaere warns that Māori should not:

> settle for mere improvements in the Pākehā system as being the ultimate goal. It is all very well to be making Pākehā law and legal institutions as Māori friendly as possible, but only so long as we do not become comfortable that we forget to aim for some more ... to remind ourselves constantly about what it is that tino rangatiratanga ultimately demands. (Mikaere 2005, p. 24)

The Treaty settlements and emergent co-governance arrangements, Mikaere and Te Aho warn, are serving to enhance the single (settler-colonial) legal order to better acknowledge tikanga "for the sake of national cohesion" rather than actually creating a "plural legal order" (Aho 2018, p. 156). Indeed, the consequences of making slight improvements to the settler-colonial legal system means that iwi interets in and responsibilities to their rohe are continuing to being undermined, with importance still given to the values and interests of the settler-state and settler society as a whole (Aho 2018; Mikaere 2011). The declaration of the Whanganui River as a legal person that owns itself and no one can assert propriety rights over it is a political compromise between Māori and Pākehā interests (Salmond 2017; Salmond et al. 2019).

Legal personhood effectively neutralised the highly politicised issue of Māori ownership of water, and meant that the river cannot be divided into units to be commodified, traded, and sold (Strang 2014). In 1990, when the Waitangi Tribunal released its inquiry report into the Whanganui River claim, the Tribunal concluded that Whanganui iwi possessed what amounted to proprietorship of the river. Iwi legally asserted their interests in their awa even though, in the words of the Tribunal, "Māori did not think in terms of ownership in the same way as Europeans. What they possessed is equated with ownership for the purposes of English or New Zealand law" (Waitangi Tribunal 1999). Heated public debates followed the release of the report, with Pākehā expressing fear that Māori ownership would restrict their entitlements to water. In response to the Crown issued statements to remind the public (and iwi) that under Aotearoa's common law no one can own water (rivers, lakes, seas) and that the Tribunal is not a court and did not determine issues of law (Aho 2018; Hickford 2018; Te Aho 2019). A similar situation occurred with regard to Te Urewara. The failed attempts of iwi to gain proprietorship preceded the use of legal personality for both Te Urewara and the Whanganui River. The use of legal personhood, legal scholar Mark Hickford argues, is a mechanism "to ameliorate any perceived anxieties as to a non-Crown actor excluding through proprietorship any third parties who might have enjoyed relatively unfettered access" (Hickford 2018, p. 168). Legal personhood is presented as something less discomforting for the dominant social group (Pākehā), which preserves public access, and ensures that the geo-entity cannot be owned by any human being or institution (but more specific by Māori). Indeed, the whole concept of legal personhood is a Western concept, Indigenous scholar Jones observes, which is not the same as Māori ontologies regarding more-than-human beings possessing their own mauri, wairua, and mana (Jones 2016, p. 98). Indeed, scholars caution such attempts to codify Indigenous concepts within Western legal orders due to the possibilities of misrecognition and the associated injustices (Coulthard 2014; Hickford 2018, pp. 168–169). Indeed, in the next chapter we highlight the limits of recognitional-based environmental justice approaches in the context of the co-governance of the Waipā River.

At an international scale, Karen Engle (writing in the context of the United Nations Declaration of the Rights of Indignous Peoples) suggests that Indigenous leaders are compromising too much in strategies that emphasise the cultural and spiritual elements of their claims and downplay claims to stronger forms of self-determination. The impact is to "reify identity and indigneous rights and displace many of the economic and political issues that initally motivated much indigenous advocacy: issues of economic dependency, structural discrimination, and lack of indigenous autonomy" (Engle 2011, p. 145). Morris and Ruru state that "just because Maori have a personified worldview, it is incorrect to assume that they will always favour non-development. Maori do not tend to ascribe to a preservation standpoint, but rather a sustainable one" (Morris and Ruru 2010, p. 49). Similarly, Māori leaders (switching between Te Ao Pākehā notions of ownership and resources and Te Ao Māori concepts of rangatiratanga and kaitiakitanga) to emphasise that their interests in freshwater; their responsibilities as kaitiaki involves a delicate balance between their capacities to maintain and enhance the hauora (health) of their awa, while also seeking economic development opportunities for iwi/hapu/whānau (Bargh 2018; Bargh and Van Wagner 2019; Jones 2016; Muru-Lanning 2012, 2016a, b). Indeed, iwi leaders argue for the Treaty to be honoured and their rangatiratanga respected, which includes their entitlements to access and use their awa for economic purposes (alongside social, cultural, and spiritual; indeed, within Te Ao Māori there is no division between domains as everything is connected, the health and flourishing of the land, water, plants, animals, spirits, and people are always interwoven) (Durie 2006; Johnston 2018; Jones 2016; Walker 1996; Walker and McIntosh 2017). Indeed, the tension remains with Aotearoa, as Māori EJ continues to be constrained by the following stipulations: firstly, Māori knowledge, tikanga, and interests in awa continues to only exist within the prescribed boundaries set by the settler-state; and secondly, in instances where Māori values, laws, and entitlement conflict with those of Te Ao Pākehā, the settler-colonial values take precedent.

In other settler societies, different forms of recognition of Indigenous interests in and rights to water are occurring through colonial legal systems. In the United States, decisions by the Supreme Court of Hawai'i are increasingly recognising Indigenous Hawai'ians (Kanaka *Kānaka 'ōiwi or Kānaka Maoli*) connections to their rivers and streams but in different ways. In the United States, a longstanding legal precedent states that all citizens possess the right to enjoy and take care of things that are common to all (under law of nature) and are recognised as "public trust doctrine" (Blumm 1988; Ede 2002; Salmond 2018). In 2000, in a legal case between Indigenous Hawai'ians and local farmers, who campaigned to restore the water to streams that had been diverted by sugar plantations, the Supreme Court of Hawai'i ruled that public trust doctrine applies to all water resources and argued that this necessitated the need to adequately protect customary Indigenous Hawai'ian rights alongside the preservation of biodiversity, scenic landscapes, and waters for all citizens. A later legal ruling by the court, in 2012, for the Four Great Waters case, expanded the public trust doctrine further, and overturned water permits awarded to two companies on the basis that the permits allowed water extraction that impacted on customary Indigenous Hawai'ian practices and the rights of ordinary citizens "public trust" interests in freshwater use (Ede 2002; Kyle 2013; Papacostas 2014). Public trust doctrines are similarly used in other countries, including India and Ecuador (where nature itself is recognised in the constitution). In India, the Supreme Court determined that public trust doctrine "imposed on us by the natural world must inform all of our social institutions" and Indian society must demonstrate "respect for plants, trees, earth, sky, air and water and every form of life" (O'Donnell 2018; O'Donnell and Talbot-Jones 2018).

In the context of Aotearoa, anthropologist Anne Salmond argues that while legislation such as Te Awa Tupua Act (Whanganui River) goes someway in recognising Māori understandings of kinship centred on whakapapa, it could be taken further still. "In the spirit of bringing "two laws" together", Salmon suggests, an Aotearoa version:

> of a public trust doctrine might recognise **both** the common-law entitle-ment of all citizens to the 'lawful enjoyment' of waterways **and** whakapapa relationships between particular Māori kin networks and ancestral springs and rivers. (Salmond 2018, pp. 189–191)

Conclusion

The emergence of hybrid institutional arrangements and changing juris-prudence, in Aotearoa, demonstrate that there are a range of different avenues being employed by which Māori mana, mātauranga and tikanga can be fostered within the context of freshwater governance and manage-ment. Other examples from around the world also attest to the opportu-nities to address the ongoing ontological dissonance within colonial laws and governance structures, particularly in the context of freshwater gov-ernance and management. Different legal and governance arrangements, from legal personhood, to the rights of Mother Nature, and public doc-trine, highlight the multiple epistemological entry points and avenues that can be taken through which legal pluralism can be enacted as a means to enable Indigenous peoples' to achieve environmental justice (Clark et al. 2018; Curley 2019; Kyle 2013; Morris and Ruru 2010; Papacostas 2014; Wilson 2020; Yates et al. 2017). Yet, while new statues, court judgements, and agreements to co-govern geo-entities (between Indigenous and settler-states) all indicate efforts to disrupt settler-colonial knowledge and political structures (as part of the decolonising process), we also note the complexities and challenges of attempting to accommo-date and reconcile multiple legal systems in the context of ongoing ineq-uitable power relations between Indigenous peoples and settler-nations.

References

Ahmad, N. B. (2019). Mask Off – The Coloniality of Environmental Justice. *Widener Law Review, 25,* 195.

Aho, L. T. (2018). Governance of Water Based on Responsible Use – An Elegant Solution? In B. Martin, L. T. Aho, & M. Humphries-Kil (Eds.), *ResponsAbility: Law and Governance for Living Well with the Earth* (pp. 143–164). London: Routledge.

Anderson, A., Binney, J., & Harris, A. (2015). *Tangata Whenua: A History.* Wellington: Bridget Williams Books.

Bakker, K., Simms, R., Joe, N., & Harris, L. (2018). Indigenous Peoples and Water Governance in Canada: Regulatory Injustice and Prospects for Reform. *Water Justice, 193–209.*

Bambridge, T. (2016). *The Rahui: Legal Pluralism in Polynesian Traditional Management of Resources and Territories.* Canberra: Anu Press.

Bargh, M. (2018). Māori Political and Economic Recognition in a Diverse Economy. In D. Howard-Wagner, M. Bargh, & I. Altamirano-Jimenez (Eds.), *The Neoliberal State, Recognition and Indigenous Rights* (pp. 293–307). Canberra: ANU Press.

Bargh, M. (2020). Challenges on the Path to Treaty-Based Local Government Relationships. *Kōtuitui: New Zealand Journal of Social Sciences Online, 1–16.* https://doi.org/10.1080/1177083X.2020.1754246.

Bargh, M., & Van Wagner, E. (2019). Participation as Exclusion: Māori Engagement with the Crown Minerals Act 1991 Block Offer Process. *Journal of Human Rights and the Environment, 10*(1), 118–139.

Bark, R. H., Barber, M., Jackson, S., Maclean, K., Pollino, C., & Moggridge, B. (2015). Operationalising the Ecosystem Services Approach in Water Planning: A Case Study of Indigenous Cultural Values from the Murray–Darling Basin, Australia. *International Journal of Biodiversity Science, Ecosystem Services & Management, 11*(3), 239–249.

Baum, A. (2019). Mni Wiconi (Water Is Life): Knowledge, Power and Resistance at Standing Rock. *Ideas from IDS: Graduate Papers from 2017/18, 9.*

Bécares, L., Cormack, D., & Harris, R. (2013). Ethnic Density and Area Deprivation: Neighbourhood Effects on Māori Health and Racial Discrimination in Aotearoa/New Zealand. *Social Science & Medicine, 88,* 76–82.

Belich, J. (1996). *Making Peoples: A History of the New Zealanders, from Polynesian Settlement to the End of the Nineteenth Century.* Auckland: Penguin Press.

Belich, J. (2013). *The New Zealand Wars and the Victorian Interpretation of Racial Conflict.* Auckland: Auckland University Press.

Bell, A. (2018). A Flawed Treaty Partner: The New Zealand State, Local Government and the Politics of Recognition. In D. Howard-Wagner, M. Bargh, & I. Altamirano-Jimenez (Eds.), *The Neoliberal State, Recognition and Indigenous Rights: New Paternalism to New Imaginings* (pp. 77–92). Canberra: ANU Press.

Bergman, I. (2006). Indigenous Time, Colonial History: Sami Conceptions of Time and Ancestry and the Role of Relics in Cultural Reproduction. *Norwegian Archaeological Review, 39*(2), 151–161.

Berry, K. A., Jackson, S., Saito, L., & Forline, L. (2018). Reconceptualising Water Quality Governance to Incorporate Knowledge and Values: Case Studies from Australian and Brazilian Indigenous Communities. *Water Alternatives, 11*(1), 40.

Bischoff-Mattson, Z., & Lynch, A. H. (2017). Integrative Governance of Environmental Water in Australia's Murray–Darling Basin: Evolving Challenges and Emerging Pathways. *Environmental Management, 60*(1), 41–56.

Bischoff-Mattson, Z., Lynch, A. H., & Joachim, L. (2018). Justice, Science, or Collaboration: Divergent Perspectives on Indigenous Cultural Water in Australia's Murray–Darling Basin. *Water Policy, 20*(2), 235–251.

Blaser, M. (2014). Ontology and Indigeneity: On the Political Ontology of Heterogeneous Assemblages. *Cultural Geographies, 21*(1), 49–58.

Blaser, M., Briones, C., Burman, A., Escobar, A., Green, L., Holbraad, M., et al. (2013). Ontological Conflicts and the Stories of Peoples in Spite of Europe: Toward a Conversation on Political Ontology. *Current Anthropology, 54*(5), 547–568.

Blumm, M. C. (1988). Public Property and the Democratization of Western Water Law: A Modern View of the Public Trust Doctrine. *Environmental Law, 19*, 573.

Brierley, G., Tadaki, M., Hikuroa, D., Blue, B., Šunde, C., Tunnicliffe, J., & Salmond, A. (2019). A Geomorphic Perspective on the Rights of the River in Aotearoa New Zealand. *River Research and Applications, 35*(10), 1640–1651. https://doi.org/10.1002/rra.3343.

Bryan, M. (2017). Valuing Scared Tribal Waters Within Prior Appropriation. *Natural Resources Journal, 57*(6), 139–181.

Castleden, H., Hart, C., Cunsolo, A., Harper, S., & Martin, D. (2017). Reconciliation and Relationality in Water Research and Management in

Canada: Implementing Indigenous Ontologies, Epistemologies, and Methodologies. In S. Renzetti & D. P. Dupont (Eds.), *Water Policy and Governance in Canada* (pp. 69–95). Cham: Springer International Publishing.

Chandler, D., & Reid, J. (2018). 'Being in Being': Contesting the Ontopolitics of Indigeneity. *The European Legacy, 23*(3), 251–268.

Charpleix, L. (2018). The Whanganui River as Te Awa Tupua: Place-Based Law in a Legally Pluralistic Society. *The Geographical Journal, 184*, 19–30.

Chuji, M., Rengifo, G., & Gudynas, E. (2019). Buenvivir. *Pluriverse–A Post–Development Dictionary, 1*, 11–113.

Clark, C., Emmanouil, N., Page, J., & Pelizzon, A. (2018). Can You Hear the Rivers Sing: Legal Personhood, Ontology, and the Nitty-Gritty of Governance. *Ecology Law Quarterly, 45*, 787.

Cochrane, R. (2014). Climate Change, Buen Vivir, and the Dialectic of Enlightenment: Toward a Feminist Critical Philosophy of Climate Justice. *Hypatia, 29*(3), 576–598.

Collins, T., & Esterling, S. (2019). Fluid Personality: Indigenous Rights and the Te Awa Tupua (Whanganui River Claims Settlement) Act 2017 in Aotearoa New Zealand. *Melbourne Journal of International Law, 20*, 197.

Cosens, B., & Chaffin, B. (2016). Adaptive Governance of Water Resources Shared with Indigenous Peoples: The Role of Law. *Water, 8*(3), 97.

Coulthard, G. S. (2014). *Red Skin, White Masks: Rejecting the Colonial Politics of Recognition*. University of Minnesota Press. Retrieved May 19, 2019, from https://muse.jhu.edu/book/35470.

Court of Appeal. Trans-tasman Resources Limited v Taranaki-Whanganui Conservation Board and Others [2020] NZCA 86, No. 86 (New Zealand Court of Appeal 2020).

Crow, S. K., Tipa, G. T., Booker, D. J., & Nelson, K. D. (2018). Relationships Between Maori Values and Streamflow: Tools for Incorporating Cultural Values into Freshwater Management Decisions. *New Zealand Journal of Marine and Freshwater Research, 52*(4), 626–642.

Curley, A. (2019). "Our Winters' Rights": Challenging Colonial Water Laws. *Global Environmental Politics, 19*(3), 57–76.

Curran, D. (2019). Indigenous Processes of Consent: Repoliticizing Water Governance Through Legal Pluralism. *Water, 11*(3), 571.

Daigle, M. (2016). Awawanenitakik: The Spatial Politics of Recognition and Relational Geographies of Indigenous Self-Determination. *The Canadian Geographer/Le Géographe Canadien, 60*(2), 259–269.

Davies, M. (2010). *Legal Pluralism*. Oxford: Oxford University Press.

Dorsett, S. (2017). *Juridical Encounters: Maori and the Colonial Courts, 1840–1852*. Auckland: Auckland University Press.

Dunford, R. (2020). Converging on Food Sovereignty: Transnational Peasant Activism, Pluriversality and Counter-Hegemony. *Globalizations*, 1–15. https://doi.org/10.1080/14747731.2020.1722494.

Durie, E. T. (1994). Custom Law: Address to the New Zealand Society for Legal and Social Philosophy. *Victoria University of Wellington Law Review, 24*, 325.

Durie, M. (2006). Measuring Māori Wellbeing. *New Zealand Treasury Guest Lecture Series, 1*.

Ede, K. C. (2002). He Kanawai Pono no ka Wai (A Just Law for Water): The Application and Implications of the Public Trust Doctrine In re Water Use Permit Applications. *Ecology Law Quarterly, 29*, 283.

Engle, K. (2011). On Fragile Architecture: The UN Declaration on the Rights of Indigenous Peoples in the Context of Human Rights. *European Journal of International Law, 22*(1), 141–163.

Forster, M. (2016). Indigenous-Environmental-Autonomy-in-Aotearoa-New-Zealand. *AlterNative: An International Journal of Indigenous Peoples, 12*(3), 316–330.

Gilio-Whitaker, D. (2019). *As Long as Grass Grows: The Indigenous Fight for Environmental Justice from Colonization to Standing Rock*. Boston: Beacon Press.

Green, A., & Hendry, J. (2019). *Non-Positivist Legal Pluralism and Crises of Legitimacy in Settler-States*. SSRN Scholarly Paper No. ID 3453251. Rochester, NY: Social Science Research Network. Retrieved June 15, 2020, from https://papers.ssrn.com/abstract=3453251.

Greensill, A. N. (2010). *Inside the Resource Management Act: A Tainui Case Study*. Thesis, The University of Waikato. Retrieved June 15, 2020, from https://researchcommons.waikato.ac.nz/handle/10289/4922.

Grosfoguel, R. (2015). Transmodernity, Border Thinking, and Global Coloniality. *Nous, 13*(9).

Harmsworth, G., Awatere, S., & Robb, M. (2016). Indigenous Māori Values and Perspectives to Inform Freshwater Management in Aotearoa-New Zealand. *Ecology and Society, 21*(4), 9.

Harris, R., Tobias, M., Jeffreys, M., Waldegrave, K., Karlsen, S., & Nazroo, J. (2006). Racism and Health: The Relationship Between Experience of Racial Discrimination and Health in New Zealand. *Social Science & Medicine, 63*(6), 1428–1441.

Healy, S., Huygens, I., & Murphy, T. (2012). *Ngapuhi Speaks: He Whakaputanga and Te Tiriti o Waitangi: Independent Report on Ngapuhi Nui Tonu Claim.* Te Kawariki & Network Waitangi Whangarei.

Hendry, J. (2019). Margaret Davies: Law Unlimited: Materialism, Pluralism, and Legal Theory. *Journal of Law and Society, 46*(1), 169–173.

Hendry, J., & Tatum, M. L. (2018). Justice for Native Nations: Insights from Legal Pluralism. *Arizona Law Review, 60*, 91.

Hickford, M. (2018). Reflecting on Landscapes of Obligation, Their Making and Tacit Constitutionalisation: Freshwater Claims, Proprietorship and "Stewardship". In B. Martin, L. T. Aho, & M. Humphries-Kil (Eds.), *ResponsAbility: Law and Governance for Living Well with the Earth* (pp. 162–182). London: Routledge. https://doi.org/10.4324/9780429467622.

Howitt, R., & Suchet-Pearson, S. (2006). Changing Country, Telling Stories: Research Ethics, Methods and Empowerment in Working with Aboriginal Women. In K. Lahiri-Dutta (Ed.), *Fluid Bonds: Views on Gender and Water* (pp. 48–63). New York: Springer.

Hunt, S. E. (2014, March 3). *Witnessing the Colonialscape: Lighting the Intimate Fires of Indigenous Legal Pluralism.* Thesis, Environment: Department of Geography. Retrieved June 9, 2019, from http://summit.sfu.ca/item/14145%23310.

Iwi Rep 6. (2020, February 14). Interview with Iwi Representative 6.

Jackson, M. (1992). The Treaty and the Word: The Colonization of Māori Philosophy. In G. Oddie & R. W. Perrett (Eds.), *Justice, Ethics, and New Zealand Society* (pp. 1–10). Oxford: Oxford University Press. Retrieved July 14, 2017, from https://philpapers.org/rec/JACTTA-4.

Jackson, M. (1993). Land Loss and the Treaty of Waitangi. *Te Ao Mārama: Regaining Aotearoa. Māori Writers Speak Out, 2.*

Jackson, M. (1995). Justice and Political Power: Reasserting Maori Legal Processes. In *Legal Pluralism and the Colonial Legacy: Indigenous Experiences of Justice in Canada, Australia, and New Zealand* (pp. 243–263). Aldershot: Averbury Ashgate.

Jackson, M. (2007). Globalisation and the Colonising State of Mind. In *Resistance: An Indigenous Response to Neoliberalism* (pp. 167–182). Wellington: Huia.

Jackson, S. (2018). Indigenous Peoples and Water Justice in a Globalizing World. In K. Conca & E. Weinthal (Eds.), *The Oxford Handbook of Water Politics and Policy.* New York: Oxford University Press.

Johnston, A. (2018). Murky Waters: The Recognition of Maori Rights and Interests in Freshwater. *Auckland University Law Review, 24*, 39.

Jones, C. (2016). *New Treaty, New Tradition: Reconciling New Zealand and Maori Law*. Toronto: University of British Columbia. Retrieved June 12, 2019, from https://books.google.co.nz/books?hl=en&lr=&id=DSLCDAAAQBAJ&oi=fnd&pg=PT5&dq=Jones+2016+New+Treaty&ots=09dWY_fMZ0&sig=vEDmsW4b2_KETAJpQfJWLcrDjRg#v=onepage&q=Jones%202016%20New%20Treaty&f=false.

Kyle, M. (2013). The Four Great Waters Case: An Important Expansion of Waiahole Ditch and the Public Trust Doctrine. *University of Denver Water Law Review, 17*, 21.

Leonard, B., Parker, D. P., & Anderson, T. L. (2020). Land Quality, Land Rights, and Indigenous Poverty. *Journal of Development Economics, 143*, 102435.

LeQuesne, T. (2019). Petro-Hegemony and the Matrix of Resistance: What Can Standing Rock's Water Protectors Teach Us About Organizing for Climate Justice in the United States? *Environmental Sociology, 5*(2), 188–206.

Maldonado-Torres, N. (2016). Colonialism, Neocolonial, Internal Colonialism, the Postcolonial, Coloniality, and Decoloniality. In Y. Martínez-San Miguel, S.-J. Ben, & M. Belausteguigoitia (Eds.), *Critical Terms in Caribbean and Latin American Thought: Historical and Institutional Trajectories* (pp. 67–78). New York: Palgrave Macmillan.

Mascarenhas, M. (2007). Where the Waters Divide: First Nations, Tainted Water and Environmental Justice in Canada. *Local Environment, 12*(6), 565–577.

McGregor, D. (2018). Mino-Mnaamodzawin: Achieving Indigenous Environmental Justice in Canada. *Environment and Society, 9*(1), 7–24.

McGregor, D., Whitaker, S., & Sritharan, M. (2020). Indigenous Environmental Justice and Sustainability. *Current Opinion in Environmental Sustainability, 43*, 35–40.

Memon, P. A., & Kirk, N. (2012). Role of Indigenous Māori People in Collaborative Water Governance in Aotearoa/New Zealand. *Journal of Environmental Planning and Management, 55*(7), 941–959.

Metge, J. (1997). Commentary on Judge Durie's Custom Law. *Custom Law Guidelines Project Paper*. Unpublished.

Mikaere, A. (2005). Cultural Invasion Continued: The Ongoing Colonisation of Tikanga Maori. *Yearbook of New Zealand Jurisprudence, 8*(2), 134.

Mikaere, A. (2011). *Colonising Myths-Maori Realities: He Rukuruku Whakaaro*. Wellington: Huia Publishers.

Morris, J. D. K., & Ruru, J. (2010). Giving Voice to Rivers: Legal Personality as Recognising Indigenous Peoples' Relationships to Water? *Australian Indigenous Law Review, 14*(2), 49–62.

Muir, B. R., & Booth, A. L. (2012). An Environmental Justice Analysis of Caribou Recovery Planning, Protection of an Indigenous Culture, and Coal Mining Development in Northeast British Columbia, Canada. *Environment, Development and Sustainability, 14*(4), 455–476.

Mulgan, R. (1997). Commentary on Chief Judge Durie's Custom Law Paper from the Perspective of a Pakeha Political Scientist. *Paper, Law Commission.* Unpublished.

Muller, S., Hemming, S., & Rigney, D. (2019). Indigenous Sovereignties: Relational Ontologies and Environmental Management. *Geographical Research, 57*(4), 399–410.

Muru-Lanning, M. (2012). Māori Research Collaborations, Mātauranga Māori Science and the Appropriation of Water in New Zealand. *Anthropological Forum, 22*(2), 151–164.

Muru-Lanning, M. (2016a). *Tupuna Awa: People and Politics of the Waikato River.* Auckland: Auckland University Press.

Muru-Lanning, M. (2016b). Intergenerational Investments or Selling Ancestors? Māori Perspectives of Privatising New Zealand Electricity-Generating Assets. In P. Adds, B. Bönisch-Brednich, R. S. Hill, & G. Whimp (Eds.), *Reconciliation, Representation and Indigeneity: 'Biculturalism' in Aotearoa New Zealand* (pp. 49–64). Heidelberg: Universiatsverlag Winter Heidelberg.

Mutu, M. (2011). *The State of Maori Rights.* Wellington: Huia Publishers.

Mutu, M. (2018). Behind the Smoke and Mirrors of the Treaty of Waitangi Claims Settlement Process in New Zealand: No Prospect for Justice and Reconciliation for Māori Without Constitutional Transformation. *Journal of Global Ethics, 14*(2), 208–221. https://doi.org/10.108 0/17449626.2018.1507003.

Mutu, M. (2019). The Treaty Claims Settlement Process in New Zealand and Its Impact on Māori. *Land, 8*(10), 152. https://doi.org/10.3390/land8100152.

New Zealand Parliament. Treaty of Waitangi Act 1975. (1975). http://www. legislation.govt.nz/act/public/1975/0114/latest/DLM435515.html? search=qs_act%40bill%40regulation%40deemedreg_Tuhoe+Settlement_ resel_25_h&p=1&sr=1. Accessed 3 July 2019.

New Zealand Parliament. Resource Management Act (1991). http://www.legislation.govt.nz/act/public/1991/0069/223.0/DLM230265.html

New Zealand Parliament. Ngāi Tahu Claims Settlement Act. , Pub. L. No. No 97 (1998). http://www.legislation.govt.nz/act/public/1998/0097/latest/ DLM429090.html. Accessed 5 July 2020.

New Zealand Parliament. Ngā Wai o Maniapoto (Waipā River) Act (2012). Retrieved April 19, 2020, from http://www.legislation.govt.nz/act/public/2012/0029/latest/DLM3335204.html.

New Zealand Parliament. Te Urewera Act 2014 (2014). Retrieved July 3, 2019, from http://www.legislation.govt.nz/act/public/2014/0051/latest/ DLM6183601.html?search=qs_act%40bill%40regulation%40deemed reg_Tuhoe+Settlement_resel_25_h&p=1&sr=1.

New Zealand Parliament. Te Awa Tupua (Whanganui River Claims Settlement) Act (2017). Retrieved April 19, 2020, from http://www.legislation.govt.nz/ act/public/2017/0007/latest/whole.html.

New Zealand Supreme Court. Takamore v Clarke. 733 NZLR (NZSC 2013).

Ngāti Maniapoto, Maniapoto Maori Trust Board, & Sovereign by Right of New Zealand. (2010, September 27). Waiwaia Accord. Unpublished.

Nursey-Bray, M., & Palmer, R. (2018). Country, Climate Change Adaptation and Colonisation: Insights from an Indigenous Adaptation Planning Process, Australia. *Heliyon, 4*(3), e00565.

O'Donnell, E. L. (2018). At the Intersection of the Sacred and the Legal: Rights for Nature in Uttarakhand, India. *Journal of Environmental Law, 30*(1), 135–144.

O'Donnell, E., & Macpherson, E. (2019). Voice, Power and Legitimacy: The Role of the Legal Person in River Management in New Zealand, Chile and Australia. *Australasian Journal of Water Resources, 23*(1), 35–44.

O'Donnell, E., & Talbot-Jones, J. (2018). Creating Legal Rights for Rivers: Lessons from Australia, New Zealand, and India. *Ecology and Society, 23*(1), 7.

Orange, C. (2015). *The Treaty of Waitangi*. Wellington: Bridget Williams Books.

Oslender, U. (2019). Geographies of the Pluriverse: Decolonial Thinking and Ontological Conflict on Colombia's Pacific Coast. *Annals of the American Association of Geographers, 109*(6), 1691–1705.

Palmer, G. (1989). The Treaty of Waitangi – Principles for Crown Action. *Victoria University of Wellington Law Review, 19*, 335.

Papacostas, C. S. (2014). Traditional Water Rights, Ecology and the Public Trust Doctrine in Hawaii. *Water Policy, 16*(1), 184–196.

Parsons, M., Nalau, J., Fisher, K., & Brown, C. (2019). Disrupting Path Dependency: Making Room for Indigenous Knowledge in River Management. *Global Environmental Change, 56*, 95–113.

Premauer, J. M., & Berkes, F. (2015). A Pluralistic Approach to Protected Area Governance: Indigenous Peoples and Makuira National Park, Colombia. *Ethnobiology and Conservation, 4.*

Prendergast, J. Wi Parata v The Bishop of Wellington (1877). 1 NZLRLC 14 (17 October 1877) 3 NZJurRp (NS) 72 (SC) (Supreme Court Wellington 1877). Retrieved June 5, 2019, from http://www.nzlii.org/cgi-bin/sinodisp/nz/cases/NZJurRp/1877/183.html?query=title(Wi%20Parata%20near%20Bishop%20of%20Wellington.

Rangitāiki River Forum. (2015). Te Ara Whanui o Rangitāiki – Pathways of the Rangitaiki. Bay of Plenty Regional Council.

Robinson, D. F., & Graham, N. (2018). Legal Pluralisms, Justice and Spatial Conflicts: New Directions in Legal Geography. *The Geographical Journal, 184*(1), 3–7.

Rose, D. B. (2004). *Reports from a Wild Country: Ethics for Decolonisation.* UNSW Press.

Ruru, J. (2009). The Common Law Doctrine of Native Title Possibilities for Freshwater. In *Indigenous Legal Water Forum, Dunedin* (Vol. 27).

Ruru, J. (2012). The Right to Water as the Right to Identity: Legal Struggles of Indigenous Peoples of Aotearoa New Zealand. In F. Sultana & A. Loftus (Eds.), *The Right to Water: Politics, Governance and Social Struggle* (pp. 110–122). New York: Routledge.

Ruru, J. (2013). The Right to Water as the Right to Identity: Legal Struggles of Indigenous Peoples of Aotearoa New Zealand. In *The Right to Water: Politics, Governance and Social Struggles.* New York: Earthscan.

Ruru, J. (2018a). Listening to Papatūānuku: A Call to Reform Water Law. *Journal of the Royal Society of New Zealand, 48*(2–3), 215–224.

Ruru, J. (2018b). First Laws: Tikanga Maori in/and the Law. *Victoria University of Wellington Law Review, 49*, 211.

Ruru, J., Borrows, J., & Coyle, M. (2017). *A Treaty in Another Context: Creating Reimagined Treaty Relationships in Aotearoa New Zealand.* Toronto: University of Toronto Press.

Salmond, A. (2017). *Tears of Rangi: Experiments Across Worlds.* Auckland: Auckland University Press.

Salmond, D. A. (2018). Rivers as Ancestors and Other Realities: Governance of Waterways in Aotearoa/New Zealand. In *ResponsAbility* (pp. 183–192). Routledge.

Salmond, A., Tadaki, M., & Gregory, T. (2014). Enacting New Freshwater Geographies: Te Awaroa and the Transformative Imagination. *New Zealand Geographical Society, 70*(1), 47–55. https://doi.org/10.1111/nzg.12039.

Salmond, A., Brierley, G., & Hikuroa, D. (2019). Let the Rivers Speak. *Policy Quarterly, 15*(3). https://doi.org/10.26686/pq.v15i3.5687.

Samuel, S. (2019). *Witsaja iki, or the Good Life in Ecuadorian Amazonia: Knowledge co-Production for Climate Resilience.* International Labour Organization. Retrieved June 10, 2020, from https://cgspace.cgiar.org/handle/10568/105603.

Shah, S. H., & Rodina, L. (2018). Water Ethics, Justice, and Equity in Social-Ecological Systems Conservation: Lessons from the Queensland Wild Rivers Act. *Water Policy; Oxford, 20*(5), 933–952.

Sieder, R., & Barrera, A. (2017). Women and Legal Pluralism: Lessons from Indigenous Governance Systems in the Andes. *Journal of Latin American Studies, 49*, 633–658.

Smith, L. T. (2013). *Decolonizing Methodologies: Research and Indigenous Peoples.* New York: Zed Books.

Stephenson, J. (2001). Recognising Rangatiratanga in Resource Management for Maori Land: A Need for a New Set of Arrangements. *New Zealand Journal of Environmental Law, 5*, 159.

Strang, V. (2014). The Taniwha and the Crown: Defending Water Rights in Aotearoa/New Zealand: Defending Water Rights in Aotearoa/New Zealand. *Wiley Interdisciplinary Reviews: Water, 1*(1), 121–131.

Te Aho, L. (2019). Te Mana o te Wai: An Indigenous Perspective on Rivers and River Management. *River Research and Applications, 35*(10), 1615–1621.

Thompson-Fawcett, M., Ruru, J., & Tipa, G. (2017). Indigenous Resource Management Plans: Transporting Non-Indigenous People into the Indigenous World. *Planning Practice & Research, 32*(3), 259–273.

Tobias, J. K., & Richmond, C. A. M. (2014). "That Land Means Everything To Us as Anishinaabe...": Environmental Dispossession and Resilience on the North Shore of Lake Superior. *Health & Place, 29*, 26–33.

Todd, Z. (2014). Fish Pluralities: Human-Animal Relations and Sites of Engagement in Paulatuuq, Arctic Canada. *Études/Inuit/Studies, 38*(1–2), 217–238.

Todd, Z. (2016). An Indigenous Feminist's Take on the Ontological Turn: 'Ontology' Is Just Another Word for Colonialism. *Journal of Historical Sociology, 29*(1), 4–22.

Vickery, J., & Hunter, L. M. (2016). Native Americans: Where in Environmental Justice Research? *Society & Natural Resources, 29*(1), 36–52.

Waikato River Authority. (2016). *Waikato River Authority Annual Report 2016.*

Waitangi Tribunal. (1983). *Report of the Waitangi Tribunal on the Motunui-Waitara Claim (Waitangi Tribunal).* Wellington: Department of Justice.

Waitangi Tribunal. (1999). *The Whanganui River Report (Wai 167)*. Wellington: Legislation Direct.

Waitangi Tribunal. (2009). *Urewera Report–Part I*. Wellington: Legislation Direct.

Waitangi Tribunal. (2018). *Te Mana Whatu Ahuru: Report on Te Rohe Pōtae Claims Pre-Publication Version Parts I and II*. Wellington: Unpublished.

Walker, R. (1996). *Nga pepa a Ranginui (The Walker Papers)*. Auckland: Penguin Group.

Walker, R., & McIntosh, T. (2017). Kāwanatanga, Tino Rangatiratanga and the Constitution. In *New Zealand and the World* (Vol. 1, pp. 201–219). World Scientific. https://doi.org/10.1142/9789813232402_0013.

Watene, K. (2016). Valuing Nature: Māori Philosophy and the Capability Approach. *Oxford Development Studies, 44*(3), 287–296.

Watts, V. (2013). Indigenous Place-Thought and Agency Amongst Humans and Non Humans (First Woman and Sky Woman Go on a European World Tour!). *Decolonization: Indigeneity, Education & Society, 2*(1) Retrieved May 16, 2020, from https://jps.library.utoronto.ca/index.php/des/article/view/19145.

Weir, J. K. (2009). *Murray River Country: An Ecological Dialogue with Traditional Owners*. Aboriginal Studies Press.

Whaanga, H., Wehi, P., Cox, M., Roa, T., & Kusabs, I. (2018). Māori Oral Traditions Record and Convey Indigenous Knowledge of Marine and Freshwater Resources. *New Zealand Journal of Marine and Freshwater Research, 52*(4), 487–496.

Wheen, N. R., & Hayward, J. (2012). *Treaty of Waitangi Settlements*. Wellington: Bridget Williams Books.

Whyte, K. P. (2011). The Recognition Dimensions of Environmental Justice in Indian Country. *Environmental Justice, 4*(4), 199–205.

Whyte, K. P. (2017). The Dakota Access Pipeline, Environmental Injustice, and U.S. Colonialism. *Red Ink: An International Journal of Indigenous Literature, Arts, & Humanities, 19*(1) Retrieved May 29, 2020, from https://ssrn.com/abstract=2925513.

Whyte, K. P. (2018). Settler Colonialism, Ecology, and Environmental Injustice. *Environment and Society, 9*(1), 125–144. https://doi.org/10.3167/ares.2018.090109.

Wildcat, D. R. (2013). Introduction: Climate Change and Indigenous Peoples of the USA. *Climatic Change, 120*(3), 509–515.

Williams, E. K., Watene-Rawiri, E. M., & Tipa, G. T. (2018). Empowering Indigenous Community Engagement and Approaches in Lake Restoration:

An Āotearoa-New Zealand Perspective. In *Lake Restoration Handbook* (pp. 495–531). New York: Springer.

Wilson, N. J. (2019). "Seeing Water Like a State?": Indigenous Water Governance Through Yukon First Nation Self-Government Agreements. *Geoforum, 104*, 101–113.

Wilson, N. J. (2020). Querying Water Co-Governance: Yukon First Nations and Water Governance in the Context of Modern Land Claim Agreements. *Water Alternatives; Montpellier, 13*(1), 93–118.

Wilson, N. J., & Inkster, J. (2018). Respecting Water: Indigenous Water Governance, Ontologies, and the Politics of Kinship on the Ground. *Environment and Planning E: Nature and Space, 1*(4), 516–538.

Winter, C. J. (2018). The Paralysis of Intergenerational Justice: Decolonising Entangled Futures. Retrieved January 11, 2020, from https://ses.library.usyd.edu.au/handle/2123/18009.

Winter, C. J. (2019). Does Time Colonise Intergenerational Environmental Justice Theory? *Environmental Politics*, 1–19. https://doi.org/10.1080/09644016.2019.1569745.

Yates, J. S., Harris, L. M., & Wilson, N. J. (2017). Multiple Ontologies of Water: Politics, Conflict and Implications for Governance. *Environmental Planning D: Society and Space, 35*(5), 797–815.

Youatt, R. (2017). Personhood and the Rights of Nature: The New Subjects of Contemporary Earth Politics. *International Political Sociology, 11*, 39–54.

Zwarteveen, M. Z., & Boelens, R. (2014). Defining, Researching and Struggling for Water Justice: Some Conceptual Building Blocks for Research and Action. *Water International, 39*(2), 143–158.

7

Transforming River Governance: The Co-Governance Arrangements in the Waikato and Waipā Rivers

Since the commencement of formal British colonisation of Aotearoa New Zealand in 1840, the settler-state took deliberate efforts to exclude Māori tribes' (iwi) knowledge, values, and decision-making authority over their ancestral lands and waters. Indeed, settler-colonialism, as a structure and a process, was (and still is) premised on the suppression of other ways of knowing and being and the introduction and promotion of Western knowledge, laws, worldviews, social norms, and modes of life. In the Waipā River, as we demonstrated in earlier chapters in this book, individual settlers and government agencies undertook a wide array of activities that directly aimed to radically remake Māori waterscapes, which included the systematic clearance of vegetation, draining wetlands, lowering river and lake levels, destroying eel weirs (pā tuna), building flood levees, using waterways as waste disposal sites, and introducing new biota to supplant native biota (Park 2002; Parsons and Nalau 2016; Parsons et al. 2017, 2019; Williams et al. 2018). All these actions were premised on the suppression and marginalisation of not only Māori bodies and entire communities, but also Māori knowledge, legal and governance

The original version of this chapter was revised: The incorrect information with reference to The Waikato River Authority text has now been revised and updated. The correction to this chapter is available at https://doi.org/10.1007/978-3-030-61071-5_12

© The Author(s) 2021, corrected publication 2022
M. Parsons et al., *Decolonising Blue Spaces in the Anthropocene*, Palgrave Studies in Natural Resource Management, https://doi.org/10.1007/978-3-030-61071-5_7

systems, cultural practices, and ways of life. The consequences of the processes of dispossession, violence, and marginalisation were shown through in Māori experiencing multiple forms of environmental injustice (inequitable distribution of environmental harms, lack of participatory parity, and failure to recognise Māori identities, knowledge and values).

In this chapter, we review how Ngāti Maniapoto is seeking to address the environmental injustices related to their river (Te Awa o Waipā) through new co-governance mechanisms which reassert Māori authority over and knowledge about wai (water) and awa (rivers). We highlight the different avenues by which Māori groups are transforming approaches to the freshwater governance and management and the implications for addressing freshwater degradation in the Anthropocene. As we documented previously in Chap. 6 the emergence of new legislation and resulting, co-governance arrangements (introduced since the mid-2010s) are providing Māori iwi with greater influence in relation to day-to-day operations as well as planning and policy changes regarding river governance and management (New Zealand Parliament 1991, 2010a, 2017; Ngā Wai o Maniapoto (Waipā River) Act 2012; Rangitāiki River Forum 2015; Waikato River Authority 2011).

In this chapter, we go in-depth to examine the ways (and the extent to which) formal recognition of Indigenous knowledge systems within environmental governance and reconciliation are achieving environmental justice (EJ) with a particular focus on Treaty settlement agreements between the New Zealand Crown (Crown) and Māori iwi groups. We focus, in particular, on the practical realities of implementing and operationalising the co-governance framework established through legislation passed following a Treaty settlement between the Crown and Ngāti Maniapoto (Ngā Wai o Maniapoto (Waipā River) 2012) using EJ and the dimensions outlined in Chap. 2 as an analytical lens (distributive, procedural and recognition). We are interested in determining whether the Indigenous-state co-governance model established for the Waipā River enables Ngāti Maniapoto to exercise their mātauranga (Māori knowledge), tikanga (customary laws), rangatiratanga (chiefly authority and sovereignty), and priorities in formal governance and decision-making in a way that addresses the environmental injustices experienced by Ngāti Maniapoto. Before we proceed to our examples from Aotearoa, it is important to situate our research within the context of broader scholarship on Indigenous freshwater governance, management, and justice.

Water 'Rights' and 'Responsibilities': Water Co-Governance and Justice

Indigenous peoples, for whom freshwater is a matter of the highest importance, are typically excluded from colonial water governance frameworks (Arsenault et al. 2019; Behn and Bakker 2019; Wilson 2020). Indeed, in settler-colonial states (such as Aotearoa NZ, Canada, the United States and Australia), where the Indigenous peoples comprise the minority of the total national population and colonisation is ongoing rather than a historical period, the knowledges, values, and management practices of Indigenous peoples' remain side-lined in favour of Western knowledge and Eurocentric environmental governance and management approaches (Coombes 2006; Pulido 2017; Veracini 2010, 2011). As a consequence of this marginalisation, shared water governance arrangements are being advocated as a way in which Indigenous peoples can be included in decision-making processes regarding waterways, and to address (redress) the historical and contemporary exclusion of Indigenous peoples' knowledges, values, and practices from water governance and management regimes (Parsons et al. 2017; Poelina et al. 2019; von der Porten et al. 2015; Simms et al. 2016).

Indigenous freshwater governance scholarship demonstrates Indigenous peoples' rights to self-determination include decision-making authority based on Indigenous laws, ontologies, and epistemologies to protect freshwater for all types of life (human and more-than-human entities) as well as generations (past, present and future) (Boelens 2014; McGregor 2014; Morgan and Te Aho 2013; Wilson and Inkster 2018). Although the deliberate refusal by settler-states to acknowledge Indigenous water rights and responsibilities are at the heart of many of the environmental injustices experienced by Indigenous peoples, the implications of power disparities that advantage and normalise settler-colonial ways of thinking and acting with regard to water, including governance arrangements, requires further exploration (Simms et al. 2016; Wilson 2020). Many scholars draw attention to the important distinction between settler-colonial legal frameworks underpinned by 'rights' over or to water, and Indigenous legal and governance frameworks centred on responsibilities and duties to water as a living entity (Castleden et al. 2017; Charpleix 2018; Jackson 2018; Poelina et al. 2019). Whereas, legal frameworks based on 'rights' implies an entitlement to own, use and manage water, Indigenous frameworks emphasise duties and responsibilities for and about water, and acknowledge that people are living because of (and their lives, livelihoods, and sense of self are all entangled with) water (McGregor 2015; Robison et al. 2018). Indigenous responsibilities for water encapsulate the maintenance and protection of water, including its quality and quantity, to enable and enhance the health and wellbeing of human and more-than-human beings (including biological and metaphysical entities). These responsibilities are intergenerational (Johnston 2018). Researchers who explore the ontological politics of freshwater governance demonstrate that injustices are linked to the repeated imposition of settler conceptualisations of water as a resource and commodity for people to exploit, own and manage (as we outline in previous chapters) (McGregor 2014; Parsons et al. 2019; Salmond 2017; Wilson and Inkster 2018). Such epistemological and ontological violence marginalised Indigenous authority and knowledge, and its expression, within legal, governance, and management systems, in ways that negatively impacted the health and wellbeing of Indigenous peoples (Barber and Jackson 2015; Berry et al. 2018; McLean 2014; Wilson et al. 2019).

Water governance performs a socio-cultural function that shapes and regulates the management and development of water resources as well as the provision of water services to society to ensure water resources are kept in a "desirable state" (Pahl-wostl 2017; Pahl-Wostl et al. 2011). A water governance system, therefore, is an interconnected assemblage of social, cultural, political and legal components that enacts the role of water governance, and which embraces actors and their institutions. As an interdependent set of institutions (formal laws, professional practices, social values and norms), a water governance regime is the main structural feature of a water governance system. The water governance arrangements that exist within settler-nations are situated on a continuum ranging from Indigenous-led to colonial-led governance systems, with co-governance or shared governance occupying a middle ground (Kotaska 2013). Shared or joint governance arrangements are created to replace the adversarial, exclusionary and top-down modes of water governance and policy-making and are being implemented around the world (not only in the context of Indigenous-state relations) to bring private and public stakeholders together in collaborative decision-making processes (von der Porten et al. 2015; von der Porten and de Loë 2013; Wilson 2020). Co-governance, in the context of settler societies, requires that both parties (Indigenous and the settler-state) share authority on a state-to-state basis and that Indigenous peoples explicitly agree to share authority over their ancestral water bodies with non-Indigenous peoples (Kotaska 2013; Muller et al. 2019; Simms et al. 2016; Wilson 2020). Many scholars forcefully argue that the creation of such collaborative governance and management arrangements are critical to the advancement of Indigenous capacities to manage their water resources effectively, and improve inter-jurisdictional catchment management (Jackson 2018; Memon and Kirk 2012; Parsons et al. 2017; Poelina et al. 2019; Tsatsaros et al. 2018).

While there is a growing body of scholarship outlining the establishment of co-governance arrangements, there has been less attention given to examining the practical day-to-day realities of how Indigenous-state co-governance arrangements operate (Bakker et al. 2018; Muru-Lanning 2016; Simms et al. 2016; Wilson 2020). Although scholars often report the hesitancy of settler-state governments to share decision-making authority with Indigenous peoples as a constraint to co-governance (Bakker et al. 2018; Simms et al. 2016), existing power disparities also marginalise Indigenous legal and governance regimes by prioritising and normalising Western (settler) ontologies and epistemologies (including the types of governance) (Kotaska 2013; Simms et al. 2016; Tipa and Welch 2006).

Connecting scholarship on Indigenous-settler-state freshwater co-governance to EJ is critical to understanding the injustices faced by Indigenous peoples as a consequence of governance and management regimes. EJ scholarship, as we outline in Chap. 2, provides a three-dimensional account that can be employed to understand better the (in)justice implications of emergent Indigenous/settler-state river co-governance regimes. Schlosberg (2004) created a "trivalent concept of justice" that includes three types of justice: distributive (the allocation of environmental risks and benefits or entitlements); procedural (how decisions are made, what procedures are used to make decisions, and who shapes decisions); recognitional (what or who are valued or not) (Schlosberg 2004, p. 521) (see Fig. 7.1). Each of these dimensions overlaps and are bound together through socio-cultural, economic and political processes. However, as Indigenous scholars including Whyte (2016), McGregor (2014), and Winter (2018) caution, environmental justice for Indigenous peoples include particular configurations that necessitate that scholars consider Indigenous environmental justice as consisting of Indigenous sovereignties, knowledges, legal and governance systems. The linkage of co-governance and EJ literatures necessarily involves, as Wilson recently writes, "adapting environmental justice frameworks to acknowledge Indigenous water rights, responsibilities and authorities, as well as recognising the conflicts' sources and understandings of the jurisdiction in Indigenous and colonial legal orders" (Wilson 2020, p. 95). Accordingly, we highlight the interlocking dimensions of (in)justices in the Indigenous

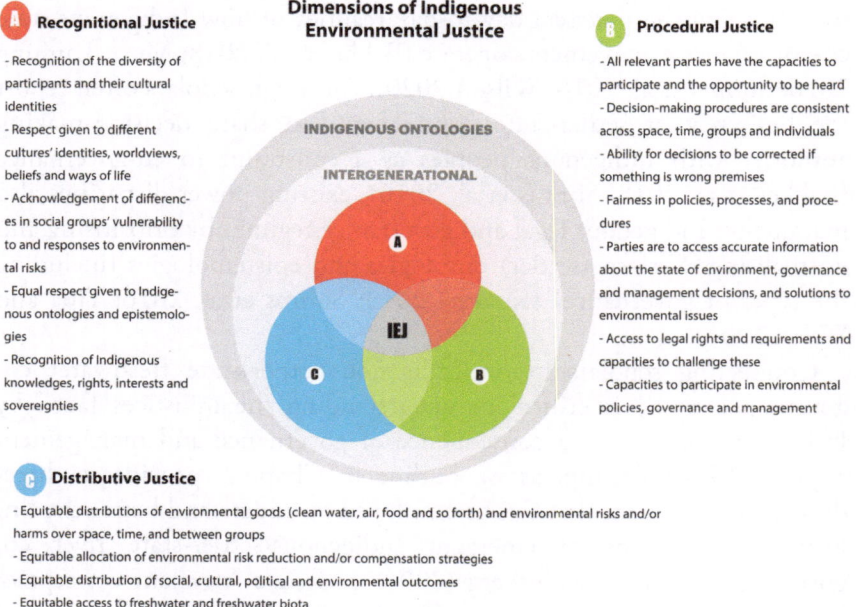

Fig. 7.1 Different dimensions of justice

and settler-state freshwater co-governance regime, and the different sources of authority that flow from different legal orders (Indigenous and settler state) within such a justice framework. In what follows, we apply this framework to examine the implementation of co-governance regimes for the Waikato and Waipā Rivers.

Treaty Settlements and Reconciliation

In 1994, the Office of Treaty Settlements (OTS) was established as a separate agency (located within the Ministry of Justice and entirely distinct from the Waitangi Tribunal) to negotiate with individual iwi (and sometimes larger pan-iwi groupings) about legal-financial reparations packages that acknowledge and sought to address the Crown's failures to honour the Treaty and as a means to reconcile with Māori (discussed further in the works of Treaty scholars including Jones and Hickford (Hickford and Humphries-Kil 2018; Jones 2016, pp. 21–22; Te Aho 2015; Wheen and Hayward 2012). A range of 'Treaty settlements' started to emerge from the mid-1990s (with the Waikato-Tainui Raupatu Settlement) and continuing into the 2020s. Treaty settlements include a formal apology from the Crown for historical and contemporary injustices against a particular iwi, financial reparations to the iwi (monetary payments and return of Crown landholdings), and the introduction of new legislation (Jones 2016; Williams et al. 2018). Ruru argues that settlements (including deeds and subsequent legislation) provide a catalyst for transforming legal education and public understandings of law by recognising Māori rights and interests and for incorporating aspects of tikanga. For instance, the Deed of Settlement that contributed to the Ngāi Tahu Claims Settlement Act (1998) included aspects of tikanga and embedded in the statute are pūrākau (traditions and stories) of the whenua (the land). More recent Treaty settlements, including those with Waikato-Tainui and Ngāti Maniapoto, include specific provisions for Māori iwi to co-govern and co-manage culturally significant sites, including rivers, lakes and national parks. And, most notably, are settlements that recognise the legal personality of Indigenous ancestors (the forest of Te Urewera and the river of Whanganui) (New Zealand Parliament 2014,

2017). We now turn our attention to recent Treaty settlements concerning the Waikato and Waipā Rivers and how these settlements and resulting legislation are exemplars of both legal and ontological pluralism.

Treaty Settlements, Legislation, and Co-Governing and Co-Managing the Waikato and Waipā Rivers

The Waipā River Deed of Settlement, reached by way of direct negotiations between Ngāti Maniapoto and the Crown, was signed on 27 September 2010. Ngā Wai o Maniapoto (Waipā River) Act 2012 (hereafter the Waipā River Act) gives effect to the Deed. The Ngāti Maniapoto Deed and legislation built on earlier Treaty settlements with another iwi (Waikato-Tainui, Raukawa, Te Arawa, and Ngati Tūwharetoa) and the Crown and resulting legislation regarding the Waikato River, which acknowledges Māori interests in and authority over waterbodies. The three acts (known as the River Acts) acknowledged the importance of the Waikato and Waipā Rivers and catchment to the five River iwi groups. The acts emphasise the need to protect and restore the river and its tributaries, with the Waipā River Act also emphasising the care and protection of the mana tuku iho o Waiwaia (in contrast to Waikato-Tainui, who identify the Waikato River as a tupuna (ancestor). Waiwaia is a taniwha who acts as a kaitiaki of Ngāti Maniapoto and is the essence and wellbeing of the Waipā River. The mana tuku iho means the ancestral authority handed down from generation to generation in respect of Waiwaia (New Zealand Parliament 2010b, 2012; Waikato-Tainui Raupatu Claims (Waikato River) Settlement Act 2010).

The River Acts formally acknowledged the historic and ongoing relationships iwi possess with the Waikato and Waipā Rivers and their tributaries, and collectively determine the architecture and mechanisms to enable co-management and co-governance across the extent of these two catchments (see Fig. 7.2). Though similar, there are differences between the two Waikato River Acts and the Waipā River Act; in particular, the Waikato River Acts refer only to "co-management" whereas the Waipā

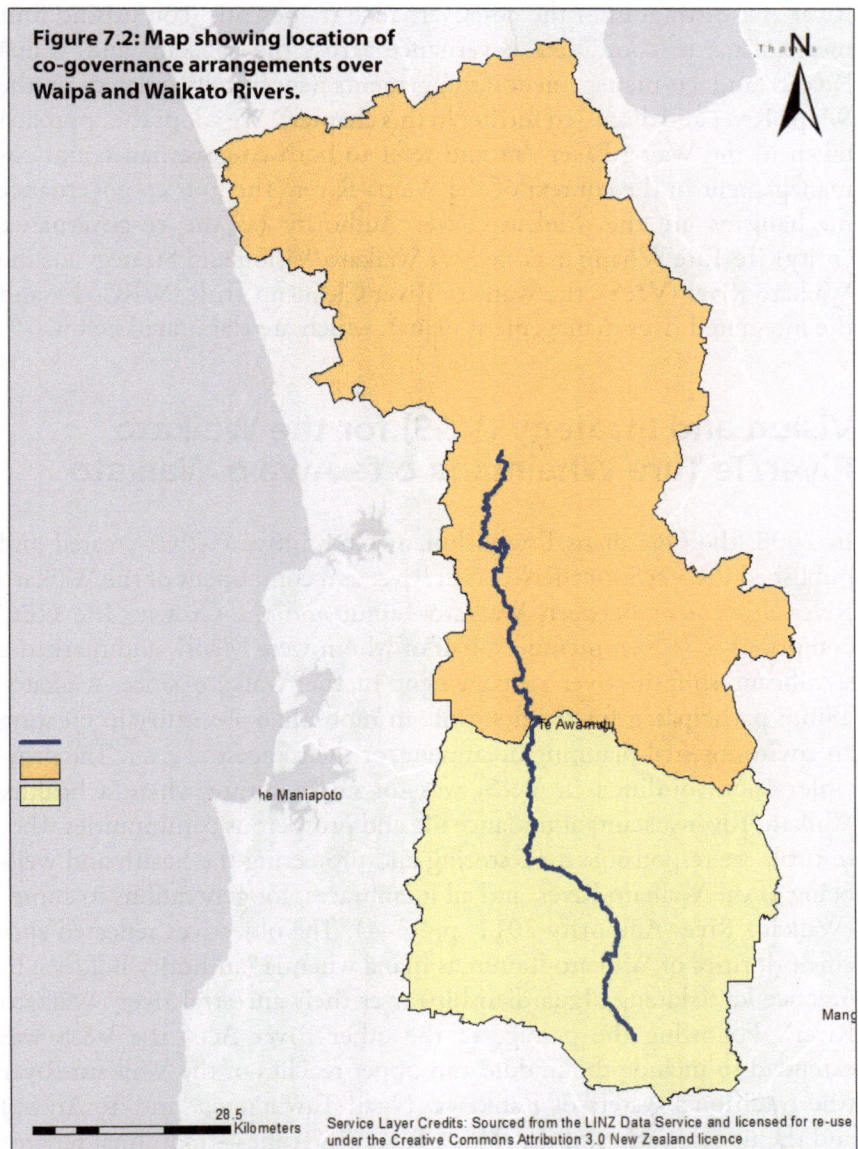

Figure 7.2: Map showing location of co-governance arrangements over Waipā and Waikato Rivers.

Te Awamutu

he Maniapoto

Mang

28.5 Kilometers

Service Layer Credits: Sourced from the LINZ Data Service and licensed for re-use under the Creative Commons Attribution 3.0 New Zealand licence

Fig. 7.2 Map showing the location of co-governance and co-management arrangements over Waipā and Waikato Rivers. (Source: Authors' own)

River Act distinguishes the co-governance framework (comprising four mechanisms to coordinate governance across the Waikato and Waipā Rivers) and co-management arrangements (specifically related to the Waipā River and discussed further in this chapter). We adopt the approach taken in the Waipā River Act and refer to both co-governance and co-management in the context of the Waipā River. The four co-governance mechanisms are the Waikato River Authority (as the co-governance entity), Te Ture Whaimana o te Awa Waikato/Vision and Strategy for the Waikato River (V&S), the Waikato River Clean-up Trust (WRCuT), and the integrated river management plans), which are elaborated below.

Vision and Strategy (V&S) for the Waikato River/Te Ture Whaimana o te Awa o Waikato

In 2008, the Guardians Establishment Committee (GEC) created and published the V&S for the Waikato River as a component of the Waikato River Settlement between Waikato-Tainui and the Crown. The GEC comprised of sixteen members, half of whom were Māori, and marked a significant shift in river management in the Waikato since Waikato-Tainui participated for the first time in more than a century in creating an environmental planning document for their ancestral river. The principle vision, outlined in V&S, was for an "a future where a healthy Waikato River sustains abundance life and prosperous communities who, in turn, are responsible for restoring and protecting the health and well-being of the Waikato River, and all it embraces, for generations to come" (Waikato River Authority 2011, pp. 3–4). The objectives reflected specific priorities of Waikato-Tainui as mana whenua (authority holders) to practice kaitiakitanga (guardianship) over their ancestral river (Waikato River). Following the passage of the other River Acts, the V&S was extended to include the middle and upper reaches of the Waikato River (the traditional waters of Raukawa, Ngati Tūwharetoa, and Te Arawa) and the upper reaches of the Waipā River (the rohe—traditional territories of Ngāti Maniapoto).

The V&S is included as a schedule in each of the River Acts and is identified by parliament as the direction setting document for the Waikato and Waipā Rivers (and its catchments) (Waikato River Authority 2011). The V&S applies to the entirety of the Waikato River's 11,000 km² catchment (from Huka Fulls to Te Puuaha o Waikato) as well as the Waipā River catchment. With the passing of the River Acts, the V&S was deemed to be part of the Waikato Regional Policy Statement (a mandatory planning document prepared under the Resource Management Act—RMA) and to prevail over all policy or planning documents that are inconsistent with it; this includes the national policy statements that are produced under it, such as the National Policy Statement on Freshwater Management (NPSFM) (Ministry for the Environment 2017). The Waikato Regional Council (WRC) (the body responsible for freshwater management for the Waikato Region under the RMA), underwent a plan change to The Waikato Regional Plan (a plan prepared to implement the Regional Policy statement) to further ensure it gives effect to the V&S (Fig. 7.3) (Waikato Regional Council 2020; Waikato River Authority 2011)

Waikato River Authority (WRA)

The Waikato River Authority (WRA) is the co-governance entity established for the Waikato River following the passing of the Waikato-Tainui Raupatu Claims (Waikato River) Settlement Act 2010, with iwi membership expanded with the passing of subsequent legislation (New Zealand Parliament 2010b, 2012). The WRA is also the sole trustee for WRCuT, which is a trust for charitable purposes that provides funding for restoration within the Waikato and Waipā river catchments to achieve the V&S. The WRA oversees the contestable funding rounds and makes decisions regarding allocation. At the time it was created, the WRA was hailed as a new era of co-governance and co-management in Aotearoa (discussed in Chap. 8) (Te Aho 2010, 2015). In line with the bi-cultural partnership that underpinned the Treaty as well as the Treaty settlements, the composition of the WRA comprises of an equal number of government (five) and River iwi representatives (one from each of the five River iwi) (Forster 2016). Representatives of local government included the

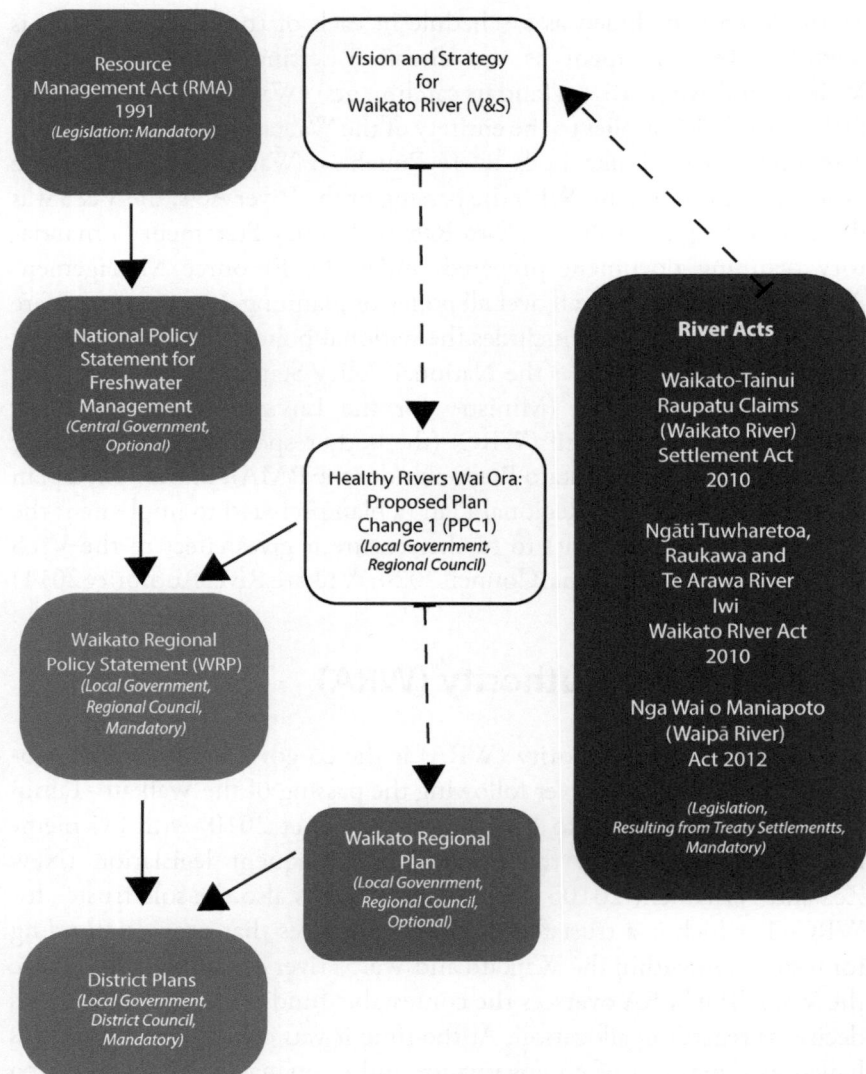

Fig. 7.3 The relationships between Aotearoa's national legislation and Waikato's local legislation framework. (Source: Authors' own)

Waikato Regional Council as well as territorial authorities (Waikato District Council, Otorohanga District Council, Waipā District Council, Hamilton City Council), and central government (Minister for the Environment appointed co-chairperson); each of the five River iwi (Waikato-Tainui, Ngāti Tuwharetoa, Raukawa, Te Arawa, Ngāti Maniapoto) appointed their representatives and decided on a shared iwi co-chairperson.

At the time of its creation, the WRA embodied co-governance far more than other government-led consultative models of governance already operating in Aotearoa such as the Canterbury Water Management Strategy (Memon and Skelton 2007; Nissen 2014; Pirsoul and Armoudian 2019). The WRA formally promotes a kaitiakitanga-based approach to river management and is focused on restoring and enhancing the mauri (life force), mana (power, authority and prestige), and health of the Waikato River and its tributaries (Waikato River Authority 2011). The WRA is intended to ensure greater Māori participation and decision-making authority within freshwater management processes (including water extraction and discharges into the waterways).

Assessing the Implementation of Co-Governance Arrangements

The creation of co-governance arrangements between the Crown and iwi, in regard to the Waikato and Waipā Rivers, represents an important shift in freshwater governance; however, the implementation of these arrangements are criticised by representatives of Ngāti Maniapoto (see Fig. 7.4: a map showing jurisdiction boundaries for co-governance of Waipā River). However, they acknowledged that beneficial changes had occurred as a consequence of the Waipā River Act and the formal recognition of Ngāti Maniapoto connections with and authority over their ancestral river, the co-governance arrangements are not living up to their expectations. Iwi acknowledges that they have experienced gains as a result of the co-governance and co-management arrangements; however, the majority of gains occurred in the context of co-management rather than co-governance arrangements (Muru-Lanning 2016; Stevens 2013). The

Iwi Waikato and Waipa River Co-Management Areas

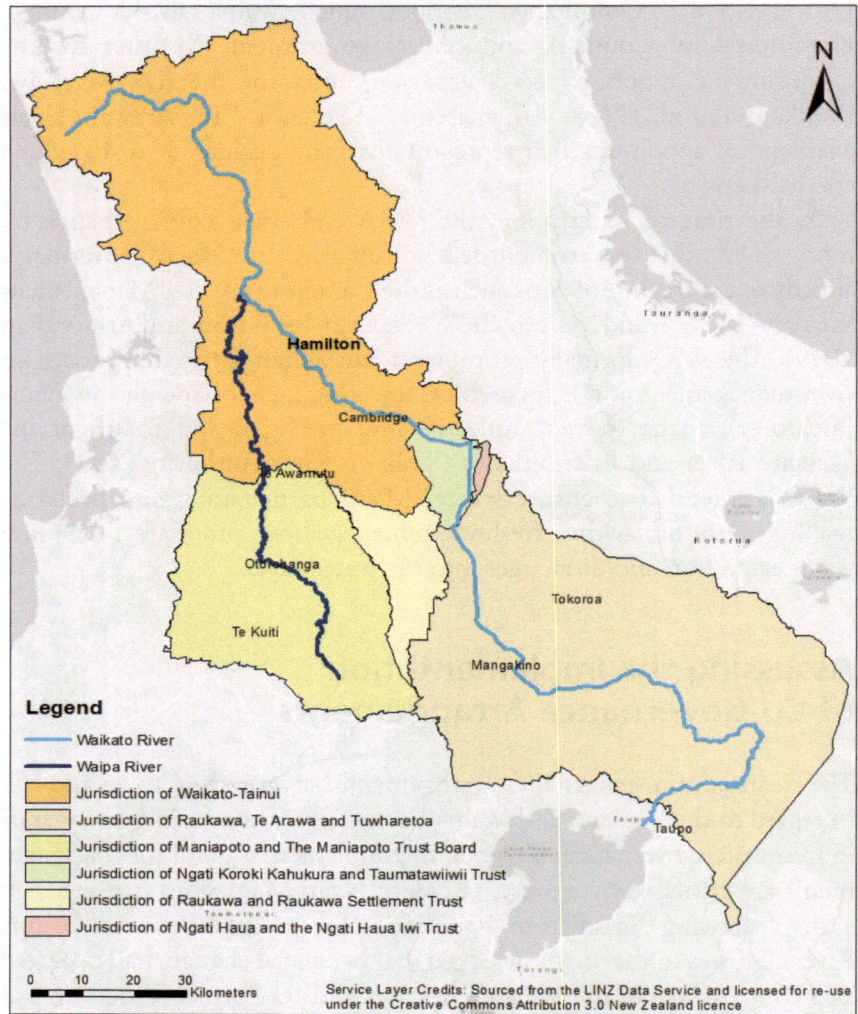

Fig. 7.4 Map showing jurisdictional boundaries of co-governance over the Waipā River. (Source: Authors' own)

critiques are organised into three broad categories: distributional; recognitional; and procedural.

Distributional (In)Justices: Lack of Resources and Capacities

Ngāti Maniapoto iwi members highlight the distributive (in)justices of freshwater governance with regard to the Waipā River (and more broadly those of other iwi within the Waikato River catchment) is a consequence of the settler state's limited acknowledgement of iwi authority (rangatiratanga), tikanga (laws), and responsibilities (as mana whenua—spiritual authority holders and kaitiaki—environmental guardians). A key query that iwi members raised within their discussions is who should possess jurisdiction and authority to make decisions about the water and land within the Waipā River catchment. From their perspective, their whakapapa (genealogy), tikanga, and centuries of governance and management of their rohe, as well as their more recent agreements with the Crown (the Deed of Settlement, Waipā River Act, and co-governance and co-management agreements), give them decision-making authority (which they now agree to share with another iwi as well as the settler-state). They possess this authority by way of the responsibilities and rights that flow on from their status as mana whenua under Māori tikanga as well as their position as Treaty and co-governance partners conferred under the settler-state laws. Moreover, various legislation (RMA and Waipā River Act), environmental plans and strategies (V&S and Waikato Regional Plan), and co-governance arrangements (WRA and WRCuT) are meant to give iwi the power to shape how their awa is governed, planned for, and managed, yet there are barriers to their capacities to influence decision-making processes.

Ngāti Maniapoto iwi members report on the numerous challenges they face in their capacities to translate their aspirations to restore and protect their awa as a living entity (as well as those human and more-than-human beings that they are kin to). They spoke of the lack of adequate money and time to allow them to participate in new co-governance and co-management processes. One member of iwi explains how the "WRC is forever asking" him and other members of Ngāti Maniapoto to participate in meetings, give expert advice to the council, and undertake other management and restoration tasks. However, the WRC regularly inform him and other members of his whānau (extended family), hapū (sub-tribes), and iwi that they (the council) cannot afford to pay them for any of their work. They simply say "good luck … Oh, we [the WRC] can't resource that" they expect "us [to work] for free and stuff". (Māori Business Owner 1 2019). Money, an iwi restoration manager declared, "that's a big barrier, is actually being able to get funding" (Māori Business Owner 1 2019).

Many iwi representatives spoke about the multiple burdens, they and others faced in their day-to-day life and how this impeded their capacities to meaningfully participate in co-management and co-governance processes as well as enact on-the-ground works to restore its mauri and that of its human and more-than-human communities. They spoke about how they and other members of their iwi held down paid work, as well as multiple other (unpaid) jobs including caring for their whānau, hapū and iwi (be it within their homes, on their marae—meeting complex for each hapū, urupā—cemetery, and wāhi tapu—sacred sites) as well as within official governance and management forums. As one research participant, who works as a scientist alongside her iwi in their restoration efforts, reports that the "hardest thing or the biggest barrier is that our kaitiaki [guardians] are time poor because they're across so many different kaupapa [initiatives]". It is not that kaitiaki "lack motivation" but that "they have so much other stuff to do … look after their whānau and keep a job so that they can support themselves" (Scientist 2 2019). Another interviewee describes how government efforts to support iwi and engage with Māori "sort of stops … at the regional level" and resourcing (be it from WRC, WRA, or district councils) are not filtering down to local level (especially marae, hapū, and whānau) (Māori Business Owner 1 2019).

> So yeah, there's definitely more work needed in that area of actually recognising what co-management means—and getting some money and resourcing done so that people are actually doing the work there. (Māori Business Owner 1 2019)

The council officials, iwi representative, complain, simply assume that all five iwi (named in the River Acts) that are part of the current freshwater co-governance and co-management arrangements can fund themselves. However, "we [Ngāti Maniapoto] don't even have that much [money] at Maniapoto because we're still [have yet to reach full Treaty] settlement" with the Crown (unlike the other River iwi). "There's definitely an inequity in there" between Ngāti Maniapoto, other iwi, and councils (Māori Business Owner 1 2019).

Without adequate financial resources, the process of decolonising freshwater co-governance arrangements and challenging the hegemony of settler-colonial rule is not possible. They report how they face substantive barriers in accessing the grants for river restoration (administered by the WRCuT) because they lack sufficient funds and trained staff members to apply for and win the contestable funding; competing against more well-resourced and larger organisations with more employees (such as Fonterra and local government bodies). Iwi also notes they do not possess sufficient financial resources to pay lawyers to hold the Crown (and its agencies) to account when it fails to uphold its end of the deal (such as launching legal action against the government or other organisations who breach environmental laws and regulations) (Iwi Rep 2 2020; Iwi Rep 4 2020; Iwi Rep 5 2019; Iwi Rep 8 2019).

Efforts to address environmental justice require attention to how resources are distributed amongst groups, not just in how environmental risks and goods are distributed across geographical and temporal domains. Financial and other capacity constraints undermine Ngāti Maniapoto (and other iwi groups') abilities to participate fully in and influence decision-making processes regarding their awa. Often, the activities surrounding co-governance and co-management of the Waipā River (including the Waikato Regional Plan change as well as resource consent applications) require that iwi and hapū organisations invest substantial amounts of their time (including staff and volunteer hours) and money (to pay staff); this includes Maniapoto Māori Trust Board (MMTB) staff as well as outside experts (including scientists, and lawyers) who are contracted to ensure that they fulfil their responsibilities as kaitiaki, and try to maintain (or improve) the mauri of the awa and the communities (human and more-than-humans) whose lives (their mauri and wairua—spiritual integrity) are dependent on it. The lack of financial parity between iwi and settler-state (not to mention private companies and other stakeholders) forces Ngāti Maniapoto to make decisions about where it is best to invest their constrained resources (time, personal, and money) and to gauge if it is worthwhile to be involved in all decisions made about their taiao (environment). While Ngāti Maniapoto is still at an early stage of the implementation of its co-governance and co-management agreements (as the Waipā River Act is less than eight years old), and there is a lot more possible.

Furthermore, once Ngāti Maniapoto and the Crown reach a final Treaty settlement, iwi hope that they will achieve further. However, there is no question that constrained resources are slowing down and impeding Ngāti Maniapoto iwi's ability to implement what they deem to be equitable, effective, and just co-management arrangements. As numerous scholars note, marginalised social groups (including Indigenous communities) require processes that give them some "locus of control over their destinies as part of a recognition of identity and place" that extends beyond words on paper (be it the Waipā River Act, Deed of Settlement and V&S) (Adger et al. 2011, p. 21). As philosopher Nancy Fraser argued, marginalised social groups need to gain participatory parity, with participation directly tied to recognition, and both linked to the distributional equity (Fraser and Honneth 2003).

Scholarship from Aotearoa and around the globe highlights that the equitable resourcing of Indigenous peoples can support efforts to build trust, enhance legitimacy and increase involvement in collaborative processes, making planning processes more procedurally inclusive, and in doing so address environmental injustices (Brink and Wamsler 2018; Denny and Fanning 2016; Harmsworth et al. 2016). Insufficient resourcing poses serious ramifications for the outcomes of collaborative freshwater management planning processes more generally, but especially in the context of Indigenous-state shared co-governance and co-management agreements (Cradock-Henry et al. 2017; Memon and Kirk 2012; Woldesenbet 2018). In the Canadian context, scholars including Roburn, Tr'ondëk Hwëch'in and Nadasdy critique co-governance agreements that expand First Nation powers to include self-government but do not provide the First Nation government with an increase in funding to build capacity and ensure the Indigenous state possesses sufficient financial, technical, and human resources necessary to successfully implement their expanded powers (Nadasdy 2017; Roburn and Hwëch'in 2012). First Nations rely on funding provided by federal and provincial governments through a variety of different projects and programmes, which frequently are only short-term funds and prone to being cut or reallocated whenever there is a change in government priorities. Accordingly, those First Nation's funds are highly unstable (compared to settler-colonial governments) (Wilson 2020). First Nations with self-governing agreements face a situation of expanding powers but even more financial insecurity (Nadasdy 2017, pp. 31–37). While the situation in Aotearoa is different from that of Canada, especially in the fact that Ngāti Maniapoto does not hold self-governing powers, financial insecurity and the precariousness of being reliant on a government project and/or grant funding are similarly experienced by Māori institutions like First Nations in Canada. The lack of money and financial stability presents implications for Māori formal institutions such as mandated tribal authorities (such as the MMTB) as well as informal institutions (hapū and marae, and whānau) capacities to engage in environmental planning and governance processes.

Our research indicates that inequalities in resourcing are contributing to inequities in iwi capacities to participate in the freshwater co-governance and co-management arrangements. In particular, representatives from iwi and hapū struggled with financial and time constraints created by the council designed planning processes (involving multiple working groups and hearing processes), as well as the Eurocentric framing of discussions, knowledges (centred on scientific and technical expertise), and governance and management approaches. Māori groups often, Harmsworth et al. (2016) previously identified, encounter such barriers (lack of resources and capacities) which limit their abilities to engage and influence freshwater management decision-making (Harmsworth et al. 2014, 2016). Yet, Ngāti Maniapoto iwi representatives maintain that lack of adequate resourcing is not the most critical factor that currently impedes their capacities to engage in and influence the new co-governance regime (Māori Business Owner 1 2019).

The Waipā River Act (and the other two River Acts), WRA, and V&S ostensibly hold the potential to disrupt entrenched settler-colonial water governance systems and management practices, particularly given the extent to which statutes and plans now afford legal recognition to tikanga, iwi values, and aspirations. However, the persistence of settler-colonial structures, including the legal order and institutional arrangements, undermines the possibilities for enacting transformative changes to the settler-colonial status quo. As one NGO worker reflects:

> I think councils [need to be] brave enough to give things a go and to relinquish some of the power that they feel like they hold. [But there is] seemingly a long way [to go] … under [the current] co-management/co-governance situation … council staff get paid to do [water] monitoring all the time [but] why aren't our kaitiaki [iwi members working to undertake river restoration] being paid on the other side of that equation? They're expected to do it voluntarily.

Once again questions about the lack of access to resources (or the continuation of distributive injustice) feature heavily in iwi accounts.

Procedural and Recognitional (In)Justices: Iwi Involvement in Planning Processes

The explicit recognition of Māori rights and interests in water, within the new legislation and co-governance regime, means little if the decision-making procedures in place do not allow for equitable and inclusive processes that allow for Māori interests to not only be recognised but also incorporated into policies, plans, and actions (Māori Business Owner 1 2019). At present, iwi members report that many central government agencies and local government authorities still treat iwi like they are just another stakeholder group, rather than Treaty and co-governance partners. Rather than decision-making authority being shared between state-to-state/iwi-government, iwi members report how the WRC continues to run its environmental planning processes (consisting of workshops, hearings, hui—meeting, planning meetings) as exercises in extended public consultation (so-called collaborative planning processes) with different stakeholder groups. Iwi are treated as little more than stakeholders by the regional council ("the same as Beef + Lamb, the same as Fonterra, the same as anyone else") even though they are both mana whenua and "iwi partners" (Māori Business Owner 1 2019).

One Ngāti Maniapoto participant, who was involved in the process of creating the new Waikato Regional Plan Change, reflected on his feelings about the (supposedly) inclusive and collaborative planning process:

> I mean yes, it's nice that we're starting to see working groups being formed to consult on [Regional Council's Regional] Plan changes, but it still seems like … a real tick in the box [bureaucratic] kind of thing for me; but it's progress, it's something, but …
>
> I found out [the community working group] not just [an iwi] working group, there [are] farmers and all of these other people as well sort of thing, so it's not giving [the River iwi] the mana to [they] deserv[e.] We [Ngāti Maniapoto and other River iwi] are co-partners, we are co-managers, we are more than the farmer down the road and the fruit grower and everything else. We are equal to the council and the government if we are [to] true governance partners and true co-management partners then we should have more than just being treated like everyone else … [The working group] was meant to be our opportunity [as mana whenua] to have our say and really it was just like [something that looked] nice [on paper,] but this is still going [to be] council [telling us their decision] because [they] need to meet the[ir] deadlines and we've [the River iwi] got to do [support] this and [the council already] decided this and [they had] already told the public. (Māori Business Owner 1 2019)

The above quote highlights the problems associated with recognition-based justice, as we previously discussed in Chap. 2, as despite the Deed of Settlement, V&S, and River Acts iwi perspectives remain marginalised within local government planning processes. In Aotearoa, like other settler-nations, it is the settler-state and its institutions who decide what Māori groups are recognised and how they are recognised. Such state-based definitions and solutions allow for existing unequal power relations between Indigenous and non-Indigenous peoples to effectively remain unchallenged. Accordingly, as scholars Coulthard and Schlosberg argue, recognition alone is not sufficient to guarantee just outcomes for Indigenous peoples (Coulthard 2014; Schlosberg 2013; Schlosberg and

Carruthers 2010). Environmental justice requires a clear focus on translating recognition (of Māori as mana whenua as well as Treaty and co-governance partners) into equitable political participation (procedures) and the achievement of distributional equity.

Such political participation requires that procedures are not just one that replicates those of the settler-state, but instead include Māori preferences about and ways of organising participatory processes (Māori Business Owner 1 2019). Iwi members express the importance of more in-depth planning procedures that are more in line with the decision-making processes run within Māori governance systems that centre on korero (talking) over days, weeks, and months (and sometimes years) about an issue, which does not just involve leaders and/or experts, but include all iwi members (who want to participate) (Mahuika 2010; Webster and Cheyne 2017). These typically involve local-level (flax-roots) discussions held on marae whereby iwi, hapū, and whānau members can discuss and debate the issue before reaching a consensus-based decision (Iwi Rep 7 2019). The new co-governance institutions and arrangements should allow iwi to be able to choose what modes of governance suit them best (to choose between Te Ao Māori or Te Ao Pākehā or occupy the "middle ground" between cultures) (O'Malley 2013). However, at present many iwi representatives felt that the co-governance arrangements are solely designed to accommodate Pākehā ways of knowing and doing rather than those of Māori.

Wider scholarship highlights that such shortcomings in collaborative or joint arrangements are commonplace (Denny and Fanning 2016; Pirsoul 2019a, b; von der Porten et al. 2015; von der Porten and de Loë 2013). Bureaucratism and box-ticking, Pirsoul and Armoudian (2019) observe, during consultative or collaborative governance and planning mechanisms mean that Indigenous peoples experience procedural inclusion in formal processes "without any substance of authentic power" (Pirsoul and Armoudian 2019, p. 4827). The concept of "wallpaper democracy"—whereby people focus on talking about the style and colour of the wallpaper without discussing the substance, arrangements or structures more generally—could be evidence in some of the procedures that are occurring within the Waikato context (Bächtiger and Parkinson 2019; Pirsoul and Armoudian 2019). Such occurrences highlight how power imbalances between partners can emerge from co-governance arrangements, which is also noticeable elsewhere in Aotearoa, where the Crown shows a tendency to treat iwi as junior partners (Morgan and Te Aho 2013; O'Sullivan 2007; Pirsoul and Armoudian 2019; Stevens 2013). An obvious example was the 2012 legal challenges made by Māori iwi against the Crown's partial sale of its government-owned hydroelectric and geothermal energy companies (including the Mighty River Power Company that operated on the Waikato River). The Treaty settlements for Waikato-Tainui, Raukawa, Ngati Tuwharetoa, and Te Arawa, from the perspective of iwi, included them being given the first right of refusal to purchase any Crown-owned assets (including energy companies) before the Crown offered them for sale on the open market. However, this did not occur, and the Crown instead sold it to private buyers (non-Māori) (Jones 2012; Muru-Lanning 2016; Strang 2014).

Procedural and Recognitional (In)Justices: Critiques of the WRA

The lack of transparency in the WRA decision-making processes is of particular concern to iwi members. One scientist acknowledges that while the co-governance and co-management arrangements are important for Ngāti Maniapoto, she holds "mixed feelings about the WRA" as an institution (Scientist 2 2019). The leadership of the WRA, in her view, is "quite underhanded and ... personality-driven, not kaupapa driven". The kaupapa (principle) for the WRA is "very clearly laid out in the vision and strategy of the Waikato and Waipā rivers"; however, it is not being translated into actions (Scientist 2 2019). There are some "big influential personalities that are leading that organisation", which means they are "ticking the boxes but not necessarily genuinely trying to give effect to the vision and strategy at all times" (Scientist 2 2019). Furthermore, since the WRA is not subject to the legislation governing freedom of information (such as the Official Information Act 1982 and the Local Government Official Information and Meetings Act 1987) its "trustees do not have to make minutes of their meetings public or publicise where meetings are held and can hold meetings behind closed doors" (Pirsoul and Armoudian 2019, p. 4827).

The closed nature of the WRA and its failure to share the information with iwi members raises concerns about what influence the settler-state, powerful interest groups (such as political parties, lobby groups and businesses), and individuals have on how the institution operates. As one scientist involved in river restoration states: "I don't know who and why this [the allocation of funding is occurring] but I still feel like it's being influenced more heavily from [other groups and] … there are complicated organisational and political relationships" (Scientist 2 2019). Indeed, while she acknowledges "the mahi [work] they [the WRA] are funding" for whānau to do is "amazing", she argues that "there's just a huge imbalance of how [the WRA] are investing the money which is intended for iwi". The overall leadership of the WRA is "kind of dictating the tone of how that funding is spent", irrespective of its policies and regulations. Interviewees expressed ambivalence about the proportion of funding awarded to non-iwi-led projects, which reflected a more general desire for iwi and hapū to exercise kaitiakitanga (environmental guardianship) over their awa as a way of redressing historical and contemporary inequities (Iwi Rep 5 2019; Iwi Rep 6, 2019; Māori Business Owner, 2019). Many of our interviewees, both Māori and non-Māori, query how the WRA makes its decisions, noting that the co-governance entity is not truly representing the interests of iwi members; instead, it may be guilty of perpetuating the distributional and procedural injustices that it is designed to redress (NGO Rep 1 2017).

The work of Māori scholar Muru-Lanning (who iwi is Waikato-Tainui) rearticulates the criticisms made by our research participants about the co-governance structure of the WRA being largely a Western model of governance rather than a Māori or hybrid governance structure (Muru-Lanning 2012a, b, 2016; Pirsoul and Armoudian 2019). Muru-Lanning (2012a) describes it as an inherently western institutional arrangement wherein "appointed representatives mak[e] formal statutory decisions on behalf of the various groups" and, thus, "a model or way of viewing the river which is foreign to most Māori and one in which they cannot easily participate" (Muru-Lanning 2012a). Indeed, the operations of the co-governance entity currently left much to be desired for iwi members we interviewed (Iwi Rep 5 2019; Māori Business Owner 1 2019). While interview participants agreed that the kaupapa (purpose and objectives) of the WRA remains correct, they also had concerns that the "the practical application of the [Waikato] River Authority is another matter" (Iwi Rep 5 2019) particularly in light of challenges arising from increasing pressure on the Waikato and Waipā Rivers (including lag effects associated with landuse activities). In marked contrast, local government officials we interviewed consider that the new co-governance arrangements are working exceeding well and see no reason for any changes to existing processes or institutional arrangements (Local Government Rep 1 2018; Local Government Rep 2 2019). These differences in perspectives highlight the extent to which settler-colonial knowledge, values, and governance systems continue to be the taken for granted norm and that those in positions of power (and settler privilege) are not able to see their ontological and epistemological blind spots, and in doing so fail to appreciate other ways of knowing and engaging with water

Recognitional (In)Justice: Ngāti Maniapoto Ontologies and Epistemologies

Ngāti Maniapoto iwi representatives criticised the current freshwater co-governance arrangements for not adequately reflecting Māori ontologies, epistemologies and governance systems. As one iwi member noted:

> If I was to draw a picture, I'd be drawing a picture saying this is *Papatūānuku* [Earth Mother]. This is *Ranginui* [Sky Father]. … If it's about caring for everything that exists between these two [*Papatūānuku* and *Ranginui*] and including them, … You can't take a *Pākehā* process and put them [*Papatūānuku* and *Ranginui* into it]—you can't… it wouldn't work. It … [is] our *mana*. Our sovereignty and our space. [The] lens … says [that] *Papatūānuku's* here … It's… [a] Māori worldview. (Iwi Rep 8 2019)

Iwi members describe the refusal (or inability) of non-Māori, specifically Pākehā council staff, consultants employed by councils and private developers to 'consult' with iwi, to seek to understand Māori ways of knowing water. Some Pākehā, iwi members, observe, remain steadfast in their narrow view of water as a resource (to be used and managed by humans for their benefit) (Iwi Rep 2 2020; Iwi Rep 5 2019; Iwi Rep 7 2019); although, some Pākehā (often scientists) perceive water as a combination of elements (H_2O) that is tied to the health of both human and ecological communities but struggle to comprehend water worlds that are not premised on Western scientific knowledge and practices (Scientist 1 2017). The limitations of current co-governance arrangements, many iwi members describe, as due to the incapacity of Pākehā to embrace plural ontologies and epistemologies of water, and their lack of respect shown to wai (water) as a living entity as well as the other more-than-human entities that dwell within their waterscapes and landscapes. For Ngāti Maniapoto, the Waipā River is a living being, she (as the river is female) is the tupuna of the iwi, who possesses her own mauri and wairua (Iwi Rep 5 2019).

Elsewhere in this book, we detail the meaning of rivers as more-than-human-entities for iwi, and the implications of different ontologies of water for how water is (and was historically) governed and managed in Aotearoa. In particular, we build on the work of other scholars to highlight how the lack of respect given to rivers as entities who possess their own mauri, wairua, and mana translates into water governance and decisions that continue to favour the settler-colonial status quo (Blaser 2014; Sundberg 2014). Iwi members highlight how their abilities to harvest freshwater resources (most notably indigenous biota including freshwater eels/tuna) remains severely limited due to poor water quality and a low number of aquatic fauna and flora living in the awa, and how their attempts to hold councils to account through planning processes remains restricted (Iwi Rep 2 2020; Iwi Rep 4 2020; Iwi Rep 5 2019; Iwi Rep 8 2019).

Iwi members argue that to understand the purpose of the River Acts and the WRA requires that decision-makers comprehend the "Maori component" of the River Acts, which includes a preamble written in Te Reo Māori (the Māori language) and Māori terms (such as mana, mauri, and wairua) used throughout each of the three River Acts. Similarly, the recent changes made to the Waikato Regional Plan include references to mātauranga and tikanga. Thus, it is important that officials (be it Māori or non-Māori) involved in co-governance initiatives ("it doesn't matter who's sitting across the table" as one iwi representative stated) "always hold the true intent" of the meanings of the terms. All parties should be working to protect and restore the mauri and wairua of the awa as outlined in the statutes and V&S (Iwi Rep 5 2019). Iwi members spoke of being tired and frustrated that they are always responsible for explaining not only Te Ao Māori and tikanga but also legal documents (such as the Treaty of Waitangi, Treaty settlements, and the Waipā River Act) to officials and stakeholders, and that the settler-state needed to ensure it educated its staff on such matters.

Water pollution, which we discussed earlier in Chap. 5, continues to be a problem despite the new co-governance and co-management arrangements. Local councils continue their earlier practices of discharging sewage into the waterways, despite Māori opposition to water-based waste disposal. District councils continue to use the practice (which is authorised by the WRC and supported by the judgement of the courts) because such water-based waste disposal is, according to the dominant knowledge, values and management approaches of Te Ao Pākehā, both cost-effective and safe. Yet, from the perspective of Ngāti Maniapoto, the disposal of waste into water is a deeply offensive and completely unacceptable practice which breaches their laws and threatens the health and wellbeing of both human and more-than-human communities (Amohanga et al. 1997; Hauauru Ki Uta Regional Management Committee 2012; Unknown Author 1996; Waitomo District Council 2011); any human waste (treated or untreated) that enters water diminishes the mauri, wairua, and mana of that water as well as all those who are connected to that water (be it mana whenua, the taniwha waiwaia, or flora and fauna). Such ontological politics of water are similarly found in other settler-colonial contexts. In the context of the Yukon (Canada), where a range of water co-governance agreements exist between Indigenous First Nations and the settler-state, scholars including Wilson and Inkster (2018) note how Indigenous views of the water as a living entity is continually disregarded by the settler-state (Wilson 2019; Wilson et al. 2019; Wilson and Inkster 2018). The disrespect of water (as a living entity) is shown, as the work of Wilson (2020) highlights, in the "water licensing decisions that prioritise industrial water use over First Nation relationships to water and over the current or projected impacts of the water licenses" (Wilson 2020, p. 108); which parallel the ongoing decisions made in Aotearoa (by central and local governments as well as the courts) to prioritise local councils' use of rivers, lakes, and seas for waste disposal over Māori relationships to water. While numerous scholars advocate for approaches that employ ontological pluralism to address power disparities between Indigenous and non-Indigenous peoples in colonial contexts, there remains a disconnect between theory and practice (Ahmad 2019; Blaney and Tickner 2017; Blaser 2014; Grosfoguel 2015; Howitt and Suchet-Pearson 2006; Maldonado-Torres 2016). The ontological conflicts evident in the Waipā River of Aotearoa (paralleling those in Yukon Canada) highlight the problems associated with trying to translate ontological pluralism (even when embedded within statutes and co-governance agreements) from paper into on-the-ground (more specifically in planning forums and court decisions) actions.

The lack of knowledge amongst government officials extended not only to their failure to comprehend Māori worldviews and values but also to their limited understanding of Waikato and Aotearoa histories. Iwi members note that few government officials seemed to know anything about Te Tiriti o Waitangi (the Treaty of Waitangi) nor of Māori experiences of violence and dispossession as a consequence of the Crown's failure to honour the Treaty. Without this historical knowledge, officials remained ignorant about what the Treaty principles (see Chap. 6) are as well as the purpose of the Treaty settlements and the new co-governance arrangements. Ngāti Maniapoto iwi members (in line with Te Ao Māori) perceive everything as connected, with the past events (particularly those experienced by one's ancestors) of critical importance to guiding the decisions that current and future generations make; thus, "everything [is based on] relationships" between living beings (human and more-than-human) (Iwi Rep 5 2019). In contrast, iwi members report that many of those members representing the settler-state in co-governance arrangements (in line with Te Ao Pākehā) only focus on the present and "only look as far as the legislation [establishing the WRA], so they only go back to … the signing of the legislation in 2010". However, to understand the legislation and the purpose of the WRA, as one iwi member argues:

"you've got to look like way before then [to the Treaty to understand the] context". The lack of historical understanding means that in the "practical application of the mechanism [of co-governance] sometimes [is] get[ting] lost in translation and [it is currently] fall[ing] short of the expectations of iwi" (Iwi Rep 5 2019).

Mark Hickford's work on Treaty settlements suggests that there are often difficulties reconciling the "complexities of co-existence" following Treaty settlements when the reality of pluralities (of multiple worlds) are operationalised. There is an "interpretive risk" when "strangers to the processes of negotiating [Treaty settlements] end up interpreting what was agreed at earlier moments in time and constructing different ways of understanding those concepts captured in legislation and deed of settlement" (Hickford and Humphries-Kil 2018, p. 170). Yet, these interpretive dangers, we argue, are more prevalent for non-Māori than Māori iwi given the extent to which large numbers of iwi members typically participate in Waitangi Tribunal and Treaty settlement processes (as demonstrated most recently within the Te Rohe Potāe Waitangi Tribunal inquiry wherein hundreds of individuals and whānau made written and oral submissions to the inquiry as well as participating in oral histories collected by researchers). The majority of iwi representatives we interviewed had been involved in either the Tribunal inquiry and/or the Treaty settlement process in some way, and so they are no strangers to the process or documents. Rather, it is government officials and stakeholders who are the strangers, and it is iwi members who are constantly left with the task of trying to explain the kaupapa (purpose) that rests behind co-governance arrangements (Iwi Rep 5 2019). Yet, we argue, that this seemingly never-ending task imposed on Māori as the educator and/or translator to non-Māori about not only Te Ao Māori but also the shared histories, legal frameworks, and institutions that are the building blocks of modern Aotearoa (the Treaty, Treaty settlements) is simply rearticulating colonial oppression; once again, the power and entitlements are unfairly distributed to maintain the status and to privilege of Te Ao Pākehā and the settler-state over Te Ao Māori and Māori communities.

Māori scholars Nēpia Mahuika and Graham Hingangaroa Smith, drawing on theorist Paulo Freire's theory of transformative praxis, contend that both oppressor (settlers) and the oppressed (Indigenous peoples) can be liberated through the praxis of reflection and action that involves a process of consciousness-raising (Mahuika 2009; Smith 1997, 2015). Freire maintains that: "Liberation is a praxis .... [which] cannot be unfold[ed] in isolation or individualism, but only through fellowship and solidarity; therefore it cannot unfold in the antagonistic relations between the oppressors and the oppressed" (Freire 1986, p. 79). Mahuika argues, in the context of how histories of Aotearoa are written, that the "transformation of the 'nation' is not a process or dream that can only be realised by Māori alone" (Mahuika 2009, p. 143). He calls on Pākehā historians to educate themselves (about Māori histories, tikanga, and knowledges) and in doing so adopt the process of reflecting on and taking actions that transform how they research, conceptualise, and write national and local histories in a way that does not reinforce colonial narratives, stereotypes and injustices against Māori (Mahuika 2009, 2015). Indeed, we argue that such a cyclical process of learning, critically reflecting, and transforming one's actions are similarly necessarily for non-Māori (be it council officials, consultants, board members, or other stakeholders) involved in freshwater governance and management to ensure that the litany of environmental injustices experienced by Māori are not repeated in the future. We are not arguing, however, that recognition and actions by Māori (and other Indigenous peoples) to resist, contest, reflect, and take actions (which includes actions to reassert their knowledges and sovereignties) are not critical; but rather that in settler-colonial societies, wherein multiple cultures now live, the importance of all people being able to practice the act of "two-eyed seeing" (Bartlett et al. 2012), walking between worlds (Salmond 2017), or existing in the pluriverse (Conway and Singh 2011; Hutchings 2019; Oslender 2019) is a fundamental part of environmental justice. Such acts of thinking and walking in and between water worlds (of Te Ao Pākehā and Te Ao Māori) cannot be ones that only Māori are expected to perform, all those who are involved in water governance in Aotearoa should be expected to walk the ontological and epistemological tightrope as a means to address current and avoid future injustices.

Conclusion

Ngāti Maniapoto iwi members argue that the ontological pluralistic visions, knowledge, and values (incorporating both Te Ao Māori and Te Ao Pākehā—the worlds of Māori and Pākehā) that underpin water co-governance agreements (as with the Treaty more than 150 years earlier) are not being translated into practice. Instead, the implementation of co-governance arrangements continues to be situated within the world of Pākehā and Western systems of governance, and in doing so, Ngāti Maniapoto ontologies and epistemologies continue to be marginalised. Yet, the results of our research highlight that research participants who whakapapa to the Waipā River strongly emphasise how changing to co-governance arrangements, in favour of more local-level (iwi- and hapū-centred) decision-making, should enable them to fulfil their aspirations to restore the health and wellbeing of their awa in ways that are more in touch with their knowledge, values, and governance structures. Flax-roots initiatives can, we suggest, ensure that "deliberative functions of establishing mutual respect and creating inclusive" and equitable decision-making processes are achieved, which includes redressing social and environmental injustices experienced by iwi (Pirsoul and Armoudian 2019, p. 4630). Yet, since environmental justice is plural rather than single (recognitional, procedural, and distributional justice), actions required to address the continued misrecognition (or failure to recognise) of Ngāti Maniapoto interests, their mātauranga and tikanga, as well as the distributional inequities, and lack of procedural inclusion they continue to experience. While explicit recognition of Indigenous ontologies and epistemologies within the legislation, government plans, and co-governance agreements are important steps towards addressing injustices, recognition alone is not enough to guarantee just outcomes. Further and ongoing changes are needed to ensure distributive equity and fairness in decision-making processes, and provide Indigenous peoples with a locus of control over their lives, livelihoods, and ancestral territories. Within the context of freshwater co-governance arrangements, the extension of iwi authorities decision-making powers over their awa and resources, which includes not only the right to be consulted but also the capacities to make decisions (including vetoing developments that are against tikanga) are critical steps to address the environmental injustices experienced by Māori.

References

Adger, W. N., Barnett, J., Chapin Iii, F. S., & Ellemor, H. (2011). This Must Be the Place: Underrepresentation of Identity and Meaning in Climate Change Decision-Making. *Global Environmental Politics, 11*(2), 1–25.

Ahmad, N. B. (2019). Mask Off – The Coloniality of Environmental Justice. *Widener Law Review, 25*, 195.

Amohanga, J., Ormsby, R., & Ormsby, M. (1997). *Further Response on Behalf of Nehenehenui Regional Management Committee Submission by Jacqui Amohanga, Rachael Ormsby, and Massey Ormsby, 24 November 1997. 60 41 51A, Volume 2.* Hamilton: Waikato Regional Council. Unpublished.

Arsenault, R., Bourassa, C., Diver, S., McGregor, D., & Witham, A. (2019). Including Indigenous Knowledge Systems in Environmental Assessments: Restructuring the Process. *Global Environmental Politics, 19*(3), 120–132.

Bächtiger, A., & Parkinson, J. (2019). *Mapping and Measuring Deliberation: Towards a New Deliberative Quality.* Oxford: Oxford University Press.

Bakker, K., Simms, R., Joe, N., & Harris, L. (2018). Indigenous Peoples and Water Governance in Canada: Regulatory Injustice and Prospects for Reform. *Water Justice.*

Barber, M., & Jackson, S. (2015). Remembering 'the Blackfellows' Dam': Australian Aboriginal Water Management and Settler Colonial Riparian Law in the Upper Roper River, Northern Territory. *Settler Colonial Studies, 5*(4), 282–301.

Bartlett, C., Marshall, M., & Marshall, A. (2012). Two-Eyed Seeing and Other Lessons Learned Within a Co-Learning Journey of Bringing Together Indigenous and Mainstream Knowledges and Ways of Knowing. *Journal of Environmental Studies and Sciences, 2*(4), 331–340.

Behn, C., & Bakker, K. (2019). Rendering Technical, Rendering Sacred: The Politics of Hydroelectric Development on British Columbia's Saaghii Naachii/Peace River. *Global Environmental Politics, 19*(3), 98–119.

Berry, K. A., Jackson, S., Saito, L., & Forline, L. (2018). Reconceptualising Water Quality Governance to Incorporate Knowledge and Values: Case Studies from Australian and Brazilian Indigenous Communities. *Water Alternatives, 11*(1), 40.

Blaney, D. L., & Tickner, A. B. (2017). Worlding, Ontological Politics and the Possibility of a Decolonial IR. *Millennium, 11*(1), 293–311.

Blaser, M. (2014). Ontology and Indigeneity: On the Political Ontology of Heterogeneous Assemblages. *Cultural Geographies, 21*(1), 49–58.

Boelens, R. (2014). Cultural Politics and the Hydrosocial Cycle: Water, Power and Identity in the Andean Highlands. *Geoforum, 57*, 234–247.

Brink, E., & Wamsler, C. (2018). Collaborative Governance for Climate Change Adaptation: Mapping Citizen–Municipality Interactions. *Environmental Policy and Governance, 28*(2), 82–97.

Castleden, H., Hart, C., Cunsolo, A., Harper, S., & Martin, D. (2017). Reconciliation and Relationality in Water Research and Management in Canada: Implementing Indigenous Ontologies, Epistemologies, and Methodologies. In S. Renzetti & D. P. Dupont (Eds.), *Water Policy and Governance in Canada* (pp. 69–95). Cham: Springer International Publishing.

Charpleix, L. (2018). The Whanganui River as Te Awa Tupua: Place-Based Law in a Legally Pluralistic Society. *The Geographical Journal, 184*, 19–30.

Conway, J., & Singh, J. (2011). Radical Democracy in Global Perspective: Notes from the Pluriverse. *Third World Quarterly, 32*(4), 689–706.

Coombes, A. E. (2006). *Rethinking Settler Colonialism: History and Memory in Australia, Canada, New Zealand and South Africa*. Manchester: Manchester University Press.

Coulthard, G. S. (2014). *Red Skin, White Masks: Rejecting the Colonial Politics of Recognition*. University of Minnesota Press. https://muse.jhu.edu/book/35470. Accessed 19 May 2019.

Cradock-Henry, N. A., Greenhalgh, S., Brown, P., & Sinner, J. (2017). Factors Influencing Successful Collaboration for Freshwater Management in Aotearoa, New Zealand. *Ecology and Society, 22*(2) Retrieved June 20, 2020, from https://www.jstor.org/stable/26270085.

Denny, S. K., & Fanning, L. M. (2016). A Mi'kmaw Perspective on Advancing Salmon Governance in Nova Scotia, Canada: Setting the Stage for Collaborative Co-Existence. *International Indigenous Policy Journal; London, 7*(3) Retrieved June 18, 2020, from http://search.proquest.com/docview/1858128395/abstract/288C64E807CE45C6PQ/1.

Forster, M. (2016). Indigenous-Environmental-Autonomy-in-Aotearoa-New-Zealand. *AlterNative: An International Journal of Indigenous Peoples, 12*(3), 316–330.

Fraser, N., & Honneth, A. (2003). *Redistribution or Recognition?: A Political-Philosophical Exchange.* London; New York: Verso.

Freire, P. (1986). *Pedagogy of the Oppressed, Trans Myra Bergman Ramos with an "Introduction" by Richard Shaull.* Hammondsworth, Middlesex: Penguin Books.

Grosfoguel, R. (2015). Transmodernity, Border Thinking, and Global Coloniality. *Nous, 13*(9).

Harmsworth, G., Awatere, S., & Procter, J. (2014). Meeting Water Quality and Quantity Standards to Sustain Cultural Values. In *21st Century Watershed Technology Conference and Workshop, Improving Water Quality and the Environment, The University of Waikato, New Zealand* (pp. 3–7) Retrieved May 18, 2017, from https://elibrary.asabe.org/azdez.asp?AID=45188&T=2.

Harmsworth, G., Awatere, S., & Robb, M. (2016). Indigenous Māori Values and Perspectives to Inform Freshwater Management in Aotearoa-New Zealand. *Ecology and Society, 21*(4).

Hauauru Ki Uta Regional Management Committee. (2012). *Clarrie Tapara and Hauauru Ki Uta Regional Management Commitee to the Mayor, Waitomo District Council, 23 August 2012. Document Number 2252764. Waikato Regional Council Archives.* Hamilton: Waikato Regional Council. Unpublished.

Hickford, M., & Humphries-Kil, M. (2018). Reflecting on Landscapes of Obligation, Their Making and Tacit Constitutionalisation: Freshwater Claims, Proprietorship and "Stewardship". In B. Martin, L. T. Aho, & M. Humphries-Kil (Eds.), *ResponsAbility: Law and Governance for Living Well with the Earth* (pp. 162–182). London: Routledge.

Howitt, R., & Suchet-Pearson, S. (2006). Rethinking the Building Blocks: Ontological Pluralism and the Idea of 'Management'. *Geografiska Annaler: Series B, Human Geography, 88*(3), 323–335.

Hutchings, K. (2019). Decolonizing Global Ethics: Thinking with the Pluriverse. *Ethics & International Affairs, 33*(2), 115–125.

Iwi Rep 2. (2020, February 13). Interview with Iwi Representative 2.

Iwi Rep 4. (2020, February 14). Interview with Iwi Representative 4.

Iwi Rep 5. (2019, March 25). Interview with Interview Iwi Representative 5.

Iwi Rep 7. (2019, May 16). Interview with Iwi Representative 7.

Iwi Rep 8. (2019, October 9). Interview with Iwi Representative 8.

Jackson, S. (2018). Indigenous Peoples and Water Justice in a Globalizing World. In K. Conca & E. Weinthal (Eds.), *The Oxford Handbook of Water Politics and Policy*. New York: Oxford University Press.

Johnston, A. (2018). Murky Waters: The Recognition of Maori Rights and Interests in Freshwater. *Auckland University Law Review, 24*, 39.

Jones, C. (2012). *Māori Council Water Rights Case Rejected* (SSRN Scholarly Paper No. ID 2545638). Rochester, NY: Social Science Research Network. Retrieved June 19, 2020, from https://papers.ssrn.com/abstract=2545638.

Jones, C. (2016). *New Treaty, New Tradition: Reconciling New Zealand and Maori Law*. Toronto: University of British Columbia. Retrieved June 12, 2019, from https://books.google.co.nz/books?hl=en&lr=&id=DSLCDAAAQBAJ&oi=fnd&pg=PT5&dq=Jones+2016+New+Treaty&ots=09dWY_fMZ0&sig=vEDmsW4b2_KETAJpQfJWLcrDjRg#v=onepage&q=Jones%202016%20New%20Treaty&f=false.

Kotaska, J. G. (2013). *Reconciliation 'at the End of the Day': Decolonizing Territorial Governance in British Columbia After Delgamuukw*. University of British Columbia. Retrieved August 15, 2019, from https://open.library.ubc.ca/cIRcle/collections/ubctheses/24/items/1.0074235.

Local Government Rep 1. (2018, October 4). Interview with Local Government Representative 1.

Local Government Rep 2. (2019, March 25). Interview with Local Government Representative 2.

Mahuika, N. (2009). Revitalizing Te Ika-a-Maui: Māori Migration and the Nation. Retrieved July 14, 2017, from http://researchcommons.waikato.ac.nz/handle/10289/6398.

Mahuika, N. (2010). Korero Tuku Iho: Our Gift and Our Responsibility. *Te Pouhere Korero, 4*, 24–40.

Mahuika, N. (2015). Re-storying Māori Legal Histories: Indigenous Articulations in Nineteenth-Century Aotearoa New Zealand. *Native American and Indigenous Studies, 2*(1), 40–66.

Maldonado-Torres, N. (2016). Colonialism, Neocolonial, Internal Colonialism, the Postcolonial, Coloniality, and Decoloniality. In Y. Martínez-San Miguel, B. Sifuentes-Jáuregui, & M. Belausteguigoitia (Eds.), *Critical Terms in Caribbean and Latin American Thought: Historical and Institutional Trajectories* (pp. 67–78). New York: Palgrave Macmillan US. https://doi.org/10.1057/9781137547903_6.

Māori Business Owner 1. (2019, August 29). Māori Business Owner 1.

McGregor, D. (2014). Traditional Knowledge and Water Governance: The Ethic of Responsibility. *AlterNative: An International Journal of Indigenous Peoples, 10*(5), 493–507.

McGregor, D. (2015). Indigenous Women, Water Justice and Zaagidowin (Love). *Canadian Woman Studies, 30*(2–3).

McLean, J. (2014). Still Colonising the Ord River, Northern Australia: A Postcolonial Geography of the Spaces Between Indigenous People's and Settlers' Interests. *The Geographical Journal, 180*(3), 198–210.

Memon, P. A., & Kirk, N. (2012). Role of Indigenous Māori People in Collaborative Water Governance in Aotearoa/New Zealand. *Journal of Environmental Planning and Management, 55*(7), 941–959.

Memon, A., & Skelton, P. (2007). Institutional Arrangements and Planning Practices to Allocate Freshwater Resources in New Zealand: A Way Forward. *New Zealand Journal of Environmental Law, 11*, 241.

Ministry for the Environment. (2017). *National Policy Statement for Freshwater Management 2014 (Updated August 2017)*. Ministry for the Environment. Retrieved August 15, 2019, from https://www.mfe.govt.nz/sites/default/files/media/Fresh%20water/nps-freshwater-ameneded-2017_0.pdf.

Morgan, T. K. K. B., & Te Aho, L. (2013). Waikato Taniwharau: Prioritising Competing Needs in the Management of the Waikato River. In J. Daniels (Ed.), *Advances in Environmental Research*. New York: Nova Science Publishers. Retrieved January 6, 2019, from https://researchcommons.waikato.ac.nz/handle/10289/8816.

Muller, S., Hemming, S., & Rigney, D. (2019). Indigenous Sovereignties: Relational Ontologies and Environmental Management. *Geographical Research, 57*(4), 399–410.

Muru-Lanning, M. (2012a). The Key Actors of Waikato River Co-Governance: Situational Analysis at Work. *AlterNative: An International Journal of Indigenous Peoples, 8*(2), 128–136.

Muru-Lanning, M. (2012b). Māori Research Collaborations, Mātauranga Māori Science and the Appropriation of Water in New Zealand. In *Anthropological Forum* (Vol. 22, pp. 151–164). Taylor & Francis.

Muru-Lanning, M. (2016). *Tupuna Awa: People and Politics of the Waikato River*. Auckland: Auckland University Press.

Nadasdy, P. (2017). *Sovereignty's Entailments: First Nation State Formation in the Yukon*. Toronto: University of Toronto Press.

New Zealand Parliament. Resource Management Act. (1991). Retrieved August 15, 2019, from http://www.legislation.govt.nz/act/public/1991/0069/223.0/DLM230265.html.

New Zealand Parliament. Ngāi Tahu Claims Settlement Act., Pub. L. No. No 97 (1998). http://www.legislation.govt.nz/act/public/1998/0097/latest/DLM429090.html. Accessed 5 July 2020.

New Zealand Parliament. Waikato-Tainui Raupatu Claims (Waikato River) Settlement Act. (2010a).

New Zealand Parliament. Ngāti Tuwharetoa, Raukawa, and Te Arawa River Iwi Waikato River Act. (2010b).

New Zealand Parliament. Ngā Wai o Maniapoto (Waipā River) Act. (2012). Retrieved April 19, 2020, from http://www.legislation.govt.nz/act/public/2012/0029/latest/DLM3335204.html.

New Zealand Parliament. Te Urewera Act 2014. (2014). Retrieved July 3, 2019, from http://www.legislation.govt.nz/act/public/2014/0051/latest/DLM6183601.html?search=qs_act%40bill%40regulation%40deemed reg_Tuhoe+Settlement_resel_25_h&p=1&sr=1.

New Zealand Parliament. Te Awa Tupua (Whanganui River Claims Settlement) Act. (2017). Retrieved April 19, 2020, from http://www.legislation.govt.nz/act/public/2017/0007/latest/whole.html.

Ngā Wai o Maniapoto (Waipā River) Act. (2012).

NGO Rep 1. (2017, September 28). NGO Representative 1.

Nissen, S. (2014). Who's in and Who's out? Inclusion and Exclusion in Canterbury's Freshwater Governance: Canterbury's Freshwater Governance. *New Zealand Geographer, 70*(1), 33–46.

O'Malley, V. (2013). *The Meeting Place: Maori and Pakeha Encounters, 1642–1840*. Auckland: Auckland University Press.

O'Sullivan, D. (2007). *Beyond Biculturalism: The Politics of an Indigenous Minority*. Wellington: Huia Publishers.

Oslender, U. (2019). Geographies of the Pluriverse: Decolonial Thinking and Ontological Conflict on Colombia's Pacific Coast. *Annals of the American Association of Geographers, 109*(6), 1691–1705.

Pahl-wostl, C. (2017). An Evolutionary Perspective on Water Governance: From Understanding to Transformation. *Water Resources Management, 31*(10), 2917–2932.

Pahl-Wostl, C., Jeffrey, P., Isendahl, N., & Brugnach, M. (2011). Maturing the New Water Management Paradigm: Progressing from Aspiration to Practice. *Water Resources Management, 25*(3), 837–856.

Park, G. (2002). Swamps Which Might Doubtless Easily Be Drained: Swamp Drainage and Its Impact on the Indigenous. In E. Pawson & T. Brooking (Eds.), *Environmental Histories of New Zealand* (pp. 176–185). Auckland: Oxford University Press.

Parsons, M., & Nalau, J. (2016). Historical Analogies as Tools in Understanding Transformation. *Global Environmental Change, 38*, 82–96.

Parsons, M., Nalau, J., & Fisher, K. (2017). Alternative Perspectives on Sustainability: Indigenous Knowledge and Methodologies. *Challenges in Sustainability, 5*(1), 7–14.

Parsons, M., Nalau, J., Fisher, K., & Brown, C. (2019). Disrupting Path Dependency: Making Room for Indigenous Knowledge in River Management. *Global Environmental Change, 56*, 95–113.

Pirsoul, N. (2019a). The Deliberative Deficit of Prior Consultation Mechanisms. *Australian Journal of Political Science, 54*(2), 255–271.

Pirsoul, N. (2019b). Recognition and Deliberation: A Deliberative Corrective to Liberal Multicultural Policies. *Journal of Deliberative Democracy, 15*(1), 10.

Pirsoul, N., & Armoudian, M. (2019). Deliberative Democracy and Water Management in New Zealand: A Critical Approach to Collaborative Governance and Co-Management Initiatives. *Water Resources Management, 33*(14), 4821–4834.

Poelina, A., Taylor, K. S., & Perdrisat, I. (2019). Martuwarra Fitzroy River Council: An Indigenous Cultural Approach to Collaborative Water Governance. *Australasian Journal of Environmental Management, 26*(3), 236–254.

von der Porten, S., & de Loë, R. C. (2013). Collaborative Approaches to Governance for Water and Indigenous Peoples: A Case Study from British Columbia, Canada. *Geoforum, 50*(1), 149–160.

von der Porten, S., de Loë, R., & Plummer, R. (2015). Collaborative Environmental Governance and Indigenous Peoples: Recommendations for Practice. *Environmental Practice, 17*(2), 134–144.

Pulido, L. (2017). Geographies of Race and Ethnicity III: Settler Colonialism and Nonnative People of Color. *Progress in Human Geography, 42*(2), 309–318.

Rangitāiki River Forum. (2015). *Te Ara Whanui o Rangitāiki – Pathways of the Rangitaiki*. Bay of Plenty Regional Council.

Robison, J., Cosens, B., Jackson, S., Leonard, K., & McCool, D. (2018). Indigenous Water Justice. *Lewis & Clark Law Review, 22*, 841.

Roburn, S., & Hwëch'in, T. (2012). Weathering Changes: Cultivating Local and Traditional Knowledge of Environmental Change in Tr'ondëk Hwëch'in Traditional Territory. *Arctic, 65*(4), 439–455.

Salmond, A. (2017). *Tears of Rangi: Experiments Across Worlds*. Auckland: Auckland University Press.

Schlosberg, D. (2004). Reconceiving Environmental Justice: Global Movements and Political Theories. *Environmental Politics, 13*(3), 517–540.

Schlosberg, D. (2013). Theorising Environmental Justice: The Expanding Sphere of a Discourse. *Environmental Politics, 22*(1), 37–55.

Schlosberg, D., & Carruthers, D. (2010). Indigenous Struggles, Environmental Justice, and Community Capabilities. *Global Environmental Politics, 10*(4), 12–35.

Scientist 1. (2017, September 4). Interview with Scientist 1.

Scientist 2. (2019, November 7). Interview with Scientist 2.

Simms, R., Harris, L., Joe, N., & Bakker, K. (2016). Navigating the Tensions in Collaborative Watershed Governance: Water Governance and Indigenous Communities in British Columbia, Canada. *Geoforum, 73*, 6–16.

Smith, G. H. (1997). *The Development of Kaupapa Maori: Theory and Praxis* (Thesis). ResearchSpace@Auckland. Retrieved August 15, 2019, from https://researchspace.auckland.ac.nz/handle/2292/623.

Smith, G. H. (2015). Chapter Three: Equity as Critical Praxis: The Self-Development of Te Whare Wānanga Awanuiārangi. *Counterpoints, 500*, 55–77.

Stevens, M. J. (2013). Ngā Tahu and the "Nature" of Māori Modernity. In E. Pawson & T. Brooking (Eds.), *Making a New Land: Environmental Histories of New Zealand* (pp. 293–309). Dunedin: Otago University Press.

Strang, V. (2014). The Taniwha and the Crown: Defending Water Rights in Aotearoa/New Zealand: Defending Water Rights in Aotearoa/New Zealand. *Wiley Interdisciplinary Reviews: Water, 1*(1), 121–131.

Sundberg, J. (2014). Decolonizing Posthumanist Geographies. *Cultural Geographies, 21*(1), 33–47.

Te Aho, L. (2010). Attempting to Integrate Indigenous Traditional Knowledge of Waterways with Western Science: To Restore and Protect the Health and Well-Being of an Ancestral River. In *4th International Traditional Knowledge Conference 2010* (p. 328).

Te Aho, L. (2015). *The Waikato River Settlement: Exploring a Model for Co-Management and Protection of Natural and Cultural Resources*. Ka Hula Ao Center for Excellence in Native Hawaiian Law, Richardson School of Law. Retrieved January 6, 2019, from https://researchcommons.waikato.ac.nz/handle/10289/10414.

Tipa, G., & Welch, R. (2006). Comanagement of Natural Resources: Issues of Definition from an Indigenous Community Perspective. *The Journal of Applied Behavioral Science, 42*(3), 373–391.

Tsatsaros, J. H., Wellman, J. L., Bohnet, I. C., Brodie, J. E., & Valentine, P. (2018). Indigenous Water Governance in Australia: Comparisons with the United States and Canada. *Water, 10*(11), 1639.

Unknown Author. (1996). *Secretary of Te Nehenehenui Regional Management Committee to Works Consultancy Ltd, 16 July 1996, 60 41 51A, Volume 1.* Hamilton: Waikato Regional Council. Unpublished.

Veracini, L. (2010). *Settler Colonialism: A Theoretical Overview.* New York: Springer.

Veracini, L. (2011). Introducing: Settler Colonial Studies. *Settler Colonial Studies, 1*(1), 1–12.

Waikato Regional Council. (2020). *Waikato Regional Plan* (Online Version). Waikato Regional Council. Government. Retrieved June 18, 2020, from https://www.waikatoregion.govt.nz/council/policy-and-plans/rules-and-regulation/regional-plan/waikato-regional-plan/.

Waikato River Authority. (2011). *Vision and Strategy for the Waikato River.*

Waikato-Tainui Raupatu Claims (Waikato River) Settlement Act. (2010). Retrieved July 3, 2019, from http://www.legislation.govt.nz/act/public/2010/0024/latest/DLM1630002.html.

Waitomo District Council. (2011). *Minutes of Waitomo District Council, Chamber of Commerce, Te Kuiti, 28 June 2011.* Unpublished. Retrieved February 8, 2020, from http://waitomo.govt.nz/Documents/Documents/Council%20Confirmed%20Minutes/PRIMARY_n252926_v1_Council_Minutes_-_28_June_2011.pdf.

Webster, K., & Cheyne, C. (2017). Creating Treaty-Based Local Governance in New Zealand: Māori and Pākehā Views. *Kōtuitui: New Zealand Journal of Social Sciences Online, 12*(2), 146–164.

Wheen, N. R., & Hayward, J. (2012). *Treaty of Waitangi Settlements.* Wellington: Bridget Williams Books.

Whyte, K. P. (2016). Indigenous Environmental Movements and the Function of Governance Institutions. In T. Gabrielson, C. Hall, J. Meyer, & D. Schlosberg (Eds.), *Oxford Handbook of Environmental Political Theory* (pp. 563–580). Oxford: Oxford University Press.

Williams, E. K., Watene-Rawiri, E. M., & Tipa, G. T. (2018). Empowering Indigenous Community Engagement and Approaches in Lake Restoration: An Āotearoa-New Zealand Perspective. In *Lake Restoration Handbook* (pp. 495–531). Cham: Springer.

Wilson, N. J. (2019). "Seeing Water Like a State?": Indigenous Water Governance Through Yukon First Nation Self-Government Agreements. *Geoforum*, 101–113.

Wilson, N. J. (2020). Querying Water Co-Governance: Yukon First Nations and Water Governance in the Context of Modern Land Claim Agreements. *Water Alternatives; Montpellier, 13*(1), 93–118.

Wilson, N. J., & Inkster, J. (2018). Respecting Water: Indigenous Water Governance, Ontologies, and the Politics of Kinship on the Ground. *Environment and Planning E: Nature and Space, 1*(4), 516–538.

Wilson, N. J., Harris, L. M., Joseph-Rear, A., Beaumont, J., & Satterfield, T. (2019). Water Is Medicine: Reimagining Water Security Through Tr'ondëk Hwëch'in Relationships to Treated and Traditional Water Sources in Yukon, Canada. *Water, 11*(3), 624.

Winter, C. J. (2018). *The Paralysis of Intergenerational Justice: Decolonising Entangled Futures* (PhD thesis). University of Sydney, Sydney. Retrieved from https://ses.library.usyd.edu.au/handle/2123/18009.

Woldesenbet, W. G. (2018). Collaborative Governance: Assessing the Problem of Weak Cross-Sectoral Collaborations for the Governance of Addis Ababa Rivers. *Applied Water Science, 8*(4), 116.

8

Co-Management in Theory and Practice: Co-Managing the Waipā River

Co-management initiatives are intended to improve the sustainable management of environments and natural resources and foster more equitable sharing of power between the state and Indigenous peoples. However, there are still ongoing debates about who actually benefits from such co-management frameworks in practice. In this chapter, we examine whether the co-management framework for the upper catchment of the Waipā River is an instrument for transforming the historically inequitable relationships between Māori and the Crown and if Ngāti Maniapoto desires for improved river health and increased capacities to exercise their self-determination rights are being fully realised. We demonstrate that the meaning of co-management differs between iwi and state actors. Ngāti Maniapoto perceive that co-management is bound to their mana (power and sovereignty) and rangatiratanga (authority and self-determination rights) over their river (Waipā River). Local government officials, in contrast, interpret that co-management is about ensuring greater consultation with iwi and making sure that iwi are involved in formal advisory bodies that feed into local government decision-making processes. We show how Ngāti Maniapoto are seeking to establish and implement their own co-management policies and plans to protect and restore their awa

© The Author(s) 2021
M. Parsons et al., *Decolonising Blue Spaces in the Anthropocene*, Palgrave Studies in Natural Resource Management, https://doi.org/10.1007/978-3-030-61071-5_8

and disrupt the authority and knowledge claims of the settler-state's environmental management regime. Indigenous groups like Ngāti Maniapoto are simultaneously following existing settler-state policies and planning processes, and also reconfiguring or subverting them to shift freshwater management away from the Eurocentric paradigm of water as a resource and a river as a landscape feature to wai (water) as a living entity that holds its own mauri (life force) and an awa (river) as a tupuna (ancestor) and taonga (treasure).

We document the ways in which the introduction of the Resource Management Act (RMA) in 1991 went some way towards procedural inclusion of Māori interests but did recognise Māori iwi decision-making authority over their rohe (traditional lands and waters). However, the emergence of new legislation and resulting co-management arrangements (introduced since the mid-2010s) are providing Māori iwi with greater influence in relation to day-to-day operations as well as planning and policy changes about river management (New Zealand Parliament 1991, 2010a, 2017; *Ngā Wai o Maniapoto (Waipā River) Act* 2012; Rangitāiki River Forum 2015; Waikato River Authority 2011). We then consider the extent to which new legislation enacted as part of Treaty settlements fulfils its potential in providing Ngāti Maniapoto with the ability to intervene in freshwater management decisions. Before we proceed to our examples from Aotearoa, it is important to situate our research within the context of broader scholarship on Indigenous freshwater co-management.

Indigenous Co-Management of Freshwater

Co-management is an increasingly prominent arrangement in the context of Indigenous peoples and natural resource management, including freshwater. Although definitions of co-management vary (Armitage et al. 2009; Berkes 2009; Denny and Fanning 2016; Dowsley 2009; Jacobson et al. 2016), it generally is used to refer to a suite of governance and management arrangements aimed at finding solutions to pressing environmental problems (Berkes 1989; Bouma et al. 2017; Diver 2016; Dowsley and Wenzel 2008). Co-management approaches strive to adjust the relationships between the state and non-state actors to provide mutual advantages to those involved. Often described as a way of managing relationships (state, Indigenous, and interest groups) rather than managing natural resources, co-management arrangements generally involve an element of sharing decision-making power between state and non-state actors (most notably Indigenous peoples) (Natcher et al. 2005; Natcher and Davis 2007; White 2020).

Co-management involves some degree of joint decision-making about freshwater management. Yet, since "jointness" takes place on a continuum, co-management (as a term and a practice) implies a high degree of ambiguity. Co-management, thus, describes consultative arrangements that involve governments seeking to consult with community stakeholders, industry, and Indigenous peoples, but not necessarily share power with them. In Canada, for instance, such co-management institutions are extensively used within fisheries and biodiversity conservation (Dowsley 2009; Feit 2005; White 2020). Such co-management arrangements are treated as advisory bodies (to provide advice to government) with the Canadian provincial and federal government continuing to retain the final decision-making authority. On the other hand, the term co-management can also apply to arrangements that enable a large degree of community control in environmental management. In the Pacific Northwest, for instance, co-management arrangements between Indigenous nations ("treaty tribes") and the settler-state were first created in the 1970s, after court decisions upheld Indigenous peoples' treaty rights to fish salmon. Since then, co-management arrangements within the Columbia River salmon fishery resulted in joint authority between state and Indigenous peoples at all levels of decision-making (with the courts playing a key role in ensuring the meaningful participation of Indigenous peoples in such arrangements) (Diver 2009; Diver et al. 2019; Pinkerton 2018). Working through such nested institutions, Columbia River Indigenous peoples who hold treaty rights are also now shaping fisheries policy at the international level and also contributing to the sustainable management of salmon fisheries (Diver 2012).

A key debate related to co-management is whether it actually contributes towards Indigenous peoples' desires for self-determination (Barrie 2018; Diver et al. 2019; Larsen 2018; Lowitt et al. 2019; Nuttall 2018).

The term self-determination itself is highly debated amongst academics and Indigenous leaders (Daigle 2016; Durie 1998; Heinämäki et al. 2017; Rifkin 2017; Walker 1990) but here is taken to mean "Indigenous communities being able to participate meaningfully in the creation of the government institution that they live with" (Diver 2014, p. 6). One central question (for scholars and practitioners) is the extent to which co-management regimes result in more equitable sharing of decision-making authority between governments and Indigenous communities in practice. Another question is whether Indigenous peoples' involvement in co-management efforts translate into meaningful benefits for them, such as greater access to resources, capacity building for Indigenous communities, and ongoing support for restoration projects.

Given the unequal power dynamics between Indigenous peoples and government entities, strong concerns are raised about co-management functioning as a tool that co-opts or continues to exclude Indigenous interests (Castro and Nielsen 2001; Muru-Lanning 2012). On the one hand, some studies conclude that many co-management initiatives are not effective because they do not result in the meaningful divisions of responsibility and collaborations. In part, this failure is traced to state bureaucratic systems continuing to privilege (and reinforce) the position of the state (and interest groups aligned with the state), and hegemonic Euro-Western knowledge systems continue to marginalise other (Indigenous and other non-Western) worldviews, knowledges and values (Nadasdy 2007; Parsons et al. 2019; Weir 2009). On the other hand, some scholars maintain that co-management arrangements provide clear opportunities to shift institutional norms, change societal expectations about management outcomes, and contribute to policy transformations (Diver 2016; Te Aho 2015; Zurba et al. 2012). Since state institutions are not monolithic and neither are Indigenous organisations, there is the potential for co-management arrangements to be flexible and adapted to suit changing expectations, knowledges, and socio-economic, political, cultural and ecological conditions. While there are imperfections with existing co-management arrangements, there is mounting evidence that co-management arrangements can provide Indigenous peoples with the ability to develop and pursue their own environmental management and restoration initiatives that can co-exist with government-led resource management strategies (Berkes and Armitage 2010; Denny and Fanning 2016; Diver 2014; White 2020). Accordingly, the existing literature on co-management broadly addresses both the opportunities and challenges of sharing knowledge between Indigenous communities and government agencies.

The ability of government departments to accept different (multiple) ontologies and epistemologies is often hampered by how they are designed (although that differs between contexts).

The extent to which state agencies recognise Indigenous peoples' spiritual relationships with their local environments, is an indication of the disconnect between settler-state (and the dominant worldview that underpins the state) and Indigenous peoples' and their knowledge systems (Castleden et al. 2017; Howitt and Suchet-Pearson 2006; Lavau 2013). Common practises of attempting to translate Indigenous knowledge concepts into narrow categories or formats that fit within the existing knowledge systems and institutional structures of settler-state agencies often result in incomplete representations (an injustice by way of misrecognition) of complex Indigenous concepts (Arsenault et al. 2019; Hikuroa et al. 2011; Nadasdy 2007). Following Diver (2014) and Weinstein (1999), there are also concerns surrounding the "capture" of information, where the redistribution of information can lead to a lack of collective power over important natural resources, especially under open-access circumstances (such as everyone being able to access water or hunt fauna freely) (Diver 2014; Heaslip 2008; Weinstein 1999). For example, sharing in-depth community knowledge with agencies about the location of natural resources, such as flora and fauna, could result in increased harvesting by outsiders (Marsh et al. 2015; Nadasdy 2007; Ross et al. 2011).

Some scholars argue that co-management is an ongoing problem-solving process, rather than an entrenched model that involves extensive discussion, deliberation, negotiation and joint social learning within networks established for problem-solving (Berkes 2009; Zurba et al. 2012). From this standpoint, research into co-management arrangements should be directed at understanding how different management tasks are organised and allocated, and by extension concentrate on the functions and activities, rather than the structure, of the co-management regime. Such an approach to examining co-management arrangements offers the potential to highlight how the sharing of power and responsibilities is the result (not the starting point) of the process (Berkes 2009; Zurba et al. 2012).

The Resource Management Act: Recognition of Māori Interests

First introduced in 1991, the Resource Management Act (RMA) remains Aotearoa's key legislation for environmental management (Barnett and Pauling 2005; Grundy and Gleeson 1996). The purpose of the RMA is "to promote the sustainable management of natural and physical resources" (section 5) (Crow et al. 2018; Knight 2016; New Zealand Parliament 1967). It allows for greater public participation than previous legislation, which specifically provides for Māori to participate in planning processes (resource consents) (Burton and Cocklin 1996; Lowry and Simon-Kumar 2017). The RMA (1991) contains specific provisions related to Māori interests in Part II of the Act (Purpose and Principles), which instructs all persons exercising functions and powers under it. Inclusion of the Māori concepts of kaitiakitanga (environmental guardianship) and taonga (treasures) marked a significant shift in recognition of Māori interests in environmental management that was glaringly absent from earlier environmental (and other) legislation.

The RMA (1991) went some way to address the lack of procedural inclusion for Māori in government-led environmental planning and decision-making processes in that iwi (tribes) were required to be consulted about planning applications (resource consents) that could impact them; however, this required district or regional councils to determine whether the applications were sufficiently important to justify public notification and public hearing processes (Burton and Cocklin 1996; Lowry and Simon-Kumar 2017). In this way, decision-making powers remained with the settler-state (district and regional councils), and Māori capacities to shape decisions remained severely constrained. Indeed, procedural, recognition and distributive injustices (as we discussed in regard to water pollution in the Waipā River in Chap. 5) continued even after the RMA (1991) was introduced.

The inclusion of Māori interests in water and land management through legislation paralleled developments in other settler-societies. In all settler-nations, Indigenous peoples have called for greater recognition of their rights and responsibilities to (and for) their rohe as well as the legal authority to make decisions about their rohe and their people. International scholars highlight that attempts to recognise Indigenous peoples through legislation (such as the RMA) frequently disregard or redress the multiple ways in which the setter-state deliberately excluded Indigenous cultures, knowledges, and practices from environmental management regimes (Hartwig et al. 2018; Jackson and Barber 2016; McLean 2014; Poelina et al. 2019). The re-distribution of power and changes to water governance and management approaches are a fundamental part of addressing the historical and contemporary environmental injustices faced by Indigenous peoples. Thus, the creation of new collaborative governance and management arrangements are critical to the advancement of Indigenous capacities to manage their water resources effectively and improve river management (Tsatsaros et al. 2018). Next, we examine the development of co-management arrangements between the New Zealand Government (Crown) and Ngāti Maniapoto in 2012 and show how it represented an important shift in formal decision-making processes in relation to the Waipā River. In doing so, the co-management arrangements strengthen the provisions of the RMA to ensure that Ngāti Maniapoto are procedural included in freshwater management.

Giving Effect to Co-Management of the Waipā River

As the preceding chapters make clear, Ngāti Maniapoto endured ongoing and systematic exclusion from formal institutional processes and the management of the Waipā since colonisation and were affected significantly by the radical transformation of land- and waterscapes in their rohe. In 2012, co-governance and co-management of the Waipā River were formalised through the passing of the Ngā Wai o Maniapoto (Waipā River) Act 2012 (referred to as the Waipā River Act), which gives effect to the Deed of Settlement and Waiwaia Accord signed by the Crown and Maniapoto Maori Trust Board (MMTB) in 2010. MMTB, as the only existing governance entity for Ngāti Maniapoto, currently represents Ngāti Maniapoto in national matters affecting iwi Māori and is the post-settlement governance entity for Ngāti Maniapoto (Jones 2016).

In the Deed of Settlement and Waipā Act, the Crown acknowledges the social, cultural and historical significance of the enduring relationship between Ngāti Maniapoto and its river (Waipā River), their guardian Waiwaia (a taniwha a supernatural being that dwells in waters), and the mana (power) of both the Waipā River and the iwi. It also acknowledges that Ngāti Maniapoto experiences distress as a consequence of the deterioration of the health of the Waipā River. Despite acknowledgements of Ngāti Maniapoto as kaitiaki (guardians) and the relevance of tikanga (laws) within the Deed of Settlement and legislation, nowhere in either document does the concept of rangatiratanga (authority and self-determination rights) appear. Indeed, the settler-state's failure to explicitly recognise rangatiratanga, along with several other principles underpinning Ngāti Maniapoto ways of being and knowing, remained a source of continued environmental injustice for iwi.

Principles for Co-Management of the Waipa River

As a means of regulating the interactions between partners, the principles contained in the Waipā River Act (discussed earlier in Chap. 7) focus on ensuring reciprocal relationships are maintained between parties and that all parties receive benefits from these relationships. The foregrounding of place-based principles and values, including Ngāti Maniapoto expressions of kaitiakitanga and tikanga, (re)asserts Ngāti Maniapoto mātauranga (knowledge) and legal orders and traditions (tikanga) suppressed by colonisation and emphasises the situatedness (or the 'where') of law (Davies 2015). The guiding principles of interpretation within the legislation give legal recognition to the relationship between Ngāti Maniapoto, the Waipā River, and Waiwaia as well as principles and values important to Ngāti Maniapoto including their relationship with the Crown. In addition to identifying the Vision and Strategy (discussed in Chap. 7) as the primary direction-setting document for the Waipā River and Waikato River (which also asserts the authority of mātauranga and Māori values), the guiding principles strongly assert Māori concepts as the foundation for co-governance and co-management. In particular, the Waipā River Act provides guidance on the interpretation of mana, rangatiratanga, kāwanatanga (governorship), kaitiakitanga as they pertain to Ngāti Maniapoto and their relationship to the Waipā. The importance of reciprocity in maintaining relationships is expressed in the principles of te mana o te wai (the quality and integrity of the waters), nga wai o Maniapoto (the deep-felt obligation of Maniapoto to restore, maintain, and protect all waters within the Maniapoto rohe), te mana o te Waipa (which refers to the historical, intellectual, physical and spiritual relationship between Maniapoto and the Waipā River), and te mana tuku iho o Waiwaia. Moreover, principles relating to processes and procedures for working together to ensure efficient and practical outcomes emphasise partnership (in light of the Treaty of Waitangi), integration (across a number of levels and a range of agencies) and integrity (a shared commitment to act to protect the integrity of the deed) as fundamental to the co-governance and co-management framework (New Zealand Parliament 2012).

In addition to co-governance arrangements, another significant component of the Waipā River Act (as well as the other two River Acts) was the creation of co-management agreements. As Māori legal scholar Jacinta Ruru states "[t]hese are the first statutes in New Zealand to evaluate Māori to co-management roles with the Crown in regard to fresh water" (Ruru 2013, p. 311). The co-management arrangements include iwi Environmental Plans, Integrated River Management Plans, as well as the Joint Management Agreements (JMAs) Table 8.1. Unlike the JMAs under the provisions of the RMA (1991), JMAs under the Waipā River Act are mandatory. Within each JMA, parts are compulsory (such as water monitoring and enforcement, preparation and amendments of

Table 8.1 Co-management mechanisms for the Waipā River

River objectives	Clause 4.3 of the deed stipulates that Ngāti Maniapoto identify their objectives for the Māori River and that these objectives must be consistent with the overarching purpose of the deed. The River objectives are: (1) Inclusive and valued relationships between all key stakeholders; (2) Maniapoto ancestral relationship is revitalised and recognised; and, (3) Partner / River relationships are clear, maintained and focussed.
Regulations	The Act provides opportunities to make regulations consistent with the overarching purpose of the Act for the management of species and habitats of the Upper Waipa River. The *Fisheries Plan for the Upper Waipā River* was launched in 2016 in anticipation of the development of regulations with Ministry of Primary Industries, which have not eventuated.
Iwi Management Plans	The Waipā River Act reinforces the position of Iwi Management Plans as documents to be considered as required under the RMA. In 2014, Maniapoto reviewed *He Mahere Taiao Maniapoto Iwi Environment Management Plan 2007*, with co-funding from the WRA. The revised *iwi* plan, *Ko Tā Maniapoto Mahere Taiao*, was launched in 2016. It provides high-level direction setting, and describes issues, objectives, policies and actions to protect, restore and enhance the relationship of Maniapoto with the environment including their economic, social, cultural and spiritual relationships[a]
Capacity funding	Capacity (co-management) funding is provided annually on an equal basis to enable Ngāti Maniapoto (and each of the other River *Iwi*) to participate in the co-governance and co-management arrangements under their respective deeds.[b]
Crown-iwi accords	The Accords provide direct lines of communication and engagement between MMTB and Crown agencies. The Waiwaia Accord is the overarching Accord under the Maniapoto deed of settlement, with nine other Accords added as Schedules. Between September 2010 and 2014, 10 Maniapoto—Crown Accords were developed and signed by both parties. During 2015–2016, Ngāti Maniapoto developed and proposed eight Accord Implementation Plans to the Crown agencies for adoption and signoff.[c]

(continued)

Table 8.1 (continued)

Joint management agreements	Part 3 of the Act outlines the duty to make joint management agreements (JMAs) between Maniapoto and each authority. This provides clear opportunities for Maniapoto to participate in formal decision making processes.

[a]Maniapoto Maori Trust Board, *Ko Tā Maniapoto Mahere Taiao – Maniapoto Environmental Management Plan. Maniapoto Māori Trust Board* (2016).
[b]Ministry for the Environment & Te Puni Kōkiri, *Review of the Waikato and Waipa Rivers Arrangements 2016–17. Crown Report for Collective Review* (2017); *Ngā Wai o Maniapoto (Waipā River) Act* (2012).
[c]Maniapoto Maori Trust Board, *Review of the Deed in Relation to the Co-Governance and Co-Management of the Waipa River* (Maniapoto Maori Trust Board, 2017).

planning documents) and other parts are by agreement between parties (iwi and government). Instead of making multiple JMAs with different local authorities, the Maniapoto Māori Trust Board (MMTB)—the mandated representative body for Ngāti Maniapoto—entered into one collective joint management agreement with the five local authorities that possessed jurisdiction in relation to the Waipā River; namely: WRC, Waikato District Council, Waipā District Council, Otorohanga District Council, and Waitomo District Council. The single JMA was intended as an instrument to strengthen and build better and more effective partnerships and relationships across the parties. As such, the Maniapoto JMA provides a framework for local authorities and MMTB to work together to carry out the functions, duties and powers provided for and to give effect to the Waipā River Act. The Ngā Wai o Waipā Co-governance Forum was formed to determine whether the JMA is being implemented to the satisfaction of all parties and in accordance with the principles set out in the JMA. The Forum comprises equal numbers of representatives from local authorities and MMTB and meets at least annually or more frequently if necessary with secretariat support provided by WRC.

Operationalising co-Management Arrangements

Following the signing of their deed of settlement, MMTB sought to develop and implement the co-management arrangements outlined in the deed as part of their annual programme of work. Almost immediately after the Waipā River Act was passed, MMTB embarked on a number of projects focused on management and restoration of the Waipā. These projects were supported in various ways by the funding opportunities enabled through the Act (for instance, through the WRCuT—the Waikato River Clean-up Trust, annual funding as part of the settlement agreement, and funding enabled through JMAs and Accords). The Act outlines five specific co-management arrangements available to Ngāti Maniapoto in giving effect to their Deed of Settlement.

River Objectives

Clause 4.3 of the deed stipulates that Ngāti Maniapoto identify their objectives for the Waipā River and that these objectives must be consistent with the overarching purpose of the deed. The River objectives are: (1) Inclusive and valued relationships between all key stakeholders; (2) Ngāti Maniapoto ancestral relationship is revitalised and recognised; and, (3) Partner/River relationships are clear, maintained and focussed. In realising these objectives (and the Vision and Strategy—V&S) and as part of their commitment to protecting the health of the Waipā River, Ngāti Maniapoto focused on identifying and prioritising restoration projects for the Waipā through a series of marae-based wānanga between 2013 and 2014. This process built upon Ngāti Maniapoto contributions to the Waikato River Independent Scoping Study process, which had focused primarily on the Waikato River, and was partly funded by Ministry of the Environment under the co-management arrangements (National Institute of Water and Atmospheric Research Ltd 2010). *The Maniapoto Priorities for the Restoration of the Waipā River Catchment* report provides anybody who wants to enter into a research or restoration relationship with Maniapoto a really clear guide… It does set the direction that [Maniapoto] can follow, but also leaves room for flexibility. Because priorities change and those were the priorities of the time and the whanau that were involved in that process. But new whanau are getting engaged and have priorities that they also want to achieve. (Kelly R)

The Maniapoto Priorities Report articulates Maniapoto whanau aspirations, values and issues in relation to the Waipā and has synergies with other plans and policies including the WRC's *Waipā Catchment Plan* and the *Waipā Zone Management Plan* (Maniapoto Maori Trust Board 2017; NIWA 2014). The *Waipā Catchment Plan* was developed by WRC in conjunction with Ngāti Maniapoto through a collaborative process to guide the implementation of integrated catchment management activities within the Waipā River. The *Waipā Catchment Plan* identifies 100 actions points, ranging from large erosion and sediment control projects to biodiversity, to looking after peat lakes, to working with Māori landowners (WRC 2014). These action points, though not enforceable, provide the focus for WRC to undertake restoration and management within the Waipā catchment and progress is measured against attainment (Iwi Rep 1 2017; Iwi Rep 2 2020; Local Government Rep 1 2018; Local Government Rep 2 2019). Overarching the *Catchment Plan* is the *Waipā Zone Management Plan,* which sets out high-level strategies and objectives to guide management activities 'to revitalise the waters of the Waipa River and its tributaries by 2050' (WRC 2012). The *Maniapoto Priorities Report* formed the basis for the Waipā projects included in the *Restoration Strategy*.

Crown-Iwi Accords

The Waiwaia Accord is the overarching Accord signed at the time of the Maniapoto deed of settlement, with nine other Accords added as Schedules. Between September 2010 and 2014, Maniapoto developed 10 Crown Accords, which were signed by both parties. During 2015–2016, Ngāti Maniapoto developed and proposed eight Accord Implementation

Plans to the Crown agencies for adoption and signoff (MMTB 2017). The Accords provide direct lines of communication and engagement between MMTB and Crown agencies.

Regulations

The Act provides opportunities to make regulations consistent with the overarching purpose of the Act for the management of species and habitats of the Upper Waipa River. MMTB developed the *Fisheries Plan for the Upper Waipā River* with co-funding from the WRA as a planning document to provide for the protection, restoration and enhancement of the fisheries resources of the Waipā River catchment (MMTB 2017; Watene-Rawiri et al. 2015). The *Fisheries Plan* was launched in 2016 in anticipation of the development of regulations with the Ministry of Primary Industries, but these have yet to eventuate.

Iwi Management Plans

Iwi Management Plans (IMPs) are planning documents developed by recognised iwi authorities and which outline their aspirations and objectives for their rohe. Under the RMA, local authorities must keep and maintain IMPs, and local authorities shall take into account IMPs in their various planning efforts (Thompson-Fawcett et al. 2017). The Waipā River Act reinforces the position of IMPs as documents to be considered as required under the RMA. In 2014, Maniapoto undertook to review *He Mahere Taiao Maniapoto Iwi Environment Management Plan 2007*, with co-funding from the WRA. The revised iwi plan, *Ko Tā Maniapoto Mahere Taiao*, was launched in 2016 (Kowhai Consulting Ltd 2007; Maniapoto Maori Trust Board 2016). This plan provides the direction of iwi and hapū, and describes issues, objectives, policies and actions to protect, restore and enhance the relationship of Maniapoto with the environment including their economic, social, cultural and spiritual relationships.

Joint Management Agreements

Unlike the JMAs under the provisions of the RMA (1991), JMAs under the Waipā River Act are mandatory. Within each JMA parts are compulsory (such as water monitoring and enforcement, preparation and amendments of planning documents) and other parts are by agreement between parties (iwi and government). Instead of making multiple JMAs with different local authorities, the Maniapoto Māori Trust Board (MMTB)—the mandated representative body for the Ngāti Maniapoto—entered into one collective joint management agreement with the five local authorities that possessed jurisdiction in relation to the Waipā River; namely: WRC, Waikato District Council, Waipā District Council, Otorohanga District Council, and Waitomo District Council. The single JMA was intended as an instrument to strengthen and build better and more effective partnerships and relationships across the parties. As such, the Maniapoto JMA provides a framework for local authorities and MMTB to work together to carry out the functions, duties and powers provided for and to give effect to the Waipā River Act. The Ngā Wai o Waipā Co-governance Forum was formed to determine whether the JMA is being implemented to the satisfaction of all parties and in accordance with the principles set out in the JMA. The Forum comprises equal numbers of representatives from local authorities and MMTB, and meets at least annually or more frequently if necessary with secretariat support provided by WRC.

In 2017, a review was undertaken to assess the effectiveness of the Maniapoto JMA and to identify areas of potential improvement (Brough Resource Management Limited 2017; Ministry for the Environment and Te Puni Kōkiri 2017). The review found there was continuing support for the Ngāti Maniapoto JMA amongst both government and iwi, and that the JMA provided a strong legislative foundation to give effect to the government-iwi partnership. However, there were further opportunities to expand the working relationships between government and iwi in the future. The review identified possible opportunities to include community, economic and environmental projects that align to iwi aspirations as per the Ngāti Maniapoto JMA and Waipā River Act (Maniapoto Māori

Trust Board et al. 2013; New Zealand Parliament 2012). The review highlighted the benefits of holding more meetings to discuss strategic outcomes and to promote collaborative projects, with the need to augment formal meetings with informal gatherings as a way to understand and accommodate iwi aspirations for the future.

At an operational level, the review determined the Ngāti Maniapoto JMA was an effective tool, but there were limits on its effectiveness due to lack of resourcing as well as lack of clear communication between different tiers of decision-making Ngāti Maniapoto iwi. Many iwi members outside of leadership roles (within WRA, MMTB, and councils) expressed a lack of knowledge about the nature of co-management relationships. As one iwi representative informed us: "I don't know anything about co-management or co-governance or what that is" as she was focused on the "mahi" (work) at the flax-roots level rather what goes on in the offices and boardrooms (be it of MMTB, WRA, and the various local councils) (Iwi Rep 7 2019b). Ngāti Maniapoto iwi members emphasise how the co-management arrangements, as with co-governance, for the Waipā River, need to be reconfigured to fit with iwi approaches. So rather than the regional council holding a single hui (meeting) with iwi representatives—on one marae (tribal meeting area with complex of buildings) or in a board room—to discuss a resource consent application require longer and more in-depth discussion and negotiation processes need to be the norm. Iwi representatives maintain that each time an issue of freshwater management arises, (such as the resource consent applications for the Otorohanga District Council to discharge wastewater into the river which we discussed in Chap. 5), a series of hui or wānanga need to be held (which involve local hapū and iwi as well as co-management partners) to ensure that people are fully informed and consensus is built. Such community-level rather regional-level approach challenge the existing practices of western-style institutions (Iwi Rep 7 2019a). Iwi Representative 7 reported that:

> more information [needs to be] available to whānau, hāpu and iwi. [And it] actually, [needs to go] back to the whānau, hāpu and iwi, and actually ask [them] the questions around [freshwater governance and management]. Because, at the end of the day, it's actually those people who are

looking after the awa, who are down the awa all the time. You can't tell me those that are in a governance position [within the WRA and WRC] are the ones who are actually doing the mahi [work] down at awa [to restore it]. They're just making decisions. (Iwi Rep 7 2019a)

The importance of Māori modes of decision-making that emphasis community discussion (whānau, hapū and iwi) communities are about to talk about information and issues to reach a consensus (at the flax-roots level) differs from the standard (settler-state) planning processes undertaken.

Indeed, as many scholars already observe, there are problems associated with current state-based recognition of Indigenous land and water rights as the power to define what (or whom) is recognised remains vested with the settler-state (and Western cultures) rather than within those of Indigenous peoples themselves. Rights, as Corntassel and Bryce observe, are "state constructions that do not necessarily reflect inherent indigenous responsibilities to their homelands". Furthermore, discussions of land and water rights compartmentalises Indigenous "self-determination from governance and community wellbeing from homelands and relationships to the natural world" (Corntassel and Bryce 2011, pp. 152–153). By embedding themselves within the settler-state centred recognition of Indigenous rights, Indigenous communities risk reinforcing the settler-colonial status quo (replicating knowledge, governance and management approaches of the state) rather than honouring their relationships and ways of interacting with their traditional lands, waters, and human and more-than-human entities that dwell there. Accordingly, scholars including Coulthard, Corntassel and Bryce argue that approaches need to centre on Indigenous resurgence involving reconnecting Indigenous peoples to their traditional lands and waters, socio-cultural practices, languages, knowledges, and ways of governing and managing their environments (Corntassel and Bryce 2011; Coulthard 2014; McGregor 2014). Emphasis is placed of how to reclaim, restore and regenerate their relationships (and responsibilities for) their homelands through decolonising processes that transform "indigenous struggles for freedom from performance to everyday practice" (Corntassel and Bryce 2011, p. 153). Indeed, in the context of Ngāti Maniapoto, the ways in which iwi members are seeking to manage and restore their awa highlights how iwi are disrupting and moving beyond the narrowly defined state-based discourse of Māori water rights. They are practising their everyday responsibilities (as kaitiaki) to their rohe and their kin (including their whānau, their human- and more-than-human ancestors) and in doing so showing manaakitanga (respect and care for others) towards the awa, wai, plants and animals, and the Waipā River's supernatural guardian Waiwaia. In order, to enact their kaitiakitanga-based (environmental guardianship) practices of sustainably managing and using freshwater resources for their subsistence as well as economic development needs, iwi members identify how the current planning regime (WRP, JMA, and district plans) remains inadequate.

Integrated Management Plan

Each of the Treaty settlements and the resulting legislation that established the co-governance and co-management of the Waikato and Waipā Rivers allows for the development of integrated river management plans in collaboration with central and local government authorities (including WRC) (New Zealand Parliament 2010a, b, 2012). The integrated river plans are intended to allow iwi, hapū and whānau to be more directly involved in plan development as well as its implementation, and the practice of river management and restoration. As of 2020, no plans have been created; however, in late 2018 Ngāti Maniapoto notified WRC of their desire to start work to develop an Upper Waipā River Integrated Management Plan (Iwi Rep 4 2020; Iwi Rep 5 2019; Iwi Rep 7 2019a). Ngāti Maniapoto participants express the hope that the Integrated River Management Plan will provide a powerful mechanism (a "lever") by which iwi and hapū can seek to improve freshwater governance and management at a hapū- or flax-roots-level and allow for the realisation of iwi desires for the freedom (agency) to choose their own paths towards more sustainable freshwater futures (Māori Business Owner 1 2019). One iwi participant expresses his hope that the:

> Integrated Plan [would be] an amazing tool for whānau for when they want to do something to know who they go to. Then they've got something

to take with them to these people, so they don't just get fobbed off and say oh no, we don't have time for that. They can say no, but you have to make time for this. (Māori Business Owner 1 2019)

Not Trickling Down to Flax-Roots-Level

The extent to which the benefits of co-governance and co-management arrangements for the Waipā River are' trickling down' to the flax-roots-level (marae, hapū, whānau) are uneven; this raises questions about the structure of the co-governance entity, how co-governance and co-management are enacted, and who (what) are the winners and losers in the new institutional arrangements. Capacity and resourcing are key issues for the formal iwi institutions (such as the MMTB) as well as other social groupings within the iwi (including hapū, marae, whānau). Many express concern that those higher up the co-governance and co-management ladder (who sit on boards) can earn wages or salaries from their positions, whereas those lower-down do all the mahi (work) on a voluntary basis. They argue that the iwi needs to ensure that those within their iwi, hapū and whānau and hapū are taken care of.

[W]e tend to forget about those people, right down [at the grassroots level] … who are doing the [water] testing [and cultural health assessment framework which MMTB is creating and implementing with scientists]. … Those are the people who are gathering the data [for the MMTB and scientists but] they're not paid to do that. Some of them take time off [their jobs], just to attend [the wānanga and hui]. So, [we need to make] sure that they're well looked after, well resourced. If you've got a pen in your office, then, make sure you give a pen to the people down the river… Because we don't own the knowledge, we don't own mātauranga. So, rather than being a gatekeeper, we just [feed] the people with it. A good leader will always be someone who will train someone up to be better than them, and not be phased by it. [If] I could train twenty people up to be twenty times better than me, cool. They'll be awesome. (Iwi Rep 7 2019b)

Indeed, for some members of Ngāti Maniapoto, the establishment of the WRA, new legislation, and co-management plans are just not being translated into tangible changes to materially improve the wellbeing Ngāti Maniapoto as a collective group (iwi, hapū and whānau). Indeed, successful co-management needs to foster beneficial relationships within an iwi (based on the values of whanaungatanga—the centrality of kinship, whakapapa that binds the Māori world together and manaatikitanga—the process of demonstrating generosity, respect, and care for others) as well as meaningful relationships between iwi, local government, and the Crown, industries, and other stakeholders. Thus, as we stressed in our previous chapter, adequate resources need to be provided not only by the settler-state to Māori groups to address Māori disadvantages (addressing the distributive disparities faced by iwi) and in doing so "level the playing field in terms of capacity for collaboration" (Porten et al. 2015, p. 134). These duties, to ensure the distribution of adequate resources, also extend to co-governance and co-management institutions (including the MMTB) so that iwi/hapū/whānau, who face limited access to resources (be it financial, technical or human), can access support so they can participate in planning for and taking actions to manage and restore their awa.

Co-Management Strengthening Procedural Inclusion and Recognition

From 1991, as we mentioned earlier, the introduction of the RMA did go some (limited) way to recognising Māori relationships with their rohe, and provided an avenue for Māori to be procedurally included in planning processes. Likewise, the new legislation and institutional arrangements (including the WRA, the V&S, WRP and JMAs) over the last decade provides Ngāti Maniapoto with greater recognition and procedural inclusion under the planning regime governed by local government (WRC and district councils). In particular, the mandatory nature of the JMA, including the provisions of water monitoring, provides Ngāti Maniapoto with more information about the health of their awa as well as the abilities to conduct their own water testing; previously the iwi encountered substantive difficulties accessing information from district councils that restricted their capacities to participate in decision-making. Bryant's work into community participation in pollution prevention highlighted how participatory practices (fair procedures) can foster environmental justice through ensuring that:

> rules, regulations, behaviors, policies and decisions support sustainable communities where people can interact with confidence that the environment is safe, nurturing, and productive. (Bryant 1995, p. 6)

He observes that the principles of environmental justice can be achieved when a community is able to reach their full potential cultivated through: "democratic decision-making and personal empowerment ... where both cultural and biodiversity are respected and highly revered and where ... justice prevails" (Bryant 1995, p. 6). Bryant's work demonstrates how the co-management arrangements being implemented within the Waipā River can (or could potentially address) some of the environmental injustices experienced by Ngāti Maniapoto by providing them greater abilities to influence local-level environmental planning decisions.

Members of Ngāti Maniapoto note how, despite the new legislation and co-management agreements, they are disappointed that water quality of their awa remains poor, and they cannot restrict activities that negatively affect their awa (through vetoing resource consent applications). Iwi members' hopes that their rangatiratanga would be recognised and empowered through the new co-governance and co-management arrangements are yet to be realised:

> We've got the [Accords], we've got the JMAs [Joint Management Agreements between councils and iwi] they aren't working. ... We're hoping that that will give more [power] to get things done. (Māori Business Owner 1 2019)

Indeed, Ngāti Maniapoto iwi members stress that they seek to exercise their rangatiratanga not as actions of secession (that challenge state sovereignty as some Pākehā and government officials fear), but rather to ensure that the principles embedded in the Treaty more than a 150 years ago are finally honoured (including active protection of rangatiratanga and iwi-Crown partnership). In the words of one iwi representative "co-management … to me should be we make the decision jointly at the end", it is put simply an exemplar of what rangatira who signed to Treaty envisioned, the ability of Māori to choose how they would or could live (walking in one, both or between the worlds of Māori and Pākehā) (Jones 2016; Māori Business Owner 1 2019; Salmond 2017). However, one iwi member questions whether Ngāti Maniapoto actually possess (under their Deed of Settlement and Waipā River Act) any level of authority that comes close to the rangatiratanga promised to them under the Treaty, as the iwi still cannot hold governments to account for their failures to follow laws, management plans, and regulations:

> there's no … regulation of these management agreements, there's no-one checking up on them [the councils] to make sure that they're actually [doing what was promised]. They've just been put [it] into an [policy] and they're left [it] there to [carry on] do[ing] their job. (Māori Business Owner 1 2019)

In previous chapters, we documented the disproportionate distribution of environmental impacts on iwi as a consequence of freshwater degradation linked to settler colonialism, and how the lack of legal and regulatory powers to prevent governments' and individual settlers' destroying their landscapes and waterscapes added another layer to the environmental injustices they experienced. The Treaty settlements, new legislation and co-governance and co-management arrangements all explicitly acknowledge the negative effects of iwi and do provide some mechanisms (legislation, funding, co-governance and co-management) by which environmental injustices faced by Māori can be addressed. Yet, these mechanisms remain incomplete and imperfect tools. Co-management needs to create opportunities for reciprocal and meaningful relationships between iwi and the settler-state (and its various agencies). Such relationships cannot be premised, however, on the maintenance of settler status quo (whereby settler-state institutions, government officials, and interest groups control how freshwater is managed) which led to the current state of freshwater degradation (and broader challenges associated with the Anthropocene). Instead, it needs to involve a transformative decolonising shift, that allows for iwi to enact their kaitiakitanga practices in ways that accord to their ways of governing and managing freshwaters, and for iwi-state to find a new (more equitable) relationship based on shared partnership and mutual respect. Scholars highlight how Indigenous responsibilities-based management approaches are critical avenues by which Indigenous peoples' to revitalise their relationships, promote the regeneration of sustainable water, land, and food systems in communities, and transmit their knowledge, values, and practices to future generations (Corntassel and Bryce 2011). Yet, more than co-management offers the opportunities of "learning together" between different groups and within groups, and the capacities to draw on multiple knowledges, and to design management approaches that can accompany pluralistic ways of knowing the world(s) (Hopkins et al. 2019).

Unlike the experiences encountered by other marginalised populations, the circumstances that contribute to environmental injustices for Indigenous peoples differ and are (arguably) more complex: encompassing distributional, procedural and recognitional justice. Much of the difference, as we discuss in previous chapters, rest in the ways in which, under settler-colonial rule, environmental laws, governance and management approaches are underpinned by settler-colonial knowledges, values and practices that are antithetical to Indigenous peoples achieving environmental justice (Muir and Booth 2012, p. 458). Another basis of this difference rests on Indigenous ontologies that are holistic, relational, and place-based wherein socio-cultural interactions, spirituality, and ecological attributes are all interwoven together. Injustices borne by Ngāti Maniapoto (like other Indigenous peoples), as we document in previous chapters, include the exploitation and degradation of environmental resources required for subsistence, and the destruction of wāhi tapu (sacred sites) and the graves belonging to their ancestors. According to this view, Schlosberg and Carruthers observe that:

> Indigenous demands for environmental justice go beyond distributional equity to emphasize the defense and very function of Indigenous communities—their ability to continue and reproduce their traditions, practices, cosmologies, and the relationship with nature that tie native peoples to their ancestral lands. (Schlosberg and Carruthers 2010, p. 13)

As we document in the previous chapters of this book, the settler-state's laws, policies and actions (including those that radically transformed Māori landscapes and waterscapes) negatively affected their capacities to access their traditional food sources, maintain their economic livelihoods, and ensure their responsibilities as kaitiaki (environmental guardians). The inability of Māori to maintain the mauri (life force) and wairua (spiritual integrity) of their awa (to which is classified as their kin) are not only direct assaults (injustices) against the mana whenua (tribal group with authority over their rohe), but also direct assaults against the "cultural practices and beliefs" that iwi require to ensure their cultural continuance (Schlosberg and Carruthers 2010, p. 13). As Schlosberg and Carruthers (2010) point out, the survival of Indigenous peoples is directly connected to their sustainable interactions with their land and waters, and with the enactment of their laws, practices, ceremonies, and beliefs connected to their places. The need to take into account a particular groups' historical and cultural basis that is critical to achieving environmental justice for those communities. When key components of Indigenous peoples are removed, the abilities of Indigenous communities to determine their own futures are therefore also removed.

The new co-management arrangements are, in many respects, enhancing the capacity of Ngāti Maniapoto to self-identify (as an iwi and as mana whenua of the upper catchment of the Waipā River). Therefore the legislation and co-management planning tools are reversing some of the injustices caused by colonisation, which includes those that threatened the cultural continuance of iwi. Indeed, through its iwi Environmental Plans, the JMA, and the legislation, Ngāti Maniapoto are asserting their mātauranga (knowledge), tikanga (laws), and kawa (ceremonies) within freshwater management. They are also articulating their aspirations and objectives through its own policies and those created in collaboration with local government authorities support the resurgence of mana whakahaere of Ngāti Maniapoto. In the restoration projects funded by WRA, for instance, the mātauranga and tikanga of Maniapoto as a dynamic grouping of people with their own conceptualisation and commitment to intergenerational Indigenous environmental justice (which extends to include the more-than-human actors most notably the Waipā River herself and the taniwha Waiwaia himself) are being not only expressed but also acted on. The Waipā River Act, the JMA, and the other co-management agreements, thus, are enabling Ngāti Maniapoto involvement in local-level decision-making processes about their awa to a far greater extent than in the past. Formerly excluded Ngāti Maniapoto are now central actors (but not necessarily equitable Treaty partners) in the freshwater governance, management and planning about their awa.

Conclusion

Rather than seeing co-management is a single piece of legislation or an institution, this chapter highlights that successful co-management between Indigenous peoples and settler states should instead be seen as a process (as part of wider decolonising processes). It is a process that is premised on sharing decision-making responsibilities, between settler-state and Indigenous authority-holders, which involves considering not only different management plans and regulations, but also the processes wherein these plans and regulations are translated into on-the-ground actions that address the material and metaphysical health and wellbeing of Indigenous communities. A critical component of this process is the consideration of Indigenous peoples' authority (self-determination, rangatiratanga holders) and responsibilities to their traditional lands and waters (rohe or homelands) as well as to their kin-group (which includes human and more-than-human beings). No longer can Indigenous peoples be framed as just another stakeholder group or a marginalised community, instead their authority as Indigenous peoples, First Nations, or in the Aotearoa context, mana whenua needs to be the basis for equitable, effective, and sustainable co-management arrangements that take into account the recognitional, distributional and procedural components of environmental justice. The first steps towards more successful co-management partnerships between Ngāti Maniapoto and the setter-state are being made in the upper reaches of the Waipā River, and it remains to be seen if that early promise can be translated into lasting environmental just outcomes for Ngāti Maniapoto.

References

Armitage, D. R., Plummer, R., Berkes, F., Arthur, R. I., Charles, A. T., Davidson-Hunt, I. J., et al. (2009). Adaptive Co-Management for Social–Ecological Complexity. *Frontiers in Ecology and the Environment, 7*(2), 95–102.

Arsenault, R., Bourassa, C., Diver, S., McGregor, D., & Witham, A. (2019). Including Indigenous Knowledge Systems in Environmental Assessments: Restructuring the Process. *Global Environmental Politics, 19*(3), 120–132.

Barnett, J., & Pauling, J. (2005). The Environmental Effects of New Zealand's Free-Market Reforms. *Environment, Development and Sustainability, 7*(2), 271–289.

Barrie, G. N. (2018). International Law and Indigenous People: Self-Determination, Development, Consent and Co-Management. *Comparative and International Law Journal of Southern Africa, 51*(2), 171–184.

Berkes, F. (1989). Co-Management and the James Bay Agreement. In E. Pinkerton (Ed.), *Co-Operative Management of Local Fisheries: New Directions for Improved Management and Community Development* (pp. 181–182). Vancouver, Canada: University of British Columbia Press.

Berkes, F. (2009). Evolution of Co-Management: Role of Knowledge Generation, Bridging Organizations and Social Learning. *Journal of Environmental Management, 90*(5), 1692–1702.

Berkes, F., & Armitage, D. (2010). Co-Management Institutions, Knowledge, and Learning: Adapting to Change in the Arctic. *Études/Inuit/Studies, 34*(1), 109–131.

Bouma, J., Reyes-García, V., Huanca, T., & Arrazola, S. (2017). Understanding Conditions for Co-Management: A Framed Field Experiment Amongst the Tsimane, Bolivia. *Ecological Economics, 141*, 32–42.

Brough Resource Management Limited. (2017). *Effectiveness Review of the Waikato and Waipa Rivers Co-Governance and Co-Management Framework*. Report Prepared for Ministry for the Environment. Wellington: Ministry for the Environment.

Bryant, B. (1995). Pollution Prevention and Participatory Research as a Methodology for Environmental Justice. *Virginia Environmental Law Journal, 14*(4), 589–613.

Burton, L., & Cocklin, C. (1996). Water Resource Management and Environmental Policy Reform in New Zealand: Regionalism, Allocation, and Indigenous Relations. *Colorado Journal of International Environmental Law and Policy, 7*, 331.

Castleden, H., Hart, C., Cunsolo, A., Harper, S., & Martin, D. (2017). Reconciliation and Relationality in Water Research and Management in Canada: Implementing Indigenous Ontologies, Epistemologies, and Methodologies. In S. Renzetti & D. P. Dupont (Eds.), *Water Policy and Governance in Canada* (pp. 69–95). Cham: Springer International Publishing.

Castro, A. P., & Nielsen, E. (2001). Indigenous People and Co-Management: Implications for Conflict Management. *Environmental Science & Policy, 4*(4), 229–239.

Corntassel, J., & Bryce, C. (2011). Practicing Sustainable Self-Determination: Indigenous Approaches to Cultural Restoration and Revitalization Indigenous Political Actors. *Brown Journal of World Affairs, 18*(2), 151–166.

Coulthard, G. S. (2014). *Red Skin, White Masks: Rejecting the Colonial Politics of Recognition.* Minneapolis: University of Minnesota Press.

Crow, S. K., Tipa, G. T., Booker, D. J., & Nelson, K. D. (2018). Relationships Between Maori Values and Streamflow: Tools for Incorporating Cultural Values into Freshwater Management Decisions. *New Zealand Journal of Marine and Freshwater Research, 52*(4), 626–642.

Daigle, M. (2016). Awawanenitakik: The Spatial Politics of Recognition and Relational Geographies of Indigenous Self-Determination. *The Canadian Geographer/Le Géographe Canadien, 60*(2), 259–269.

Davies, A. R. (2015). *Māori and Freshwater: A Comparative Study of Freshwater Co-Management Agreements in New Zealand.* Thesis, Lincoln University.

Denny, S. K., & Fanning, L. M. (2016). A Mi'kmaw Perspective on Advancing Salmon Governance in Nova Scotia, Canada: Setting the Stage for Collaborative Co-Existence. *International Indigenous Policy Journal; London, 7*(3) Retrieved June 18, 2020, from http://search.proquest.com/docview/1858128395/abstract/288C64E807CE45C6PQ/1.

Diver, S. (2009). Towards Sustainable Fisheries: Assessing Co-Management Effectiveness for the Columbia River Basin. *Nature Precedings.* https://doi.org/10.1038/npre.2009.3754.1.

Diver, S. (2012). Columbia River Tribal Fisheries: Life History Stages of a Co-Management Institution. In *Keystone Nations: Indigenous Peoples and Salmon Across the North Pacific. School for Advanced Research Press, Santa Fe, New Mexico, USA* (pp. 207–235). Santa Fe: SAR Press.

Diver, S. (2014). *Negotiating Knowledges, Shifting Access: Natural Resource Governance with Indigenous Communities and State Agencies in the Pacific Northwest.* Berkeley: University of California.

Diver, S. (2016). Co-Management as a Catalyst: Pathways to Post-Colonial Forestry in the Klamath Basin, California. *Human Ecology, 44*(5), 533–546.

Diver, S., Ahrens, D., Arbit, T., & Bakker, K. (2019). Engaging Colonial Entanglements: "Treatment as a State" Policy for Indigenous Water Co-Governance. *Global Environmental Politics, 19*(3), 33–56.

Dowsley, M. (2009). Community Clusters in Wildlife and Environmental Management: Using TEK and Community Involvement to Improve Co-Management in an Era of Rapid Environmental Change. *Polar Research, 28*(1), 43–59.

Dowsley, M., & Wenzel, G. (2008). "The Time of the Most Polar Bears": A Co-Management Conflict in Nunavut. *Arctic, 61*, 177–189.

Durie, M. (1998). *Te mana, te kāwanatanga: The Politics of Māori Self-Determination*. Auckland: Oxford University Press.

Feit, H. A. (2005). Re-Cognizing Co-Management as Co-Governance: Visions and Histories of Conservation at James Bay. *Anthropologica, 47*, 267–288.

Grundy, K. J., & Gleeson, B. J. (1996). Sustainable Management and the Market: The Politics of Planning Reform in New Zealand. *Land Use Policy, 13*(3), 197–211.

Hartwig, L. D., Jackson, S., & Osborne, N. (2018). Recognition of Barkandji Water Rights in Australian Settler-Colonial Water Regimes. *Resources, 7*(1), 16.

Heaslip, R. (2008). Monitoring Salmon Aquaculture Waste: The Contribution of First Nations' Rights, Knowledge, and Practices in British Columbia, Canada. *Marine Policy, 32*(6), 988–996.

Heinämäki, L., Herrmann, T., & Green, C. (2017). Towards Sámi Self-Determination Over Their Cultural Heritage: The UNESCO World Heritage Site of Laponia in Northern Sweden. In A. Xanthaki, S. Valkonen, L. Heinämäki, & P. Nuorgam (Eds.), *Indigenous Peoples' Cultural Heritage* (pp. 78–103). Brill.

Hikuroa, D., Slade, A., & Gravley, D. (2011). Implementing Māori Indigenous Knowledge (mātauranga) in a Scientific Paradigm: Restoring the mauri to Te Kete Poutama. *Mai Review, 3*, 9.

Hopkins, D., Joly, T. L., Sykes, H., Waniandy, A., Grant, J., Gallagher, L., et al. (2019). "Learning Together": Braiding Indigenous and Western Knowledge Systems to Understand Freshwater Mussel Health in the Lower Athabasca Region of Alberta, Canada. *Journal of Ethnobiology, 39*(2), 315–336.

Howitt, R., & Suchet-Pearson, S. (2006). Rethinking the Building Blocks: Ontological Pluralism and the Idea of 'Management'. *Geografiska Annaler: Series B, Human Geography, 88*(3), 323–335.

Iwi Rep 1. (2017, September 29). Interview with Iwi Representative 1.

Iwi Rep 2. (2020, February 13). Interview with Iwi Representative 2.

Iwi Rep 4. (2020, February 14). Interview with Iwi Representative 4.

Iwi Rep 5. (2019, March 25). Interview with Interview Iwi Representative 5.

Iwi Rep 7. (2019a, May 16). Interview with Iwi Representative 7.

Iwi Rep 7. (2019b, June 13). Interview with Iwi Representative 7.

Jackson, S., & Barber, M. (2016). Historical and Contemporary Waterscapes of North Australia: Indigenous Attitudes to Dams and Water Diversions. *Water History, 8*(4), 385–404.

Jacobson, C., Manseau, M., Mouland, G., Brown, A., Nakashuk, A., Etooangat, B., et al. (2016). Co-Operative Management of Auyuittuq National Park: Moving Towards Greater Emphasis and Recognition of Indigenous Aspirations for the Management of Their Lands. In *Indigenous Peoples' Governance of Land and Protected Territories in the Arctic* (pp. 3–21). New York: Springer.

Jones, C. (2016). *New Treaty, New Tradition: Reconciling New Zealand and Maori Law*. Toronto: University of British Columbia. Retrieved June 12, 2019, from https://books.google.co.nz/books?hl=en&lr=&id=DSLCDAAAQBAJ&oi=fnd&pg=PT5&dq=Jones+2016+New+Treaty&ots=09dWY_fMZ0&sig=vEDmsW4b2_KETAJpQfJWLcrDjRg#v=onepage&q=Jones%202016%20New%20Treaty&f=false.

Knight, C. (2016). *New Zealand's Rivers: An Environmental History*. Christchurch: Canterbury University Press.

Kowhai Consulting Ltd. (2007). *He Mahere Taiao: The Maniapoto Iwi Environmental Management Plan for Maniapoto Maori Trust Board*. Otorohanga: Kowhai Consulting Ltd. Retrieved June 12, 2019, from www.maniapoto.iwi.nz.

Larsen, R. K. (2018). Impact Assessment and Indigenous Self-Determination: A Scalar Framework of Participation Options. *Impact Assessment and Project Appraisal, 36*(3), 208–219.

Lavau, S. (2013). Going with the Flow: Sustainable Water Management as Ontological Cleaving. *Environment and Planning D: Society and Space, 31*(3), 416–433.

Local Government Rep 1. (2018, October 4). Interview with Local Government Representative 1.

Local Government Rep 2. (2019, March 25). Interview with Local Government Representative 2.

Lowitt, K., Levkoe, C. Z., Lauzon, R., Ryan, K., & Sayers, C. D. (2019). 7 Indigenous Self-Determination and Food Sovereignty Through Fisheries Governance in the Great Lakes Region. *Civil Society and Social Movements in Food System Governance*, 145.

Lowry, A., & Simon-Kumar, R. (2017). The Paradoxes of Māori-State Inclusion: The Case Study of the Ōhiwa Harbour Strategy. *Political Science, 69*(3), 195–213. https://doi.org/10.1080/00323187.2017.1383855.

Maniapoto Maori Trust Board. (2016). *Ko Tā Maniapoto Mahere Taiao – Maniapoto Environmental Management Plan*. Maniapoto Māori Trust Board.

Maniapoto Maori Trust Board. (2017). *Review of the Deed in Relation to the Co-Governance and Co-Management of the Waipa River*. Maniapoto Maori Trust Board.

Maniapoto Māori Trust Board, Otorohanga District Council, Waikato District Council, Waikato Regional Council, Waipa District Council, & Waitomo District Council. (2013). *Joint Management Agreement*. Hamilton.

Māori Business Owner 1. (2019, August 29). Māori Business Owner 1.

Marsh, H., Grayson, J., Grech, A., Hagihara, R., & Sobtzick, S. (2015). Re-Evaluation of the Sustainability of a Marine Mammal Harvest by Indigenous People Using Several Lines of Evidence. *Biological Conservation, 192*, 324–330.

McGregor, D. (2014). Traditional Knowledge and Water Governance: The Ethic of Responsibility. *AlterNative: An International Journal of Indigenous Peoples, 10*(5), 493–507.

McLean, J. (2014). Still Colonising the Ord River, Northern Australia: A Postcolonial Geography of the Spaces Between Indigenous People's and Settlers' Interests. *The Geographical Journal, 180*(3), 198–210.

Ministry for the Environment, & Te Puni Kōkiri. (2017). *Review of the Waikato and Waipa Rivers Arrangements 2016–17. Crown Report for Collective Review*.

Muir, B. R., & Booth, A. L. (2012). An Environmental Justice Analysis of Caribou Recovery Planning, Protection of an Indigenous Culture, and Coal Mining Development in Northeast British Columbia, Canada. *Environment, Development and Sustainability, 14*(4), 455–476.

Muru-Lanning, M. (2012). The Key Actors of Waikato River Co-Governance: Situational Analysis at Work. *AlterNative: An International Journal of Indigenous Peoples, 8*(2), 128–136.

Nadasdy, P. (2007). The Gift in the Animal: The Ontology of Hunting and Human–Animal Sociality. *American Ethnologist, 34*(1), 25–43.

Natcher, D. C., & Davis, S. (2007). Rethinking Devolution: Challenges for Aboriginal Resource Management in the Yukon Territory. *Society & Natural Resources, 20*(3), 271–279.

Natcher, D. C., Davis, S., & Hickey, C. G. (2005). Co-Management: Managing Relationships, Not Resources. *Human Organization, 64*(3), 240–250.

National Institute of Water and Atmospheric Research Ltd. (2010). *Waikato River Independent Scoping Study*. NIWA Project: MFE10201 No. HAM2010-032. Hamilton: National Institute of Water & Atmospheric Research Ltd.

New Zealand Parliament. Water and Soil Conservation Act, Pub. L. No. 135. (1967). Retrieved June 16, 2020, from http://www.nzlii.org/nz/legis/hist_act/wasca19671967n135320/.

New Zealand Parliament. Resource Management Act. (1991). Retrieved June 16, 2020, from http://www.legislation.govt.nz/act/public/1991/0069/223.0/DLM230265.html.

New Zealand Parliament. Waikato-Tainui Raupatu Claims (Waikato River) Settlement Act (2010a).

New Zealand Parliament. Ngāti Tuwharetoa, Raukawa, and Te Arawa River Iwi Waikato River Act (2010b).

New Zealand Parliament. Ngā Wai o Maniapoto (Waipā River) Act (2012). Retrieved April 19, 2020, from http://www.legislation.govt.nz/act/public/2012/0029/latest/DLM3335204.html.

New Zealand Parliament. Te Awa Tupua (Whanganui River Claims Settlement) Act. (2017). Retrieved April 19, 2020, from http://www.legislation.govt.nz/act/public/2017/0007/latest/whole.html.

Ngā Wai o Maniapoto (Waipā River) Act. (2012).

NIWA. (2014). *Maniapoto Priorities for the Restoration of the Waipā River Catchment*. Wellington: NIWA.

Nuttall, M. (2018). Self-Determination and Indigenous Governance in the Arctic. In *The Routledge Handbook of the Polar Regions* (pp. 93–106). New York: Routledge.

Parsons, M., Nalau, J., Fisher, K., & Brown, C. (2019). Disrupting Path Dependency: Making Room for Indigenous Knowledge in River Management. *Global Environmental Change, 56*, 95–113.

Pinkerton, E. (2018). Legitimacy and Effectiveness Through Fisheries Co-Management. In *The Future of Ocean Governance and Capacity Development* (pp. 333–337). Boston: Brill Nijhoff.

Poelina, A., Taylor, K. S., & Perdrisat, I. (2019). Martuwarra Fitzroy River Council: An Indigenous Cultural Approach to Collaborative Water

Governance. *Australasian Journal of Environmental Management, 26*(3), 236–254.

von der Porten, S., de Loë, R., & Plummer, R. (2015). Collaborative Environmental Governance and Indigenous Peoples: Recommendations for Practice. *Environmental Practice, 17*(2), 134–144.

Rangitāiki River Forum. (2015). *Te Ara Whanui o Rangitāiki – Pathways of the Rangitaiki*. Bay of Plenty Regional Council.

Rifkin, M. (2017). *Beyond Settler Time: Temporal Sovereignty and Indigenous Self-Determination*. Duke University Press.

Ross, A., Sherman, K. P., Snodgrass, J. G., Delcore, H. D., & Sherman, R. (2011). *Indigenous Peoples and the Collaborative Stewardship of Nature: Knowledge Binds and Institutional Conflicts*. New York: Routledge.

Ruru, J. (2013). Indigenous Restitution in Settling Water Claims: The Developing Cultural and Commercial Redress Opportunities in Aotearoa, New Zealand. *Pacific Rim Law & Policy Journal, 22*, 311.

Salmond, A. (2017). *Tears of Rangi: Experiments Across Worlds*. Auckland: Auckland University Press.

Schlosberg, D., & Carruthers, D. (2010). Indigenous Struggles, Environmental Justice, and Community Capabilities. *Global Environmental Politics, 10*(4), 12–35.

Te Aho, L. (2015). *The Waikato River Settlement: Exploring a Model for Co-Management and Protection of Natural and Cultural Resources*. Ka Hula Ao Center for Excellence in Native Hawaiian Law, Richardson School of Law. Retrieved January 6, 2019, from https://researchcommons.waikato.ac.nz/handle/10289/10414.

Thompson-Fawcett, M., Ruru, J., & Tipa, G. (2017). Indigenous Resource Management Plans: Transporting Non-Indigenous People into the Indigenous World. *Planning Practice & Research, 32*(3), 259–273.

Tsatsaros, J. H., Wellman, J. L., Bohnet, I. C., Brodie, J. E., & Valentine, P. (2018). Indigenous Water Governance in Australia: Comparisons with the United States and Canada. *Water, 10*(11), 1639.

Waikato River Authority. (2011). *Vision and Strategy for the Waikato River*.

Waikato Regional Council. (2012). *Waipa Zone Management Plan*. Hamilton: Waikato Regional Council.

Waikato Regional Council. (2014). *Waipā Catchment Plan*. Hamilton: Waikato Regional Council.

Walker, R. (1990). *Ka whawhai tonu matou: Struggle Without End* (Vol. 220). Auckland: Penguin.

Watene-Rawiri, E., Kukutai, J., & Maniapoto Māori Trust Board. (2015). *He Mahere Ika: Maniapoto Upper Waipā River Fisheries Plan 2015*.

Weinstein, M. S. (1999). Pieces of the Puzzle: Solutions for Community-Based Fisheries Management from Native Canadians, Japanese Cooperatives, and Common Property Researchers. *Georgetown International Environmental Law Review, 12*, 375.

Weir, J. K. (2009). *Murray River Country: An Ecological Dialogue with Traditional Owners*. Aboriginal Studies Press.

White, G. (2020). *Indigenous Empowerment Through Co-Management: Land Claims Boards, Wildlife Management, and Environmental Regulation*. Vancouver: UBC Press.

Zurba, M., Ross, H., Izurieta, A., Rist, P., Bock, E., & Berkes, F. (2012). Building Co-Management as a Process: Problem Solving Through Partnerships in Aboriginal Country, Australia. *Environmental Management, 49*(6), 1130–1142.

9

Decolonising River Restoration: Restoration as Acts of Healing and Expression of Rangatiratanga

When one thinks through how a restored river is or could be produced within the context of historical and ongoing entanglements of Indigenous peoples and settler-colonial societies, it is critical to consider what restoration is and how it is enacted within particular environmental practices. In this chapter, we look specifically at river restoration and what constitutes a restored river (or landscape and waterscapes) in the context of the Waipā River. Over the last three decades, an ever-expanding and diverse body of scholarship on river restoration has emerged; including research from the fields of historical ecology (Beller et al. 2016; Bhatt et al. 2016; Kurashima et al. 2017; Stein et al. 2010), geomorphology (Abernethy and Rutherfurd 1998; Arnaud et al. 2015; Jacobson et al. 2011), engineering (Palmer et al. 2014), environmental management (Bhatt et al. 2016; Morandi et al. 2014; Waltham et al. 2014). The emphasis remains placed on the need to address the degradation of places, ecosystems, or keystone species through targetted restoration efforts. Across the diversity of disciplines, despite the critiques of the climax approach, the "value of a historical

The original version of this chapter was revised: The incorrect information with reference to The Waikato River Authority text has now been revised and updated. The correction to this chapter is available at https://doi.org/10.1007/978-3-030-61071-5_12

M. Parsons et al., *Decolonising Blue Spaces in the Anthropocene*, Palgrave Studies in Natural Resource Management, https://doi.org/10.1007/978-3-030-61071-5_9

perspective" continues to be noted for setting "the goals, strategies and targets" for river restoration (Beller et al. 2020). Ecological restoration is, thus, broadly framed as the "need to protect and restore both habitat remnants and modified ecosystems in management" with reference to "the value of ecosystems as cultural landscapes" (Beller et al. 2020). Yet, critical questions still need to be raised about whose "historical perspective" is being taken into account when the restoration goals, targets, and approaches are being set, and whose cultural landscapes and waterscapes are valued when decisions are being made about what is being restored and/or protected and how the restoration practices are being enacted.

Given that river restoration is predicated on acknowledgement of past mistakes (that human actions resulted in highly degraded freshwater systems), we argue that it is important to spend time critically analysing the ideas, approaches, and practices that underpin restoration projects. The intellectual underpinnings of restoration are explored in-depth by other scholars; however, the limitations of these past studies is that they only consider Western ontological and epistemological frameworks, narrating restoration projects through the experiences of Euro-Western environmentalists, and failing to destabilise nature-culture binaries (Hall 2005; Higgs 2003; Higgs et al. 2014; Hourdequin and Havlick 2016). More recent scholarship does demonstrate that ecological restoration is not the sole domain of scientific knowledge and expertise and that Indigenous and Local Knowledges can contribute towards restoration practices (Crow et al. 2018, 2020; Ens et al. 2012; Fox et al. 2017; Ratana et al. 2019; Reyes-García et al. 2019). Furthermore, the social dimensions of landscapes and waterscapes are increasingly acknowledged, and the ways in which different knowledge systems and values influence how different groups of people define restoration priorities and facilitate restoration practices (Failing et al. 2013; Fernández-Manjarrés et al. 2018; Kibler et al. 2018; Paterson-Shallard et al. 2020; Stein et al. 2010).

River restoration (both in theory and practice), however, remains largely located within the realm of the hegemonic knowledge systems, socio-cultural values, and human-environmental relations of Euro-Western cultures. Restoration practitioners (focused on restoring the functioning of ecosystems) act in the "silent interests of ecosystems" (Hall 2005, p. 11) and in doing so, rearticulate the long-standing command-and-control paradigm especially prevalent in freshwater management.

We argue that ecological restoration (specifically river restoration) is not a neutral (scientific, linear, universal) process, but one that is laden with power, authority, and ontological politics. Every time an ecological restoration project commences, a particular type of 'nature' is being expressed and enacted. Restoration practitioners, therefore, are making decisions (both conscious and unconscious) about what they want a specific place and/or ecosystem to become in the future; restoration, therefore, involves the act of envisioning or transformative changes to landscape and waterscape motivated by the desires to address environmental degradation in reference to historical reconstructions of the past (which are particular imaginative geographies that are embedded within Euro-Western knowledges and ongoing colonialism).

The Emergence of Ecological Restoration as a Field of Study and Practice

A unifying theme within early restoration scholarship was to "re-create historical associations" (Jordan and Lubick 2011, p. 2). Put simply, initial ecological restoration research and restoration projects were directed at trying to return an ecosystem (a river, a forest, an island) to a prior state (often historic and/or pre-human) (Beller et al. 2016; Humphries and Winemiller 2009; Palmer et al. 2005; Palmer et al. 2016). Such early ecological restoration work was underpinned by climax theory, coined in 1916 by plant ecologist Frederic Clements, which proposed that in the absence of external shocks or disturbances, ecosystems transition through various states till they each a stable condition (climax) (Clements 1916, 1936). In the absence of disruptions, according to the theory, ecosystems would reverse to a steady state. Within restoration activities, the theory of climax was evident in the use of 'reference ecosystem' (or a historic baseline) for which practitioners direct their efforts to return an ecosystem to a former (reference) state of being (Jordan 2003; Jordan and Lubick 2011). Later ecologists disproved climax theory on multiple occasions. Non-equilibrium ecological theories increasingly gained popularity (informed by theories of self-organisation and chaos). Such theories complicated ecological restoration by demonstrating that ecosystems are highly complex, dynamic, and uncertain, particularly in response to changes (Higgs et al. 2014; Palmer et al. 2016; Perring et al. 2015; von Wehrden et al. 2012). The singular aim of restoration ecology (as well as conservation biology and invasion biology) to return ecosystems to a historical state (often a highly idealised one) was no longer a suitable guide for ecological restoration efforts.

Recent definitions of ecological restoration generally avoid a complete commitment to entirely recreating past (pre-human) ecosystems and instead emphasise addressing environmental degradation. The Society for Ecological Restoration International, a network founded in 1987 for restoration practitioners, for instance, defines ecological restoration as:

> the process of assisting the recovery of an ecosystem that has been degraded, damaged, or destroyed … [E]cological restoration seeks to 'assist recovery' of a natural or semi-natural ecosystem rather than impose a new direction or form upon it. That is, the activity of restoration places an ecosystem on a trajectory of recovery so that it can persist and its species can adapt and evolve. (Tipa and Nelson 2017; Wehi and Lord 2017)

Other authors define ecological restoration as a "sequence of steps progressing from ascertaining the natural and anthropogenic disturbance regimes, identifying and implementing restorative measures, and monitoring key indicators to determine trajectories of the responses and the outcomes of the restoration project" (Lake et al. 2017, p. 509). Since ecosystems are dynamic entities that flux and change over time, restoration projects thus are increasingly framed as a process of transition "a continuous coming into a being of an ecosystem" (Higgs 2003, pp. 110–111). In most restoration projects, environmental scientists and management practitioners work to restore a degraded ecosystem to a historic "baseline" or criteria of prior conditions. The baseline is often a selected reference point that is defined (by scientists) as the imagined peak of ecosystem diversity and functionality, which in settler societies is often situated before European colonisation, a tendency that is critiqued in our analysis. Scholars identify three key principles for efficient and effective ecological restoration practices. Firstly, efficient restoration establishes and maintains the values of ecosystems. Secondly, effective restoration maximises beneficial outcomes at the same time as costs (in terms of time, effort, and resources) are minimised. And thirdly, successful restoration involves practitioners engaging with stakeholders and partners in a way that promotes participation and enhances people's experience of ecosystems (McDonald et al. 2018; Weber et al. 2018).

A significant amount of research and on-the-ground restoration projects are focused on riparian zones (the interface between land and water that encompasses riverbanks and channels). The majority of river restoration practices are directed at the removal of invasive plants, replanting of native vegetation, and fencing off waterways from non-native fauna (to decrease grazing pressures and effluent). Within river channels, there is also a focus on restoring habitat structure, improving fish migration pathways, and augmenting refuges for fauna (particularly in the context of drought). However, restoration scholars argue that the restoration of entire river catchments and connectivity between landscapes and waterscapes remain significant challenges. Indeed, despite small and large-scale efforts to restore freshwater systems around the globe over the last two decades (particularly in the settler-states of Australia, Aotearoa, Canada and the United States), many rivers, lakes, streams, and wetlands remain in a state of worsening environmental degradation (Davenport et al. 2010; Fernández-Manjarrés et al. 2018; Hamilton et al. 2018; Marshall et al. 2017; Paterson-Shallard et al. 2020; Saulters 2014).

Critiques of Ecological Restoration

When ecological restoration as first positioned as a new paradigm that would address environmental degradation, critiques emerged from Western philosophers. For instance in Eric Katz's *The Big Lie*, first published in 1992, restoration is critiqued for promoting human beings domination and control over nature as well as the production of restored ecosystems that were social artefacts not 'nature'. Likewise, Australian philosopher Robert Elliot claimed the restoration efforts were merely deceptions that 'faked' nature (Elliot 1997). Elliot's central thesis is that restored landscapes are technological productions presented as of equal value to so-called 'wild nature'. Both Katz and Elliot were concerned that restoration efforts subverted the goals of environmental protection in that it would be used to encourage environmental offsetting; developments could be authorised on the basis that damaged environments could be simply reproduced elsewhere.

The aim of 'historical fidelity'—returning to an 'original' or pre-disturbance—served to legitimise restoration within the natural sciences and safeguarded it from philosophical critiques (Beller et al. 2016; Stein et al. 2010). Although a commendable purpose, and maintaining their commitment to local historical place-based associations, restoration practitioners rejected charges of command-and-control or technical approach that disputed environmental preservationist thinking. Philosopher Andrew Light adopted a pragmatic viewpoint, drawing a distinction between malicious and benevolent restoration based on the intention and claims behind ecological restoration projects (Light 1994). Light supported restoration and argued that it could play a significant role in cultivating ecological citizenship (what other scholars term an environmental ethic) that posits that ethical responsibilities for nature are part of being a good citizen (Bauman and O'Brien 2019; Katz and Light 2013; MacGregor 2014). In his 2002 book, *Nature by Design*, Canadian anthropologist and restoration ecologist Eric Higgs outlined four essential qualities that enabled restoration projects to be "morally good". Firstly, a project must restore ecological integrity. Secondly, restoration must be underpinned by historical knowledge. Thirdly, a project needed to include a component of "wild design" that provided space for "nature and culture … to go wild". And fourthly, a restoration project must practise "focal restoration" that can "rebuild our concern with things that matter" (Higgs 2003, pp. 226, 285). Ecological restoration, thus, was conceptualised to be both an ecological and socially important practice that can alter human and more-than-human relationships.

The main theoretical criticisms of ecological restoration discourses were initially limited in that most remain largely within the confines of Western ontological and epistemological frameworks that rearticulate the nature-culture dichotomy. In settler societies like Australia, the United States, and Aotearoa restoration projects often continue to be premised on settler-relationships with land, water, and coasts (Connelly and Knuth 2002; Davenport et al. 2010; Moran 2010; Peters et al. 2015; Schuelke

2014). Indeed, as Callicott aptly notes, the "simple and easy understanding of the appropriate norm for ecological restoration is premised on two myths that then prevailed—the wilderness myth and the ecological-equilibrium myth" (Callicott 2002, p. 418).

In Chap. 3 of this book, we describe how the colonial discourse of an 'untamed' and 'uncultivated' nature was used to justify settler colonialism in Aotearoa, including the invasion of the Waikato and the dispossession of Māori. In different settler societies, variations of the 'wilderness myth' (terra nullius, Australia's Outback, Aotearoa's wastelands, and Canada's empty Arctic North) were used to downplay, deny or erase Indigenous peoples from their landscapes, waterscapes and seascapes (Baldwin et al. 2011; Cameron 2015; Clover and Historical Society of the Hauraki Plains 2007; Fitzmaurice 2007; Giblett 2009; Pluymers 2011; Veracini 2010). In the Australian context, for instance, the year 1788 (the start of British colonisation in Australia) is frequently employed as the baseline for restoration work with limited recognition given to Australian Indigenous peoples' thousands of years of occupation and complex ways of managing environments (Beilin and West 2016, p. 193). When environmental scientists, particularly those ascribing to the notion of a timeless stable climax ecosystem, did recognise Indigenous presence it was to criticise Indigenous peoples' for disrupting 'pristine' nature with their use of local ecosystems, use of fire, and loss of biodiversity (Flannery 2002; Head 2012).

In Australia, where the history of Aboriginal peoples' occupation stretches 60,000 years, the idea of a "balanced" and "pre-human" state is completely out of check with the socio-cultural and environmental histories (Barber and Jackson 2015; Bardsley and Wiseman 2016; Bashford 2013; Beilin and West 2016, p. 193; Langton 2006; Winter 2019). Likewise, in Aotearoa, as we and countless other scholars demonstrate, Europeans did not arrive to an empty land ('terra nullius') but to instead a country filled with waterscapes and landscapes that generations of Māori iwi (tribe), hapū (subtribe), and whānau (family) carefully created, maintained and cared for (Anderson 2003; Anderson 2002; Boswijk et al. 2005; Park 2018; Stokes 2000). Moreover, the conceptualisation of 'pristine nature', wherein European settlers arrived to settle an unoccupied and a-historical place, is forcefully challenged by Indigenous scholars (ourselves included) for its Eurocentric framing of nature.

Indigenous ontologies, as well highlight throughout this book, connect human and more-than-human communities in ongoing reciprocal (often kinship-based) relationships (de Leeuw and Hunt 2018; Todd 2014; Watts 2013; Winter 2018). From Māori worldviews, culture and nature are inextricably interwoven together to the extent that one cannot exist without the other (Clément 2017; Salmond 2014). These ontological and epistemological differences, however, feature into accounts of ecological restoration theorising or practices. Yet as we discuss throughout this book, a river for Māori (and many other Indigenous peoples) is not something that exists as a mere entity that human actors inscribe meaning on (altering and using, commanding-and-controlling, degrading and restoring). From Māori perspectives, rivers (like other geoentities) are more-than-human actors with agency, power, and a life force (as well as being the kin to particular iwi and hapū). Rivers are, therefore, from a Māori ontological viewpoint, materially and metaphysically co-constituted; simultaneously affecting and affected by others (human and more-than-human). Thus, since both these two myths (of supposedly empty undeveloped 'wilderness'/'wastelands' as well as ecosystem-equilibrium) are now largely rejected, the ethical and theoretical foundations of ecological restoration in settler societies are being increasingly uncertain and complex (Pearce 2019).

The narrative of historicising and romanticising Indigenous cultures, in wherein they are confined to "pre-history" or "traditional", remains prevalent within ecological restoration. The narrative overlooks not only Indigenous histories of land and water management, their economies, and political authority but also attempts to rob Indigenous peoples of their futures and capacities to adapt to changing circumstances (Head 2012). The idea of a historical baseline or year zero (1788 in Australia, 1840 in Aotearoa) reinforces the notion that Indigenous peoples, communities, and individuals are confined to pre-history (inherently traditional, static and unchanging), which does not allow for dynamic and fluid understandings of cultures (Bennett and van Sittert 2019; Head 2012; Head and Muir 2004; Lidström et al. 2016).

Some of the emergent ecological restoration research and projects are examining the interconnectedness of ecosystems and people, with scholars even acknowledging the presence of the plurality of ontologies and epistemologies. For instance, the idea of "eco-cultural restoration" is promoted by Dennis Martinez, founder of the Indigenous Peoples' Restoration Network. Likewise, Māori ecologist Priscilla Wehi uses the term "bi-cultural restoration" to take into account Aotearoa's Treaty partnership between two cultures (Pākehā and Māori) and the critical importance of including "cultural practices" in restoration efforts in Aotearoa (Senos et al. 2006; Wehi et al. 2019; Wehi and Lord 2017). Such work makes a significant contribution to expanding restoration discourse, but remains at the margins of the dominant restoration praxis, and, where included, does so in projects that involve a focus on Indigenous knowledge, cultural practices, or livelihoods (Reyes-García et al. 2019; Tipa and Nelson 2017; Wehi and Lord 2017). Often this research seeks to use Indigenous Knowledge to augment gaps in Western scientific knowledge (Reyes-García et al. 2019) and to identify significant cultural sites or biota that Indigenous communities want restored (Ens et al. 2012; Long et al. 2017; White et al. 2011). Indigenous knowledge (positioned as complementary to western science) is thus often framed as a tool that can be used by restoration ecologists and practitioners to better reference ecosystems (where historical data is unavailable) and ensure that Indigenous communities support restoration projects (Uprety et al. 2012). The limited spaces afforded to Indigenous peoples within restoration scholarship (confined to the realms of Indigenous ecological knowledge, cultural practices, and livelihoods) means that the field remains underpinned (seemingly unwittingly) to colonial structures, knowledges, values, and practices; and Indigenous ways of knowing and being in the world remain tokenistically referenced. Yet, a wealth of recent work from Māori scientists and social scientists demonstrates the tremendous capacity to expand socio-cultural, political, economic, and ecological thinking and practices in restoration beyond those of the West to encompass Indigenous and other peoples' ontologies and epistemologies (Carter 2019; Forster 2012; Harmsworth and Roskruge 2014; Hikuroa et al. 2011; Panelli and Tipa 2007; Tipa and Nelson 2017; Wehi et al. 2019).

Co-management and Restoration Planning

Restoration of the Waipā River involves actions and participation by a range of actors operating under a variety of mandates, seeking to achieve a range of objectives and with differing funding arrangements, including iwi and hapū, the National Wetlands Trust, and community groups. We focus, in particular, on restoration efforts of iwi arising from opportunities provided by the co-governance framework and co-management arrangements. As discussed in Chap. 5, the passing of the Waipā River Act enabled access by applicants to the Waikato Clean-up River Trust (WRCuT) for restoration projects targeting the Waipā River and tributaries. In 2017, as part of its co-management arrangements (as detailed in Chaps. 5 and 6), Ngāti Maniapoto produced a report that detailed its river management and restoration priorities for the Waipā River and its tributaries. *The Maniapoto Priorities Report* articulates the iwi's aspirations, values and issues in relation to the Waipā and links with the Maniapoto Māori Trust Board's (MMTB) other plans as well as policies developed by the Waikato Regional Council's (WRC) such as the *Waipā Catchment Plan* and the *Waipā Zone Management Plan* (Maniapoto Maori Trust Board 2017; NIWA 2014).

The *Waipā Catchment Plan* was developed by WRC in conjunction with Ngāti Maniapoto through a collaborative process to guide the implementation of integrated catchment management activities within the Waipā River. The *Waipā Catchment Plan* identifies 100 actions points, ranging from large erosion and sediment control projects to biodiversity, to looking after peat lakes, to working with Māori landowners (WRC 2014). These action points, though not enforceable, provide the focus for WRC to undertake restoration and management within the Waipā catchment and progress is measured against attainment. Overarching the *Catchment Plan* is the *Waipā Zone Management Plan*, which sets out the high-level strategy and objectives to guide management activities 'to revitalise the waters of the Waipā River and its tributaries by 2050' (WRC 2012). The *Maniapoto Priorities Report* formed the basis for the Maniapoto's Waipā restoration projects (which were included within the Waipā River Restoration Strategy).

Unlike many restoration projects, which often remove people (specifically Indigenous peoples) from the narrative, the various restoration works being undertaken by Ngāti Maniapoto are seeking to restore their cultural waterscapes and landscapes. Projects are led and managed by various layers of Maniapoto iwi/hapū/whānau, including formal iwi institutions (MMTB and Nehenehenui RMC) and informal institutions (marae-based or whānau) often in collaboration with external institutions (such as the National Institute of Water and Atmospheric Research hereafter NIWA and the Waikato River Authority hereafter WRA). Restoration projects report struggling to find sufficient and ongoing funding to support their restoration efforts. The lack of resourcing is partly a result of the limited funds available through the MMTB compared to other iwi in the Waikato region; Ngāti Maniapoto (as we outlined in Chaps. 7 and 8) are still in the process of negotiating with the central government (the Crown) for a legal and financial reparations package (known as a Treaty settlement) for historical injustices committed against the iwi by the Crown. Also, the lack of funding available to support restoration efforts is a consequence of the failures of the co-governing institution (the Waikato River Authority hereafter WRA) and its restoration funding body (the Waikato River Cleanup Trust hereafter WRCcT). The WRA was established, as we discuss in Chap. 5, as a consequence of Treaty settlements reached between five River iwi (Ngāti Maniapoto, Raukawa, Waikato Tainui, Te Arawa, and Tūwharetoa) and the Crown, and involved the Crown agreeing to share co-governance and co-management arrangements with iwi over the Waikato and Waipā rivers. The WRA administers the WRCuT which provides funding for restoration works in catchments of both rivers. However, the ways in which the WRCuT operates (specifically its process of awarding funding to restoration projects based on a yearly contestable funding round to any group doing restoration work) is heavily critiqued by our interviewees, including those who whakapapa to Maniapoto. In many respects, Maniapoto interviewees demonstrated a lack of understanding of the details of the settlement co-governance arrangements (including how WRCuT funding is awarded and to whom), and the implications of the settlement and co-governance for iwi members and those operating at the flax roots level. For these inter- viewees, their primary concerns (and criticisms) related to fairness and equity. For example, where funding is awarded to councils or industry groups, our interview participants questioned whether it was fair since these groups are generally better and more consistently resourced than iwi, hapū and whānau. For interviewees, the need to build capacity and capabilities within Maniapoto and to overcome the barriers to partici- pating in all aspects of river restoration and management (including the competing demands on people's time), compounded the desire for fair- ness and equity, and the need to take steps to ensure sustainability into the future.

Constraints on Restoration Efforts

Many interviewees spoke mahi (work) on the ground was being under- valued by their own iwi institutions (MMTB), the co-governance institu- tion (Waikato River Authority), as well as local councils. At the moment, the majority of people who are undertaking restoration projects "are doing it voluntarily on top of everything else" in their lives. In contrast, those employed in all the various institutions involved in freshwater gov- ernance and management (for MMTB, WRA, WRC, ODC) are all get- ting paid for their work (Māori Business Owner 1 2019). It was so difficult for hapū- and whānau-led restoration efforts, one iwi consultant reported, to access funding through the WRA managed WRCuT that whānau were overwhelmed by the process (Māori Business Owner 1 2019). Indeed, many Maniapoto whānau want to implement their own restoration projects but lack of "access to money" was the "big barrier" to transforming their aspirations for restoring their awa into actions. However, both scientists and iwi members expressed strong support resources to be directed to support the work and "aspirations … whatever that would be" of (Māori Business Owner 1 2019; Scientist 2 2019, p. 2). "Getting a whole heap of … submission writers" who could write fund- ing applications to the WRA, regional council, and other funding bodies "would be amazing" because at the moment whānau are "just like no, I can't even begin to grasp the concept of what that is and what that's going to take" to write a submission (Māori Business Owner 1 2019). However, if a scientist or researcher assisted them to fill out the application forms and make a submission, then they would be able to translate their aspirations for restoring their awa into on-the-ground restoration works that would benefit their whānau and future generations (Māori Business Owner 1 2019).

Another constraint to river restoration work was the lack of people able to undertake the work (which was largely unpaid). For instance, the utilisation of the Stream Health Monitoring Assessment kit (SHMAK) by hapū started in 2016; however, MMTB struggled to recruit sufficient numbers of kaitiaki who would be willing to undertake testing for them within the middle and upper parts of the Waipā River Catchment. Engagement with the project is being constrained by the fact that the majority of Maniapoto (around 80 per cent) live outside of the rohe of Maniapoto. This means that only "20 ... per cent that are left in Maniapoto ... are already highly involved in their marae, their kura [school], the kōhanga [preschool]—everything else ... that needs to be done" (Māori Business Owner 1 2019). So, for many Māori, the duties and work involved in being a "kaitiaki" (which is mostly voluntary and unpaid labour) "just seems like another thing on top of all" their own responsibilities (to their whānau, marae, hapū). The "hardest part is getting" Maniapoto (including those who live within and outside the rohe) back to their awa (Māori Business Owner 1 2019).

> Kanohi kit e kanohi [face-to-face meetings] is so crucial with Māori—like knowing who you are and what you are on about can only happen when you are standing in front of these people talking to them ... Once they're there and they realise how important this is, how amazing this is, once they've been in the water and reconnected every single person that's come has pretty much come back. It's really about getting them there in the first place to say that this is worthy of sacrificing some family time maybe or weekend time or whatever. (Māori Business Owner 1 2019)

Iwi representatives also regularly spoke of the ontological and epistemological differences between Māori and Pākehā ways of thinking and how it creates difficulties in environmental governance and management more broadly as well as in the context of restoration; which we discussed previously in Chaps. 3 and 4 in terms of two cultures (Pākehā and Māori) "talking past each other" due to differing worldviews (Metge and Kinloch 2014; Winter 2018). As Iwi Rep 5 states, Ngāti Maniapoto and other Māori people possess a "whakapapa directly to those tūpuna that were there" (who signed the Treaty, experienced colonial violence, dispossession and marginalisation), whereas the Crown (embodied by the officials of the New Zealand Government, its agencies, and local councils) "does not have that same sort of connection". Governments change, as do government officials, and the people who represent the Crown today do not represent their own tūpuna; (indeed, accordingly to Māori whakapapa it is "the Queen and her family" who should be directly representing their ancestors in Treaty Settlements and the co-governance of the Waipā). Thus, Iwi Rep 5 argues, central government (Crown) and local government officials (who adopt a Pākehā/Western worldview) do not carry with them the same sense of "intergenerational responsibility" that Māori do (Iwi Rep 5 2019). Such a worldview translates into freshwater governance, management, and restoration approaches continuing to employ short-term time planning horizons and favouring current development (benefits) over future environmental harms (costs).

Getting the Values Right

One iwi representative, who affiliates to both Waikato-Tainui and Ngāti Maniapoto iwi, spoke of the importance of ensuring that iwi values (encompasses socio-cultural, spiritual, political, economic and ecological dimensions) are translated from a conceptual framework into real-world environmental management plans, as well as on-the-ground management and restoration practices (Iwi Rep 8 2019). The post-Treaty settlement period is a "mediated, negotiated space" in which iwi, involved in the co-governance and co-management arrangements for the Waipā and Waikato Rivers, need to consider how their values can provide "the groundwork … [on which] the house [is built] up" (Iwi Rep 8 2019). The importance of thinking about what values matter and how they can be incorporated into management plans and restoration projects relates to Māori cosmology and mātauranga Māori:

> If I was to draw a picture, I'd be drawing a picture saying this is Papa-tū-ā-nuku [Earth Mother]. This is Ranginui [Sky Father]. How do we fit in here? How do we fit in whatever those values are, how do we make it fit in there? If it's about caring for everything that exists between these two [Papa-tū-ā-nuku and Ranginui] and including them, you can't lose [when you practice kaitiakitanga]. ... Because you can't do it the other way around. You can't take a Pākehā process and put them [Papa-tū-ā-nuku and Ranginui into it]—you can't... it wouldn't work. It ... [is] our mana. Our sovereignty and our space. [The] lens ... says [that] Papa-tū-ā-nuku's here ... It's... [a] Māori worldview. (Iwi Rep 8 2019)

The Maniapoto Cultural Health Assessment Framework is one such framework that seeks to translate Ngāti Maniapoto values into restoration plans and projects.

The Maniapoto Cultural Health Framework was designed through a collaborative co-design process between scientists from NIWA (National Institute of Water and Atmospheric Research), staff at the Maniapoto Māori Trust Board (MMTB), and hapū of Ngāti Maniapoto hapū (Māori Business Owner 1 2019). The framework was, as one interviewee noted, about identifying "what matters to us" as Maniapoto and how "to get it [the awa] to a restored state" (Māori Business Owner 1 2019). Central to this was what Maniapoto considered to be the goals of restoration:

> We use words like make right ... what do they [our whānau] consider to be the right state of [the awa] and some of it's not even [about] restoration. Some of them aspire for it to be better than what they've ever known it to be. (Māori Business Owner 1 2019)

The co-design process involved wānanga with different hapū throughout Maniapoto tribal boundaries and included:

> whānau doing ... brainstorms ... [which involved them] literally just put[ing] it all down on the paper—what matters to you and what is the right state of that or a good state—what in your mind is good enough for

[the awa]. It was just literally pages of writing all over it. [It was] unclear [what the NIWA scientists were] taking that away [from it] using the[ir] scientific western minds [but they] picked the things that are similar [to whanau], … they [had] made tables, … So they're doing the scientific stuff with our whānau's knowledge and then they come back to the whānau with what they've done and [then they asked] is this right? Does this resonate with you? Does this matter to you? … So, it was like redesigning the [water] monitoring [sheets] to fit what our whānau wanted to know from the monitoring. … So [NIWA scientists] added smell, because smell is important to our whānau when it comes to swimming in the water. (Māori Business Owner 1 2019)

The framework, thus, is specifically designed for the iwi and not intended to be a universal "Māori assessment framework" but specifically one that specifically caters to "what matters to the whanau", hapū and Ngāti Maniapoto iwi. It is based on their histories, their relationships with their wai (water) and awa, and their ways of doing things. Māori scientists, employed by NIWA, provided Ngāti Maniapoto with "some examples" of work they did with other whānau, hapū, and iwi around the country such as "one whānau … doing drinking water … then another … just tuna", which allowed Ngāti Maniapoto to see aspects that could be incorporated into their framework. Yet, scientists make it clear to Ngāti Maniapoto hapū and whānau involved in co-design process that they (the scientists) are "really aware of the fact that we [Māori] are all different" and even though "there is a lot of similarities between us as Māori … iwi and hapū" hold different priorities and aspirations (Māori Business Owner 1 2019). Hāpu members who participated in the co-design workshops spoke about the experience in overwhelmingly positive terms. Dr Erica Williams, a Māori scientist employed by NIWA, singled out for repeated praise for her role in co-design process not only because of her "abundance of knowledge" and openness but also for making it explicitly clear that the framework was co-designed by Ngāti Maniapoto and was to benefit them (the whānau of Maniapoto); the framework was not the property of scientists, councils or others, and was to add in Ngāti Maniapoto efforts to enact their visions of what restoration was (Māori Business Owner 1 2019).

The process of creating the Maniapoto Cultural Health Assessment Framework "sort of went backwards" by giving "the whānau the tools and then said how do you want to use them and why do we need to use them" (Māori Business Owner 1 2019). The first wananga was held at Kahotea marae "the whānau sat around and just did big brainstorms" to discuss all the things negatively affecting the awa. Then employees from Maniapoto Māori Trust Board (MMTB) and NIWA identified three key issues that different groups of whānau all identified in their brainstorming sessions ("drinking, swimming and tuna"). The "second wananga", held later that same year, involved whānau sitting down to consider what matters to them:

> This is what an acceptable state of this [awa] is for us. Obviously, something that came through was tuna—all Māori care about tuna and kai. Having lots of tuna is important to us. Having lots of tuna to supply for the poukai [ceremonial gathering in support of Kīngitanga—the Māori king movement] for instance … where the [Māori] king goes around to all [the] different mare. So, for some marae that really support the Kīngitanga [the Māori King], … it's important to have tuna at that time of year… It's not just about having heaps … of tuna so we can eat them all the time, there's actually a cultural significance to why we need this at this time of year. So that's what we've been working on [how to restore tuna numbers] this year. (Māori Business Owner 1 2019)

However, some important issues identified by whānau were not included within the framework. For instance, Rereahu (hapū of Ngāti Maniapoto) wanted to include "birds … everything really … the whole habitat" not just the water and the tuna (freshwater eels). Likewise, many whānau and hapū wanted sites of cultural significance (wāhi tapu) included in the framework (Māori Business Owner 1 2019). However, the staff from MMTB and NIWA who co-designed project came to the joint decision that the inclusion of sites of significance would be beyond the scope of the cultural assessment framework and it would be better to concentrate first on the restoration of rivers and wetlands; restoration works include those that aim to replant native trees in areas cleared in the late nineteenth and early twentieth centuries (see Figs. 9.1, 9.2, and 9.3).

CARVING OUT A HOME IN THE BACKBLOCKS: A BUSH-CLEARING SCENE ON THE TE KUITI-MOKAU ROAD. AUCKLAND.

Fig. 9.1 Area of Te Rohe Potāe cleared of forest in 1911. (Source: AWNS 19110907 3 2, Auckland City Libraries, Auckland, New Zealand)

It would require an entirely new framework for sites of cultural significance that would require tools that allowed for historical investigations as well as ways to assess the wairua (spiritual integrity) of wāhi tapu (something not measurable by scientific studies) (Māori Business Owner 1 2019).

Representatives from different hapū are now using the framework throughout Ngāti Maniapoto rohe, some of whom spoke about their experiences using the framework.

The realities of implementing the framework involved trying to translate the theory into practices (Iwi Rep 6 2020). Some groups were in the process of identifying sites and "doing those measurements" as well as starting work to clean-up "their oxbows" (crescent-shaped lakes that lie alongside a river). Others were seeking to address invasive species along the riverbanks and river channels (Iwi Rep 4 2020).

TWO BRIDGES ON THE UPPER REACH OF THE MIMI RIVER.

Fig. 9.2 Pākehā farming household standing beside an unnamed tributary of the Waipā River in 1901. Note the absence of vegetation due to deliberate actions to log and burn the Indigenous flora to be replaced by pastures. (Source: AWNS 19010419 4 3, Auckland City Libraries, Auckland, New Zealand)

Defining Restoration

One Ngāti Māniapoto interviewee wanted, through his restoration work, to return the Waipā River to "the point of where it was ... a beautiful freshwater [system] with an abundance of native species". These actions should, he hopes, contribute positively to the wellbeing of his whānau, hāpu and iwi:

> So when you're thinking of restoration you're thinking about—so some of our whānau—and restoration is in their mind is getting it back to what they had as [children] because they're older and then their kids can't have that [same experiences] but then some whānau are like no, it's even better than that. We want better than what we were putting up with back then. (Māori Business Owner 1 2019)

Fig. 9.3 Restoration efforts are primarily focused on rivers, lakes, and wetlands. The photograph shows one of the numerous peak lakes, located in the middle and lower reaches of the Waipā River catchment, which volunteers have replanted with native plants and sought to reintroduce endemic fauna. (Source: Dennis Parsons (Photographer))

The word restoration "doesn't adequately cover everything that comes" up for when whānau discuss what they want. Indeed, "restoration is [about] making right" the relationship between tangata whenua and their awa. "I suppose when you translate [the phrase] make right [into te reo Māori] it's kia tika. Kia tika is a big thing for you know, [it means to] make right, make correct—kia tika. Whereas if you were going to use restoration, [the phrase you would use would] probably be whakahoki mai which is to make something good" (Māori Business Owner 1 2019). There are some slight differences in the meaning of terms between the different languages (English and Māori), paralleling the earlier differences in the Treaty/Te Tiriti, which also highlights how iwi-led restoration efforts are underpinned by Māori understandings of what restoration is and how it should be practised.

Iwi Rep 7 considers that restoration is about "whakaoho mauri, that's to re-awaken the mauri that is in that space. Mauri is another word for being … [and] whakaoho … means to awaken" (Iwi Rep 7 2019b). River restoration, from this perspective, is focused on "reawaken[ing] space" which includes the wellbeing of all things. Another related word is "whakatō mauri. … to awaken whakatō is to grow … [so] that's to grow the mauri" of the awa (river), repo (wetlands), tuna (eels), and human beings (Iwi Rep 7 2019b). Her approach to restoration is "very holistic" and she is "thankful for the likes of NIWA" for allowing her the opportunity to be involved in a co-design process that incorporates both mātauranga Māori and western scientific knowledge (Iwi Rep 7 2019b). By being able to work with scientists as well as mātauranga Māori experts, she and other iwi representatives employed as restoration practitioners are able to maintain a balance and draw on the best of both worlds (Te Ao Māori and Te Ao Pākehā).

> [W]hat I just said was, we need to hold onto what our tūpuna had spoken about, in terms of looking after the awa, because if the awa is paru, then our people are not okay. That's what's going to bring wellbeing to our people, is when our river is clean and when our river is flowing and abundance of [kai] and when our people return back to the awa. (Iwi Rep 7 2019b)

Iwi-Led Restoration Projects: Enacting Kaitiakitanga

Five of our interviewees work for an iwi-led river restoration social enterprise and hold the official job title of "Kaitiaki" and are restoration practitioners who are involved in growing plants used in restoration works (Kaitiaki 1 2020; Kaitiaki 2 2020; Kaitiaki 3 2020; Kaitiaki 4 2020; Kaitiaki 5 2020). All are Māori wahine (women) narrate their experiences raising seeds, growing plants, and planting them as alike to motherhood; three of whom are Ngāti Maniapoto, and two are from different iwi but whose partners and children whakapapa to Ngāti Maniapoto. "We are mothers to every plant in this nursery", Kaitiaki 2 muses (Kaitiaki 2 2020). A view supported by her colleague, Kaitiaki 1 "I have 408,000 children. That's how many plants we've got on [growing]", which is framed in terms of aroha (love) and koha (gifts) to Papa-tū-ā-nuku (see Fig. 9.4) (Kaitiaki 1 2020). They raise the question: "If we aren't kaitiaki, who's going to be? Who's going to fix it? Even just a little step like this [raising plants] can make a huge difference and keep growing". The noted that for them being a paid Kaitiaki did not mean that they felt their work (as environmental guardians) stopped at the end of their working hours. Instead, their paid work made them realise that kaitiaki is a way of life: "Going home as kaitiaki, eating breakfast as kaitiaki, taking a shit as kaitiaki. All of the above kaitiaki" (Kaitiaki 4 2020). As Kaitiaki 5 similarly argues: "You're kaitiaki here, and you're kaitiaki at home … [While we are] paid for the work, but that's not the driving cause [for doing] the work" (Kaitiaki 5 2020). Instead, the driver, she maintains, is on the responsibilities to one's kin, including future generations. "There's going to be something done [to address the environmental problems] so that we leave something behind for our children, and our children's children" (Kaitiaki 5 2020). The intergenerational dimension of river restoration is key to how they conceptualised not only restoration but also environmental justice. Issues of climate change, biodiversity loss, environmental degradation, and pollution all bring attention to the longer-term future, and questions about time and intergenerational responsibilities, and river restoration practitioners frame their work in terms of their intergenerational obligations to their atua (gods), tūpuna (ancestors), and their kin (both human and more-than-human, past, living and future generations), which includes the rivers and lands in which they live and work.

As we demonstrated in previous chapters, Ngāti Maniapoto like other iwi were unable to practice kaitiakitanga over their rohe for generations as a consequence of settler-colonialism. Yet, the injustices against Indigenous peoples were not just confined to acts of violence and dispossession, but also policies and strategies that sought to exclude and marginalise Māori knowledge and values. As a result of settler-colonial encroachment of their rohe, being Maniapoto (or Māori, or Indigenous) today means constantly engaging in a struggle to reclaim, re-know, and reassert one's identity and one's "relational, place-based existence by challenging the ongoing, destructive forces of colonization" (Corntassel and Bryce 2011, p. 152). One kaitiaki recalls how she:

Fig. 9.4 Plants growing in the Pūniu River Care nursery. (Source: Melanie Mayall-Nahi (Photographer))

grew up around here and grew up in the [Puniu] river but when I ... started schooling away from here [at] Te Awamutu Intermediate and then going on to high school, [I] spent less time around [the river and] Spent less time connected. ... I'm a little bit tainted by ... Pākehā ... schooling. Like from intermediate onwards I was total immersion in English... For me, ... [a Māori] understanding [of the world is] like a different mind state from what [is taught within the mainstream education system] ... Usually, people follow the pathway of school and then either work or university if you decide to do that. But not a lot of focus [within the education system is] on the environment and protecting your whenua [land]. (Kaitiaki 3 2020)

In her role as a Māori restoration practitioner came to the realisation that she had learnt a great deal outside of formal (Pākehā) education, particularly as a child with her whanau on the marae (tribal meeting house) and in her interactions with her awa, and that this knowledge was not subordinate to that of Te Ao Pākehā. The settler-colonial logic was premised on violence (to humans and ecosystems), dispossession, and in some instances both "genocide and ecocide". From the perspective of difference Indigenous peoples around the globe, the loss of species was (and is still) the loss of kin, of culture, of knowledge, of relationships, and of modes of living. In the present-day Indigenous people express feelings of shame that they cannot speak their own language (Te Reo Māori), that they do not know their own knowledge (mātauranga) and laws (tikanga), and cannot employ the practices their ancestors used to maintain safe and sustainable environments. Indeed, in Australia and North America, many Indigenous peoples report being groups are forced to adopt (and adapt) the Indigenous knowledges of other peoples because of the scale of loss associated with "genocide and ecocide" (involving the killing of entire tribal groups, the forced removal of Indigenous peoples to government institutions, missions, and reservations, and policies of biological and cultural assimilation) (Barta 2008; Campbell 2007; Ellinghaus 2009; Kelm 1999; Parsons 2010; Rose 2004, p. 35). In these instances, ecological restoration projects can provide an avenue by which Indigenous people can acknowledge their losses, feelings of shame for being somehow 'not authentically Indigenous' (with many Māori referring to themselves as 'potatoes'—brown on the outside and white on the inside) (Bell 2014;

Bird 1999; Corntassel and Bryce 2011; Kukutai 2013; McCormack 2012; Wanhalla 2015).

As Mohawk scholar Gerald Taiaiake Alfred writes "colonialism is best conceptualized as an irresistible outcome of a multigenerational and multifaceted process of forced dispossession and attempted acculturation—a disconnection from land, culture, and community—that has resulted in political chaos and social discord" within Indigenous communities and the collective dependency on the settler-state (Alfred 2009, p. 56). Yet, in the small acts of restoring a section of a riverbank or wetland interviewees spoke they formed (revived, restored) connections with their awa, whenua, and other more-than-human kin and in doing so revitalised their own confidence in themselves (as individuals and members of their iwi). Others spoke of feeling "proud about" their work as an official job title as 'Kaitiaki' (who are paid to undertake the practices of kaitiakitanga) and allowed them to reconnect with their own whakapapa, histories, and knowledge (irrespective of what hapū or iwi they belonged to). The physical actions of growing and planting, of removing weeds and sowing seeds helped them rebuilt the relationships (with one another, their ancestors, and the non-human worlds) that were/are disrupted by colonialism.

The practices of tending to a plant, replanting a riverbank, and restoring a wetland, also allowed for restoration practitioners to maintain and enhance their connections to their ancestors, their knowledge, and their ethics (centred on their tikanga). The Ngāti Maniapoto elders, one Kaitiaki noted, always acknowledge the shared histories and ongoing relationships between her iwi and Ngāti Maniapoto, which included her "great ... grandfather [who] came here [to the Waikato Region] and fought for Maniapoto" during the Waikato Wars (1863–1864). Such recognition of inter- and intra-iwi connections, stories and histories (of reciprocity between groups, of resistance to colonial oppression, and of cultural continuance) was situated at the heart of individuals' accounts of why restoration projects were positive (not a "cultural" benefit that was secondary to the ecological benefits of restoration, inter-personal and inter-species relationships were situated at the heart of restoration). Those Māori employed within Ngāti Maniapoto-led restoration projects, but whose whakapapa was to other iwi, felt their efforts to restore the Waipā awa (the ancestor of Ngāti Maniapoto) were not only appreciated by Ngāti Maniapoto but also allowed them opportunities to learn about mātauranga and tikanga. Further, their involvement in Maniapoto-led restoration efforts meant that they sought to renew their whakapapa connections to their own whanau, hapū, iwi, and awa (Kaitiaki 1 2020). As one Kaitiaki states: "I feel like [the work has] given me the tools so that I can be confident about my own pa. And [sing the] karanga.[1] Get up and do my pepeha[2]" (Kaitiaki 2 2020).

One of our interviewees, who is involved in the co-governance and co-management of the Waipā River, described how his professional work is indelibly shaped by his identity as Ngāti Maniapoto and his iwi's history (Iwi Rep 5 2019, p. 5). In particular, he cited the importance of cultural continuation and resilience of his iwi (Ngāti Maniapoto) in the face of multiple social and environmental injustices. He recounted the history of how his tūpuna (ancestors), after the 1863–1864 Waikato War, were marginalised and experienced ongoing physical and emotional traumas as a consequence of colonial violence (to people, land, water, and biota); but they were able to survive through a commitment to unity and collective action (kotahitanga), which was underpinned by the remembering and enacting the principles and values "brought over from Hawaiki" to Aotearoa by their ancestors (Iwi Rep 5 2019). In the present-day, the efforts of Ngāti Maniapoto to enhance the health of the Waipā River is narrated by these same principles. Most notably, the principle of "wairua" (spiritual integrity) which is:

> really about identity and not forgetting that we are mana atua [sacred spiritual power from the gods], mana tangata [power of the people] and we should all remember … the stories of how we as an iwi survive and … we have a whakapapa directly to those tūpuna [ancestors] that were there, so we are the generations that they ensured our survival. (Iwi Rep 5 2019)

---

[1] A karanga is part of the cultural protocol of a powhiri (welcome ceremony). Karanga involve the exchange of calls between senior women; either to welcome (if hosts) their visitors or to acknowledge their hosts (if visitors) onto a marae (the meeting place of a hapū).

[2] A pepeha is an introductory story in which a person introduces themselves through discussing places and people, including one's whakapapa.

Even for those Māori who do not whakapapa to them Waipā River or its tributaries but instead married into the Ngāti Maniapoto iwi, they still considered themselves to be guardians (kaitiaki) of Waipā and Pūniu awa because of their responsibilities to care for their children and grandchildren. The water is the lifeblood of their Ngāti Maniapoto children and so they take the role of kaitiaki seriously:

> we're the guardians of Pūniu awa. We're trying to restore the water. Replenish and get it back to what it used to be. And that's what a kaitiaki is to me, a guardian. It's someone who looks after something or someone. Or anything. My children whakapapa back to the Pūniu. That's their awa and I couldn't love my job anymore. I'm doing this for my kids, and for my partner and all of his family. They're so proud of us [river restoration practitioners] and our mahi [our work]. (Kaitiaki 1 2020)

Restoration practices are also about re-establishing relationships between Māori (as individuals and members of whanau, hapū, and iwi) and Papa-tū-ā-nuku (Earth Mother), which were disrupted by settler-colonialism:

> You know it's so deep, so much deeper than just planting plants. And putting them out on the when. It's like acknowledging Papa-tū-ā-nuku, being one with her. Cleaning her [the Pūniu River's] waterways. Not just for her, but for our children to come, for generations to come after that. (Kaitiaki 2 2020)

These duties of care go beyond cost-benefit analysis, scientific studies, and accounts of the present-day or near-future and encompasses longer time frames and infinite future generations.

Restoration projects are situated as part of wider efforts by iwi, hapū, whānau, and other groups to maintain and strength Māori culture, knowledge and tikanga, including the Māori language (Te Reo Māori), the identities and practices of iwi. As Cherokee scholar Clint Carroll writes, that Cherokee cultural revitalisation efforts are a crucial step in creating "sovereign landscapes", which he defines as "spaces where environmental governance and management take place on Indigenous terms and in Indigenous ways, however complex and multifaceted they may be" (Carroll 2015, p. 33). Critical to producing sovereign landscapes, Carroll argues, that this involves "fashioning modes of environmental governance" that are more in line with Indigenous values and perspectives towards more-than-human worlds (Carroll 2015, p. 33). To facilitate this, many iwi representatives suggested that Ngāti Maniapoto's co-governance, co-management, and restoration projects efforts need to shift focus from one based of Western framings of rights to one focused on responsibilities; the intergenerational responsibilities to ensure that health and wellbeing of future generations, including the capacities of the next generations to exercise rangatiratanga (authority and power) and practice kaitiakitanga according to the tikanga of Maniapoto. Interviewees from Ngāti Maniapoto repeatedly articulated how their involvement in ecological restoration projects and river co-management planning were motivated by their multiple responsibilities to their rohe, including included land, water, biota, and past/present/future generations.

Such a perspective of intergenerational environmental justice (which includes for both human and more-than-human communities) is similarly observed in other Indigenous societies (Alfred 2015; Leonard et al. 2013; Norgaard et al. 2018; Nursey-Bray 2016; Nursey-Bray and Palmer 2018; Turner and Clifton 2009; Winter 2018). The importance of thinking "seven generations into the future" is, for instance, the key principle guiding Haudenosaunee Nations (one of the Indigenous peoples of the US) sustainable stewardship of their lands and waters (Brookshire and Kaza 2013; King 2006, p. 449). Many members of Indigenous Karuk people express concern about declining numbers of salmon in their rivers and how climate change is likely to make it even more difficult for future generations to engage in salmon fishing, which is not an important source of healthy protein for Karuk families but also is considered an essential part of their identity as Karuk (such concerns are similarly expressed by different Indigenous peoples within the settler-states of Canada and the United States in regard to salmon and other fish and fauna species) (Denny and Fanning 2016; Diver 2012; Norgaard et al. 2018; Todd 2014). In the Waipā context, Ngāti Maniapoto articulate their anxieties that already diminished and degraded stocks of native flora and fauna will further decline in the future, and that future generations would no longer be able to harvest from their mahinga kai (food gathering sites). However, they also articulated their hopes that restoration projects would help to improve the quality and quantity of freshwater biota; and in doing so heal the interwoven physical and spiritual traumas experienced on Ngāti Maniapoto as a collective group (iwi, hapū, whānau) but also those endured by individuals, and the traumas infected by colonialism and degradation on their more-than-than human kin (including the river and water).

An unhealthy river not only causes aquatic flora and fauna to be unwell but also causes suffering for tangata whenua who are kin to the river and the more-than-human entities who dwell within waters of the river and its tributaries. As Panelli and Tipa, writing in the context of Māori well-being, demonstrate notions of reciprocity and the interconnectivity of individuals and wider social, ecological, and metaphysical entities means that the health and wellbeing of individuals cannot be simply be separated from the socio-cultural units in which they live (Durie 1998; Panelli and Tipa 2007). Like many other Indigenous cultures, the wellbeing of iwi Māori (referring to people who identify with a specific tribal group) is predicated on consideration of both individual and collective experiences, which includes self, whānau, hapū and iwi (Panelli and Tipa 2007). Accordingly, an understanding the Ngāti Maniapoto health and wellbeing is situated within the health and wellbeing of their rohe, and associated resources (Durie 1998; Panelli and Tipa 2007; Tipa and Teirney 2006). Stories of the healing qualities of "pristine waters" of the Waipā waterscapes (it's rivers, streams, wetlands, lakes and springs) are recounted by current generations of Ngāti Maniapoto. So too do they tell the stories of their ancestors watching on, protesting against, and sometimes participating in actions that radically changed the taiao (environment) and resulted in its "the mauri (life force) of the river [being] degraded" (Iwi Rep 5 2019). As a young person, one of our interviewees recalled how his wife told him: "well once the river's healed our people will be healed and that will set the way forward. That will help us" (Iwi Rep 5 2019). At the time, he dismissed what his wife told him, thinking to himself "that's just bullshit basically"; but now "I've learnt [actually] working in this space [of river restoration], everything's connected" (Iwi Rep 5 2019). As he now realises: "It's about wisdom really and it's very true. If you're drinking polluted water, or water that's been treated with chemicals, you're not going to be 100 per cent right you know?" Indeed, the process of establishing co-management arrangements, contributing to river planning documents, and engaging in restoration work, many within Maniapoto report renewed appreciation for their mātauranga. As we discussed earlier in Chap. 2, mātauranga is premised on a holistic way of thinking wherein human and more-than-human are constituted and co-constituted through kinship relationships (whakapapa) that are place-based and intergenerational.

For Ngāti Maniapoto, like their neighbours Waikato-Tainui (whose rohe is to the north and includes the lower and middle reaches of the Waipā River as well as the Waikato River), when people are unwell, they traditionally would go to the awa to sprinkle themselves and swim in the waters of the Waipā to be blessed, healed, and relax. In 2020 local Māori recount how they still go swimming in the local waterways of the Waipā and Pūniu Rivers and still anoint themselves with water to receive the blessing of their ancestors (see Fig. 9.5) (Kaitiaki 2 2020; Kaitiaki 5 2020). In the context of river restoration work, iwi representatives narrate how spiritual dimensions (the "wairua of it") underpin the work they do:

> water is life … Rivers are the blood of the land … We get nurtured by the land and it all interconnects. We're all [are all interrelated]—life and death [are] really connected closely to us all, so if we don't stop [the destruction], if we don't look after our taiao or our waterways we will inevitably perish, or get really sick … It's just like if we don't look after our people, … the same thing [happens]. We just create a desperate situation. (Iwi Rep 5 2019)

Others recount how they no longer swim in the waterways due to pollution but continue to rely on the awa to maintain their physical and mental wellbeing.

> I use the river to reset. Like so that's my… when the head is heavy, I come down to the river by myself. Hopefully, there's no one down here. And you know spend time by myself and think, process. (Kaitiaki 3 2020)

Fig. 9.5 People swimming in the Waipā River in 2019. Note this stretch of river (like much of the river catchment) still does not include any riparian plantings and cows graze right beside the river. The Waikato contains the most amount of cattle (1.9 million dairy cows and 480,000 beef cattle) of any region in Aotearoa, which equates to 5.11 cows per person. Unsurprisingly the Waikato region also generates the highest greenhouse gas emissions in the country chiefly as a consequence of cattle (who produce methane gas) and energy production. (Source: Karen Fisher (Photographer). (Statistics New Zealand 2020))

Some other interviewees record that they continue to swim in the waters of the Waipā and Pūniu Rivers, even though the water tastes "paru" (dirty, filthy), filled with algae, silt, and sediment. They also harvest foods (tuna and freshwater mussels), but they taste of dirt and sometimes sewage when they eat them (Kaitiaki 1 2020; Kaitiaki 2 2020; Kaitiaki 4 2020). All the iwi restoration practitioners expressed the hope that their efforts to restore the Waipā and its tributaries would ensure that their descendants would enjoy greater capacities to swim, harvest kai (food), and even drink the waters of the Waipā without fear of E.coli and other infectious diseases.

Grief and Hope

A significant (oftentimes) unacknowledged part of restoration work involves people addressing the difficult emotions attached to recognising environmental loss and degradation. As environmental historian Lillian Pearce writes: "Those who forge strong bonds with a place open themselves up and to confronting the overwhelming realities of ecological decline, extinction and the wider impacts of climate change" (Pearce 2019, p. 269). Pearce, writing from the perspective of Australians (most of whom were White Australians) engaging in efforts to restore their lands, acknowledges that the affective experiences of restoration work different between cultural settings. Yet, Indigenous peoples lived realities of environmental degradation and destruction (of ecocide) are not just confined just to the immediate present but instead are intergenerational and interconnected with their experiences of colonial dispossession. Indigenous peoples globally live with the legacies of environmental dispossession (the processes that reduced their abilities to access resources from their local environments) and environmental injustices that have radically affected their relationships with lands and waterways that sustained their livelihoods and social, cultural, economic, and spiritual well-being for generations (Tobias and Richmond 2014).

Many interviewees reflected on their own whānau (family) histories of loss (alienation from their land, language, homes, food sources, and ways of life). One Ngāti Maniapoto iwi representative recalled how his mother and whānau were made "homeless in Ōtorohanga because of the river diversion" when their land was compulsorily acquired by the local government (Waikato River Authority) under the Public Works Act as part of flood management scheme for the township instituted after a large flood event in 1958 Many Māori who lived alongside the waterways were left "[h]omeless, landless, resource-less, economy-less [sic] … from the river diversion in Ōtorohanga" and many migrated away in search of work (Māori Business Owner 1 2019). The decision to divert the river part of a wider Otorohanga Flood Management Scheme that involved the re-engineering of the channels of the rivers and streams that surrounded Otorohanga, the drainage of remaining areas of wetland, the removal of trees along the riverbanks, and the construction of large flood levees (District Commissioner of Works 1958; McLeod 1964). All these works involved the government compulsory acquiring ("taking") of Māori land, which included whānau land as well as the area reserved as a Māori urupā (cemetery). Such actions contributed to further environmental dispossession of Ngāti Maniapoto, whereby people's homes, cultivations, wāhi tapu (sacred sites), and mahinga kai (food gathering sites) were destroyed (paralleling actions decades earlier when the wetlands were drained as discussed in Chap. 3).

Such historic experiences of environmental dispossession continued to impact people's day to day lives in the 2010s, with interviewees talking about how they were not able to harvest the resources and prepare the foods that their ancestors did (Māori Business Owner 1 2019). They spoke of their own lives (childhood and adulthood) as well as those of their parents and previous generations (grandparents, great-grandparents, and so on) were all connected with the taiao and how cumulative impacts slowly eroded their relationships with the wai and disrupted the mauri and wairua (spiritual integrity) of all Ngāti Maniapoto kin (human and more-than-human). As a child, one interviewee, recalls: "we could walk along the river. You could head the birds. You could hear the river...you could feel the wairua in the river. Now it's just a trickle", but before (before drainage, river realignment, and flood levees) the water used to rush down it (Iwi Rep 4 2020). The interviewee also describes the smell of the river, and how smell is an indicator of health. If the water smelt a particular way people knew it was healthy and people could swim, drink, and harvest food from it, with the smell of the Waipā now described as paru and foul-smelling (an indication to iwi of its polluted status). In addition to harvesting food, water, medicines and other materials from their waterscapes, Maniapoto also used rivers and streams to preserve foods. In particular, one interviewee recalls how she and other kuia (grandmothers) used to put corn into baskets and place it in the river (attached it to poles) and leave it there, in the rushing waters, until the corn was fermented (a local delicacy which is no longer prepared or eaten) (Iwi Rep 4 2020). They spoke of how their ancestors used to harvest tuna from the lakes, rivers, and streams of the Waipā River catchment during the early-to-mid twentieth century using pā tuna (eel weirs), but they could not so not only due to the limited numbers of tuna living in the waterways but also because of drainage of wetlands (outlined in Chap. 3) and various legal restrictions placed on their abilities to access waterways and built physical structures (eel weirs) in waters that they did not possess the legal authority (under settler-state legal order) to do so; even if tikanga recognise them (Ngāti Maniapoto) as mana whenua which gave them the authority. Iwi representatives recall being able to see where they tied the hīnaki (woven basket used to catch tuna) to the bottom of the river because the water was so clear and "there were times when you could walk along [the riverbanks] and the water would shine with all the [fish]" (see Figs. 9.6 and 9.7) (Iwi Rep 4 2020). However, "doesn't do that anymore" as the water (Iwi Rep 4 2020). Indeed, as we describe in Chap. 5, the Waipā River and its tributaries and now filled with sediment, algae blooms, and pollutants from towns, factories, and farms meaning that few aquatic fauna can survive in it; its waters are heavily critiqued by scientists, iwi, and other local residents for its unappealing looks (being a murky brown with an occasional dash of green from algae) and smell (mud mixed with effluent). In the past, one interviewee recounts, the clear and "clean" waterways were filled with fish and other aquatic life (Iwi Rep 6 2020). The interviewee recalls: "[you'd] put a hīnaki in [the river and] it would come out full [with tuna] for any gatherings". One could "feed the [entire] whānau" (family) and "all the marae" (hapū-based meeting complex) with tuna caught in a single hīnaki. However, in the present-day "you can't do it, … I think the last time we put a hīnaki in would have been probably four, five years ago and it was in there for a week. We were lucky to get two" (Iwi Rep 4 2020).

Hobb wrote applied the psychological concept of "stages of grief" (denial, anger, bargaining, depression and acceptance) to people's relationships with environmental degradation and changes (Hobbs 2013). Yet, unlike the death of a person, the losses of environments (of animals, plants, waterways) are uncertain, chronic, and diffuse. The practical and metaphorical lessons of grieving, Hobbs and other scholars suggest are helpful to extend to think about how different cultures respond to environmental changes (Cunsolo and Ellis 2018; Dawson 2015; Hobbs 2013; Pearce 2019). Although Hobbs principally examines the loss of ecosystems and species, research by Pearce demonstrates that "loss can

Fig. 9.6 Photograph of one type of hīnaki. (Source: McDonald, James Ingram, 1865–1935. Photographs. Ref/ PA1-q-257-71-1. Alexander Turnbull Library, Wellington, New Zealand)

Fig. 9.7 Another type of hīnaki, Whanganui River area (Located south of the Waipā River). (Source: McDonald, James Ingram, 1865–1935. Photographs. Ref/ PA1-q-257-72-2. Alexander Turnbull Library, Wellington, New Zealand)

come in much more complex and connected ways, and the communal experience of loss can resurrect a more ethical, relational and collective grief" (Pearce 2019, p. 261). Pearce describes her own experiences of seeing images of environmental destruction of the forests in Tasmania (Australia) where she once lived (as a consequence of wildfire) and the sobering realisation that the landscapes she took bushwalks through as a young adult would not be the one's her daughter encountered. Similarly, van Dooren examines how the experience of mourning can help people more conscious of their relationships with more-than-humans and inspire a caring responsibility (van Dooren 2014). Van Dooren encourages making space for reflection, more and actions, likewise Pearce and Hobbs suggest that the grieving process is a fundamental part of restoration works as a way to "collectively grieve the possibilities and relationships and ways of being in the world that go along with species and ways of life" (van Dooren 2014; Hobbs 2013; Pearce 2019, p. 263). Yet, all of these discussions of grief and loss (sometimes referred to in terms solastalgia) are very much framed from the vantage point of Euro-Western intellectual traditions, cultures, and histories.

Many Euro-Western scholars, as we discuss in Chap. 1, situate the Anthropocene as a recent arrival (mid-twentieth century) that is radically changing environments and in doing so placing their ontological security and their homes on shaking ground (Steffen et al. 2011, 2015). However, for Indigenous peoples and Indigenous allies start of the Anthropocene begun hundreds of years earlier when European colonisation commenced, and their homelands were invaded. Indigenous peoples are not just suddenly recognising the environmental changes are occurring (as we highlight in Chaps. 3, 4, and 5 settler-colonialism in Aotearoa involved deliberate actions to alter Māori landscapes and waterscapes since the 1840s radically) nor are they just experiencing loss and grief (Davis and Todd 2017; Whyte 2017). Indigenous peoples, as Whyte argues, already know what it feels like to experience loss of relationships, ways of life, and worlds, they continue to live with the loss and grief associated caused by (and still causing) colonialism (and the multiple environmental injustices) (Whyte 2017, 2018). Moreover, unlike Western ontologies, Indigenous ontologies (like Te Ao Māori) are already premised on reciprocal relationships between human and more-than humans, kinship, and intergenerational responsibilities of care, and so experiences of the earth-shattering traumas and losses associated colonisation were (and are) not needed for Indigenous peoples to adopt a "position of ontological plurality that takes seriously inter-species relationships" which Western environmental management and restoration practitioners are encouraged to embrace (Pearce 2019, p. 269). Principles of caring and responsibilities for the wellbeing of other species are already woven into mātauranga Māori and tikanga (as we highlight in Chaps. 2 and 4), and for Māori restoration practitioners in the Waipā restoration work was not about only about grief but also about hope.

Hope emerges from amidst the eco-violence, uncertainty, and losses of the Anthropocene, when Indigenous responsibilities (intergenerational environmental justice embracing both human and more-than-humans) are not only recognised but also transformed into actions. It involves the reciprocal processes of healing relationships between people and rivers through changing how rivers are governed, management, and interacted with, which involves shifting away from the ontological and epistemological privileging of Western cultures and knowledges (Fox et al. 2017; Te Aho 2010, 2019). The process and outcomes of restoration efforts, Māori restoration practitioners maintained, should be directed at the process of healing (the waters, biota, and themselves) because "Healthy waters [means] healthy people" (Kaitiaki 5 2020).

Iwi representatives who worked in restoration projects drew direct links between their own engagement in restoration efforts and what they deemed to be the ultimate goal of restoration (and what would be considered successful restoration): the need to improve, enhance and maintain the mauri of wai, whenua (land), and taonga (treasures) for future generations. Those taonga included the birds who live in native forest lining the rivers as well as within the wetlands. As one interviewee recalls:

> When we were small, you could walk along [the riverbank] and hear the birds, so we have hope with the, some of the native plants that we've put back along the river, that all those birds would come back. (Iwi Rep 4 2020)

Maniapoto Māori Trust Board wants to conduct a count of kererū (New Zealand Pigeon *Hemiphaga novaeseelandiae*) so that the iwi has a rough idea of the number and distribution of kererū in their rohe (Baranyovits 2017; Bell 1996; Schotborgh 2005). Kereū are considered by Ngāti Maniapoto (and many other iwi) to be a taonga (formerly eaten by Māori but now protected under law), whereas ecologists refer to them as a keystone species. Kererū are significant for the survival of a large number of forest species and are the major or only seed dispersers for more than 60 tree and shrub species in Aotearoa (Bell 1996; Carpenter 2019). Accordingly, many iwi restoration practitioners would like to expand their restoration efforts from riparian areas and wetlands to encompass the whenua more and include creating more habitat for kererū.

Future restoration efforts iwi participants also spoke about included those aimed at providing more suitable habitats for tuna. One proposed idea is the creation of "tuna hotels", which is already used by another iwi, and involves the introduction of "white plastic pipes" into riverbanks. The tuna hotels are "put ... in the side of a fast-flowing awa so that tuna can go in there and rest" (Iwi Rep 6 2020). Along similar lines, the Waipā Rerenoa Restoration Project is "looking at putting rock structures in, just to slow the awa down" so that "the tuna can hide behind th[e] rock structures" (Iwi Rep 4 2020). At the moment, the water in the awa as it runs through the township of Otorohanga runs too swiftly for native aquatic fauna, with straight sides, and no vegetation. As a consequence of the ongoing engineering works undertaken at Otorohanga, to drain the wetlands and mitigate flood risk (discussed earlier in Chap. 4), the Waipā "River has sped up a lot", and so there are no longer any "resting spots, even for fish" (Iwi Rep 6 2020). Thus, the restoration works planned by hapū involve actions "to slow the river down and to have areas for the habitats [of] fish [and] tuna" (Iwi Rep 4 2020). The conceptualisation of restoration work contributing positively, even if only incrementally, to future generations motivated individuals and collectives engagement in existing and support for larger-scale restoration projects within their rohe.

Hope lies in the realms of imagining and dreaming, language and discussing, and translating the abstract into the practical (be it within one's home, backyard, riverbank, wetland, forest, rohe, marae, office or community). Restoration is about hope. It is about acknowledging the long-silenced histories of Indigenous environmental dispossession and ecological destruction, and in doing so, challenging the curated settler-colonies histories peaceful settlement and nation-building. By recognising (in-laws, policies, and plans) the coupled impacts of colonialism on people and ecosystems, the new co-governance and co-management arrangements are facilitating hopefulness amongst Ngāti Maniapoto. Especially amongst those working to enact restoration on the ground, and despite the issues, individuals and groups are facing with lack of sufficient funding to support their restoration efforts they express a renewed sense of hope for the future. Iwi Representative 7 talked about her huge optimism about the future after talking to young people and seeing their "potential. I see this light ... the kids that are a blank canvas. Some of our rangatahi [younger generations] I see put up some beautiful posts [on Facebook] and I think you're in a good space" (Iwi Rep 7 2019a). Indeed, many spoke of their hope that future generations will be willing to continue to step up and embrace Maniapoto ways of being (tikanga and mātauranga) and adapt to changing situations. Ngāti Maniapoto already possess long, rich histories of stepping up, be it by taking leadership roles (within Kīngitanga and the creation of Rohe Pōtae), or resisting colonial intrusions (during the Waikato Wars, seeking to maintain rangatiratanga within Rohe Pōtae) and launching petitions and legal campaigns in the face of injustices. Restoration practices are just another way Ngāti Maniapoto are stepping up and taking actions to address environmental degradation and achieve more justice futures grounded in their mātauranga (knowledge), tikanga (laws), mana (power and sovereignty), and rangatiratanga (authority and power). The act of stepping into the muddy, polluted waters of the Waipā and Pūniu to plant raupo, standing amongst seemingly impregnatable thorn-covered gorse bushes (*Ulex europaeus*) and valiantly to remove them as one's arms become covered in scratches, may not seem like being part of any sort of healing process. However, small scale restoration works are grounded expressions of hope, of healing, and of authority, which are situated in particular places, histories, cultures, and relationships. Hope, environmental historian Alison Pouliot writes:

> is one of those intangible things that begs for logic and for something less tangible; something based in emotions and belief. Hope arises from the capacity to feel and care. It relies on past experiences to project into the future. (Pouliot 2016, p. 339)

Hope comes from turning up in a cold and foggy winter's day and stands with mud coming up to one's knees to plant flax and manuka (see Figs. 9.8 and 9.9) beside Lake Ngāroto, Pūnui or Waipā Rivers (or one of the numerous lakes or waterways within the catchment). It is displaying acts of caring to others amidst all the uncertainties (of the Anthropocene, of everyday life). Every so other these practical acts of grounded hope are meet with a non-monetary gift in the form of the return of tuna swimming in a stream or the sound of a flock of kererū cooing in the trees. Every returned species, every section of whenua (land) and repo (wetland) replanted, is a promise fulfilled by kaitiaki (to themselves, their ancestors, children and grandchildren, and their kin), a reciprocal

Fig. 9.8 Manuka replanted around Lake Rotopiko (one of the peat lakes within the Waipā catchment). The lake, including a small area of wetlands and forest, was restored by volunteers under the direction of the National Wetlands Trust. The restoration efforts were funded, in part, through a grant received by the WRA (through the WRCuT). (Source: Dennis Parsons (Photographer))

message that communicates both the benefits and responsibilities of environmental stewardship for future generations.

Yet, the hope provided by river restoration is (and needs to be) realistic and based on the reality of local contexts and conditions. Restoration, therefore, involves the revival or creation of relationships between people and taiao (environment)—their landscapes and waterscapes—which are based on knowledge of how environmental conditions are constantly changing, and so too social conditions are changing. One iwi representative noted that the iwi needed to create more opportunities within its various institutions (be they iwi-, hapū-, and co-led) such as the MMTB and WRA so that Maniapoto youth are mentored and trained so that they can take up leadership roles within river governance, management and restoration projects (Iwi Rep 7 2019a). The passing of experiential

Fig. 9.9 Established trees and shrubs at a wetland restoration project (Lake Rotopiko) within the Waipā River catchment. This photograph shows newly established plantings of tī kōuka (cabbage trees or *Cordyline australis*) and Harakeke (flax or *Phormium tenax*) as well as older trees (remnant stands of Dacrycarpus dacrydioides or kahikatea) that escaped being felled or burnt during the 1860s–1960s efforts of landowners' and governments to removal all indigenous vegetation from the catchment. (Source: Dennis Parsons (Photographer))

knowledge, including of restoration, to younger generations within the iwi is considered a critical part of the cultural continuance (its mana) and wellbeing of both Maniapoto (iwi/hapū/whānau) and its taiao (environment). The educational component is, many iwi representatives note, currently missing from the ways in which environmental governance, management, and restoration is being framed within Maniapoto's own institutions (as well as within its various collaborations with other iwi, councils and scientific bodies). An iwi representative noted the iwi needed to develop a clearer "succession plan" so that younger generations within the iwi, hapū and whānau could learn from the older generations about how to care for the awa, the whanau, and each other. A successive plan would mean that when individuals with particular knowledge and skills departed Maniapoto (due to death, retirement, or changing jobs), their knowledge was passed onto the next generation (Iwi Rep 7 2019a). Iwi participants also emphasised the need to educate younger generations in mātauranga, tikanga, kaitiakitanga as well as western knowledge so that youth are fully equipped to walk between and exist in multiple worlds (Māori and Pākehā). Since restoration efforts are expected to be ongoing (stretching forward for at least the next half-century) iwi interviewees emphasised the importance of future generations taking up the mandate of protecting the awa (kaitiakitanga). By understanding the overlapping and ongoing processes of environmental degradation, restoration, and regeneration, we argue, one starts to appreciate and envision how to implement more meaningful, sustainable, and substantive decolonising practices. Future generations of Maniapoto (as well as other iwi and non-Māori) will map out their own pathways for restoration, which address their needs, concerns and aspirations (Corntassel and Bryce 2011, pp. 160–161). Decolonisation does not just involve legal and political performances (of Treaty settlements, co-governance arrangements, of legal personhood) aimed at recognition; it also moves into the everyday realm of socio-cultural and ecological practices where Māori plant trees on the side of a riverbank, count birds in the trees, and make an endless submission to councils demanding actions to address water pollutions. All these acts are ways Ngāti Maniapoto assert their rangatiratanga (power and authority), exercise their kaitiakitanga (environmental guardianship), and enact whanaungatanga (kinship, relationship, sense of connection gained through shared experiences and working together).

A commitment to kaitiakitanga and whanaungatanga has the possibilities to reshape and regain connections with more-than-human entities. The revitalisation of mātauranga (knowledge) and tikanga (is) is of certain importance to supporting Ngāti Maniapoto and other iwi visions of healthy spaces, beings (human and non-human), and relationships.

One iwi member, who is involved in restoration works and still harvests kai (food) from her rohe (despite her knowledge that the waters are polluted and she may become ill from consuming the good), argues that restoration is about re-establishing and maintaining her (and her wider whānau, hapū and iwi) relationships with their more-than-human kin (see Fig. 9.10). The act of being able to harvest and eat food collected

Fig. 9.10 Eel caught in the Pūnui River being prepared for cooking. (Source: Melanie Mayall-Nahi)

from her mahinga kai (food gathering sites) is an expression of her connectivity with her ancestors and all her kin:

> karakia [prayer] ... [that] addresses the kai [food] as being our tupuna
> [ancestors] and us having whakapapa [genealogy] to it. So, it calls the gods
> forth within the kai [food] and within ourselves, as being [a] relation—
> having a relationship to each other through our genealogy to the gods. It
> calls them forth and it says, come, and be our feast with us, in this feast you
> are the feast. It addresses us all as the children of Ranginui and Papatūānuku.
> Before you eat, you—whether it's—I don't know—a leaf or something—
> you sniff whatever it is in your kai and it's symbolic of releasing the [mauri]
> from the kai [food] that you have gathered for it to continue. (Iwi
> Rep 3 2020)

The ontological and epistemological differences between Western liberal and those of Indigenous intellectual traditions, as noted previously
by McGregor (2014) and Winter (2019), means that the processes of
ontological and epistemological pluralism within freshwater governance,
management, and restoration are fraught with ongoing obstacles. Indeed,
the dominant (Pākehā-centred) settler-colonial society in Aotearoa is not
always accepting of mātauranga and tikanga (at least not on the terms
that Māori find acceptable). However, in order for Aotearoa to move
towards more sustainable ways of interacting with water (governing,
managing, restoring), it is of crucial importance that we (as academics,
community members, and a nation) undertake the process of decolonising and disrupting the status quo. At the moment, we still find ourselves
in the situation where water quality in Aotearoa is poor (and worsening),
which is closely connected to the ever-increasing amounts of pollutants
are pouring into our waterways; at the same time increasing demand for
water (chiefly from irrigated agriculture) are contributing to growing
water scarcity. All of which is meaning diminishing biodiversity and
reducing capacities for communities to seek physical and emotional subsistence from the waters and biota of the rivers.

Conclusion

The current situation, as we argue in previous chapters, involves tikanga (laws) being broken (tapu of waste, the kinship obligations to care for the land and water, the sustainable use of biota for future generations) without any negative consequences on the wrongdoers (settler-state). When Indigenous "laws are broken with no resource", McGregor (2014, p. 19) observes in the Canadian First Nations context, the legal order is destabilised, which in turn causes negative consequences for Indigenous peoples. In sum, the foundations of the dominant settler-colonial legal order, and the settler society itself, stands on "shaky ground due to the ongoing and often wilful ignorance" of Indigenous laws. Settler societies' efforts to reconcile Indigenous peoples (remedying injustice through legal, institutional, and financial reparations) necessitates the recognition and empowerment of Indigenous ontologies and epistemologies in ways that include more-than-human rights, responsibilities, and justice requirements (McGregor 2014, 2018). The notion that there mutual responsibilities that are shared between human and more-than-human beings (such as people holding duties of care to water and vice versus) rests at the heart of many Indigenous ontologies including Te Ao Māori. The act of planting a tree or clearing a pathway so fish can migrate up or down a river again is a practice of kaitiakitanga (as our interviewees quoted earlier in this chapter highlight), it is also an act based on the belief that they (as tangata whenua and as kaitiaki) are responsible for caring for (in this instance restoring) the river and its flora and fauna.

The dominant discourse of environmental justice typically situates responsibility for achieving justice (and addressing injustice) securely with government (that often hold an administrative responsibility towards Indigenous peoples). However, in most instances, these arrangements do not provide adequately for environmental justice (from the perspective of Indigenous peoples). Indeed, McGregor argues that if such responsibilities (for defining what justice is and how it can be achieved) remain solely within the hands of government decision-making and within Western legal systems, it is highly unlikely that Indigenous environmental justice will ever be secured in any meaningful and lasting manner. Indeed central and local governments in Aotearoa consistently talk about sustainability and healthy rivers, but rarely take actions that create any on-the-ground improvements that contribute to better water quality, increased biodiversity, and enhanced connectivity between tangata whenua, awa, and whenua. Indeed, relying on the settler-state and Western legal orders is unlikely to achieve environmental justice or sustainability in ways that Māori or other Indigenous peoples require. Instead, this chapter highlights how restoration efforts are an avenue by which members of Ngāti Maniapoto (as well as people from other iwi) are (in the process of) restoring the balance (between them and their more-than-human kin) and in doing so restoring just (reciprocal) relationships within their rohe. It is not a short-term process, but rather a long-term (interviewees suggest it will take anywhere between 50 to 100 years to restore their awa). Restoration is an intergenerational pathway by which individuals and groups are seeking to achieve environmental justice in their own way, using their own ways of knowing and being (outside the narrow confines of Western liberal thought premised on the division of nature from culture and the upholding of the world economic order). Such an assertion does not absolve the settler-state of Aotearoa (or other colonial governments) of all the responsibilities they currently possess with respect to environmental injustices and addressing justice, but rather that the current settler-colonial system built on capitalistic exploitation regularly shows itself as being inadequate at achieving Indigenous environmental justice (as demonstrated by the widespread Indigenous opposition to mining in Aotearoa, Australia, and Canada, pipelines in Canada and the US, hydroelectric dams in Brazil and India, to name just a few) (McGregor 2014). So it is time to consider the pluralistic nature of and practices of achieving and sustaining Indigenous environmental justice. For Ngāti Maniapoto, the small act of planting a tree, building a 'hotel' for eels in the side of a riverbank, and re-establishing their connections between themselves and their tūpuna through engaging with Te Awa o Waipā (the Waipā River), are steps towards bringing about environmental justice.

References

Abernethy, B., & Rutherfurd, I. D. (1998). Where Along a River's Length Will Vegetation Most Effectively Stabilise Stream Banks? *Geomorphology, 23*(1), 55–75.

Alfred, T. (2009). Colonialism and State Dependency. *Journal of Aboriginal Health, 5*, 42–60.

Alfred, T. (2015). Cultural Strength: Restoring the Place of Indigenous Knowledge in Practice and Policy. *Australian Aboriginal Studies, 1*, 3.

Anderson, A. (2002). A Fragile Plenty: Pre-European Maori and the New Zealand Environment. In *Environmental Histories of New Zealand*. Auckland: Oxford University Press.

Anderson, A. (2003). *Prodigious Birds: Moas and Moa-Hunting in New Zealand*. Christchurch: Cambridge University Press.

Arnaud, F., Piégay, H., Schmitt, L., Rollet, A. J., Ferrier, V., & Béal, D. (2015). Historical Geomorphic Analysis (1932–2011) of a By-Passed River Reach in Process-Based Restoration Perspectives: The Old Rhine Downstream of the Kembs Diversion Dam (France, Germany). *Geomorphology, 236*, 163–177.

Baldwin, A., Cameron, L., & Kobayashi, A. (2011). *Rethinking the Great White North: Race, Nature, and the Historical Geographies of Whiteness in Canada*. UBC Press. Retrieved August 12, 2017, from https://books.google.co.nz/boo ks?hl=en&lr=&id=V8UyF3vufhsC&oi=fnd&pg=PP2&dq=Catriona+Sandil ands+Cape+Breton&ots=kFzZM2WoBe&sig=6IqcqS IIp_hYLJicdELYKVWDLlc.

Baranyovits, A. (2017). *Urban Ecology of an Endemic Pigeon, the Kererū*. PhD Thesis, University of Auckland, Auckland.

Barber, M., & Jackson, S. (2015). Remembering 'the Blackfellows' Dam': Australian Aboriginal Water Management and Settler Colonial Riparian Law in the Upper Roper River, Northern Territory. *Settler Colonial Studies, 5*(4), 282–301.

Bardsley, D. K., & Wiseman, N. D. (2016). Socio-Ecological Lessons for the Anthropocene: Learning from the Remote Indigenous Communities of Central Australia. *Anthropocene, 14*, 58–70.

Barta, T. (2008). Decent Disposal: Australian Historians and the Recovery of Genocide. In *The Historiography of Genocide* (pp. 296–322). Springer.

Bashford, P. A. (2013). The Anthropocene Is Modern History: Reflections on Climate and Australian Deep Time. *Australian Historical Studies, 44*(3), 341–349.

Bauman, W. A., & O'Brien, K. J. (2019). *Environmental Ethics and Uncertainty: Wrestling with Wicked Problems.* New York: Routledge.

Beilin, R., & West, S. (2016). Performing Natures: Adaptive Management Practice in the "Externally Unfolding Present". In L. Head, K. Saltzman, G. Setten, & M. Stenseke (Eds.), *Nature, Temporality and Environmental Management: Scandinavian and Australian Perspectives on Peoples and Landscapes* (pp. 186–203). London; New York: Routledge.

Bell, R. E. (1996). *Seed Dispersal by Kereru (Hemiphaga Novaeseelandiae) at Wenderholm Regional Park.* Thesis, University of Auckland.

Bell, A. (2014). *Relating Indigenous and Settler Identities: Beyond Domination.* London: Palgrave Macmillan UK.

Beller, E. E., Downs, P. W., Grossinger, R. M., Orr, B. K., & Salomon, M. N. (2016). From Past Patterns to Future Potential: Using Historical Ecology to Inform River Restoration on an Intermittent California River. *Landscape Ecology, 31*(3), 581–600.

Beller, E. E., McClenachan, L., Zavaleta, E. S., & Larsen, L. G. (2020). Past Forward: Recommendations from Historical Ecology for Ecosystem Management. *Global Ecology and Conservation, 21*, e00836.

Bennett, B. M., & van Sittert, L. (2019). Historicising Perceptions and the National Management Framework for Invasive Alien Plants in South Africa. *Journal of Environmental Management, 229*, 174–181.

Bhatt, J. P., Manish, K., Mehta, R., & Pandit, M. K. (2016). Assessing Potential Conservation and Restoration Areas of Freshwater Fish Fauna in the Indian River Basins. *Environmental Management, 57*(5), 1098–1111.

Bird, M. Y. (1999). What We Want to Be Called: Indigenous Peoples' Perspectives on Racial and Ethnic Identity Labels. *American Indian Quarterly, 23*(2), 1–21.

Boswijk, G., Fowler, A., & Palmer, J. (2005). *Hidden Histories: Tree-Ring Analysis of late Holocene Swamp Kauri, Waikato, New Zealand, 9.*

Brookshire, D., & Kaza, N. (2013). Planning for Seven Generations: Energy Planning of American Indian Tribes. *Energy Policy, 62*, 1506–1514.

Callicott, J. B. (2002). Choosing Appropriate Temporal and Spatial Scales for Ecological Restoration. *Journal of Biosciences, 27*(4), 409–420.

Cameron, E. S. (2015). *Far Off Metal River: Inuit Lands, Settler Stories, and the Makings of the Contemporary Arctic.* Vancouver: UBC Press.

Campbell, J. (2007). *Invisible Invaders: Smallpox and Other Diseases in Aboriginal Australia, 1780–1880.* Carlton South, VIC: Melbourne University Press.

Carpenter, J. K. (2019). *Legacy of Loss: Seed Dispersal by Kererū and Flightless Birds in New Zealand.* PhD Thesis, University of Canterbury, Christchurch.

Carroll, C. (2015). *Roots of Our Renewal: Ethnobotany and Cherokee Environmental Governance.* Minneapolis: University of Minnesota Press.

Carter, L. (2019). He Korowai o Matainaka/The Cloak of Matainaka: Traditional Ecological Knowledge in Climate Change Adaptation – Te Wai Pounamu, New Zealand. *New Zealand Journal of Ecology, 43*(3), 1–8.

Clément, V. (2017). Dancing Bodies and Indigenous Ontology: What Does the Haka Reveal About the Māori Relationship with the Earth? *Transactions of the Institute of British Geographers, 42*(2), 317–328.

Clements, F. E. (1916). *Plant Succession: An Analysis of the Development of Vegetation.* Washington, DC: Carnegie Institution of Washington.

Clements, F. E. (1936). Nature and Structure of the Climax. *Journal of Ecology, 24*(1), 252–284.

Clover, K., & Historical Society of the Hauraki Plains. (2007). *Taming of the Hauraki Swamp: Stories of Some of the Pioneers Who Lived on the Hauraki Plains and Whose Lives Have Been Influential in Converting the Area from Swamp to Farms.* Ngatea [N.Z.]: Historical Society of the Hauraki Plains.

Connelly, N. A., & Knuth, B. A. (2002). Using the Coorientation Model to Compare Community Leaders' and Local Residents' Views About Hudson River Ecosystem Restoration. *Society & Natural Resources, 15*(10), 933–948.

Corntassel, J., & Bryce, C. (2011). Practicing Sustainable Self-Determination: Indigenous Approaches to Cultural Restoration and Revitalization Indigenous Political Actors. *Brown Journal of World Affairs, 18*(2), 151–166.

Crow, S. K., Tipa, G. T., Booker, D. J., & Nelson, K. D. (2018). Relationships Between Maori Values and Streamflow: Tools for Incorporating Cultural Values into Freshwater Management Decisions. *New Zealand Journal of Marine and Freshwater Research, 52*(4), 626–642.

Crow, S. K., Tipa, G. T., Nelson, K. D., & Whitehead, A. L. (2020). Incorporating Māori Values into Land Management Decision Tools. *New Zealand Journal of Marine and Freshwater Research*, 1–18.

Cunsolo, A., & Ellis, N. R. (2018). Ecological Grief as a Mental Health Response to Climate Change-Related Loss. *Nature Climate Change, 8*(4), 275.

Davenport, M. A., Bridges, C. A., Mangun, J. C., Carver, A. D., Williard, K. W., & Jones, E. O. (2010). Building Local Community Commitment to Wetlands Restoration: A Case Study of the Cache River Wetlands in Southern Illinois, USA. *Environmental Management, 45*(4), 711–722.

Davis, H., & Todd, Z. (2017). On the Importance of a Date, or Decolonizing the Anthropocene. *ACME: An International E-Journal for Critical Geographies, 16*(4).

Dawson, C. (2015). Wai Tangi, Waters of Grief, Wai Ora, Waters of Life: Rivers, Reports, and Reconciliation in Aotearoa New Zealand. *Ecocriticism of the Global South, 93.*

de Leeuw, S., & Hunt, S. (2018). Unsettling Decolonizing Geographies. *Geography Compass, 12*(7), e12376.

Denny, S. K., & Fanning, L. M. (2016). A Mi'kmaw Perspective on Advancing Salmon Governance in Nova Scotia, Canada: Setting the Stage for Collaborative Co-existence. *International Indigenous Policy Journal; London, 7*(3) Retrieved June 18, 2020, from http://search.proquest.com/docview/1858128395/abstract/288C64E807CE45C6PQ/1.

District Commissioner of Works. (1958). *District Commissioner of Works to Commissioner of Works*, 3 March 1958, AATE A1002 5113 13/125/1, Archives New Zealand Auckland.

Diver, S. (2012). Columbia River Tribal Fisheries: Life History Stages of a Co-management Institution. In *Keystone Nations: Indigenous Peoples and Salmon Across the North Pacific. School for Advanced Research Press, Santa Fe, New Mexico, USA* (pp. 207–235). Santa Fe: SAR Press.

Durie, M. H. (1998). *Te Mana, Te Kāwanatanga: the Politics of Self Determination.* Auckland: Oxford University Press.

Ellinghaus, K. (2009). Biological Absorption and Genocide: A Comparison of Indigenous Assimilation Policies in the United States and Australia. *Genocide Studies and Prevention.* Retrieved May 17, 2020, from https://www.utpjournals.press/doi/abs/10.3138/gsp.4.1.59.

Elliot, R. (1997). *Faking Nature: The Ethics of Environmental Restoration.* London; New York: Routledge.

Ens, E. J., Finlayson, M., Preuss, K., Jackson, S., & Holcombe, S. (2012). Australian Approaches for Managing 'Country' Using Indigenous and Non-Indigenous Knowledge. *Ecological Management & Restoration, 13*(1), 100–107.

Failing, L., Gregory, R., & Higgins, P. (2013). Science, Uncertainty, and Values in Ecological Restoration: A Case Study in Structured Decision-Making and Adaptive Management. *Restoration Ecology, 21*(4), 422–430.

Fernández-Manjarrés, J. F., Roturier, S., & Bilhaut, A.-G. (2018). The Emergence of the Social-Ecological Restoration Concept. *Restoration Ecology, 26*(3), 404–410.

Fitzmaurice, A. (2007). The Genealogy of Terra Nullius. *Australian Historical Studies, 38*(129), 1–15.

Flannery, T. (2002). *The Future Eaters: An Ecological History of the Australasian Lands and People.* New York: Grove Press.

Forster, M. E. (2012). *Hei Whenua Papatipu: Kaitiakitanga and the Politics of Enhancing the Mauri of Wetlands.* Doctor of Philosophy, Massey University, Palmerston North.

Fox, C. A., Reo, N. J., Turner, D. A., Cook, J., Dituri, F., Fessell, B., et al. (2017). "The River Is Us; The River Is in Our Veins": Re-Defining River Restoration in Three Indigenous Communities. *Sustainability Science,* 1–13.

Giblett, R. (2009). Wilderness to Wasteland in the Photography of the American West. *Continuum, 23*(1), 43–52.

Hall, M. (2005). *Earth Repair: A Transatlantic History of Environmental Restoration.* Charlottesville: University of Virginia Press.

Hamilton, D. P., Collier, K. J., Quinn, J. M., & Howard-Williams, C. (2018). *Lake Restoration Handbook: A New Zealand Perspective.* Cham: Springer.

Harmsworth, G., & Roskruge, N. (2014). Indigenous Maori Values, Perspectives, and Knowledge of Soils in Aotearoa-New Zealand. In *The Soil Underfoot: Infinite Possibilities for a Finite Resource.* Boca Raton, FL: CRC Press, Taylor & Francis Group.

Head, L. (2012). Decentring 1788: Beyond Biotic Nativeness. *Geographical Research, 50*(2), 166–178.

Head, L., & Muir, P. (2004). Nativeness, Invasiveness, and Nation in Australian Plants. *Geographical Review, 94*(2), 199–217.

Higgs, E. (2003). *Nature by Design: People, Natural Process, and Ecological Restoration.* Cambridge, MA: MIT Press.

Higgs, E., Falk, D. A., Guerrini, A., Hall, M., Harris, J., Hobbs, R. J., et al. (2014). The Changing Role of History in Restoration Ecology. *Frontiers in Ecology and the Environment, 12*(9), 499–506.

Hikuroa, D., Slade, A., & Gravley, D. (2011). Implementing Māori Indigenous Knowledge (Mātauranga) in a Scientific Paradigm: Restoring the Mauri to Te Kete Poutama. *MAI Review, 3*(1), 9.

Hobbs, R. J. (2013). Grieving for the Past and Hoping for the Future: Balancing Polarizing Perspectives in Conservation and Restoration. *Restoration Ecology, 21*(2), 145–148.

Hourdequin, M., & Havlick, D. G. (2016). *Restoring Layered Landscapes: History, Ecology, and Culture*. New York: Oxford University Press.

Humphries, P., & Winemiller, K. O. (2009). Historical Impacts on River Fauna, Shifting Baselines, and Challenges for Restoration. *BioScience, 59*(8), 673–684.

Iwi Rep 3. (2020, February 13). Interview with Iwi Representative 3.

Iwi Rep 4. (2020, February 14). Interview with Iwi Representative 4.

Iwi Rep 5. (2019, March 25). Interview with Interview Iwi Representative 5.

Iwi Rep 6. (2020, February 14). Interview with Iwi Representative 6.

Iwi Rep 7. (2019a, May 16). Interview with Iwi Representative 7.

Iwi Rep 7. (2019b, June 13). Interview with Iwi Representative 7.

Iwi Rep 8. (2019, October 9). Interview with Iwi Representative 8.

Jacobson, R. B., Janke, T. P., & Skold, J. J. (2011). Hydrologic and Geomorphic Considerations in Restoration of River-Floodplain Connectivity in a Highly Altered River System, Lower Missouri River, USA. *Wetlands Ecology and Management, 19*(4), 295–316.

Jordan, W. R. (2003). *The Sunflower Forest: Ecological Restoration and the New Communion with Nature*. University of California Press.

Jordan, W. R., & Lubick, G. M. (2011). *Making Nature Whole: A History of Ecological Restoration*. Washington; London: Island Press.

Kaitiaki 1. (2020, February 4). Interview with Kaitiaki 1.

Kaitiaki 2. (2020, February 4). Interview with Kaitiaki 2.

Kaitiaki 3. (2020, February 5). Interview with Kaitiaki 3.

Kaitiaki 4. (2020, February 5). Interview with Kaitiaki 4.

Kaitiaki 5. (2020, February 5). Interview with Kaitiaki 5.

Katz, E., & Light, A. (2013). *Environmental Pragmatism*. Routledge.

Kelm, M.-E. (1999). *Colonizing Bodies: Aboriginal Health and Healing in British Columbia, 1900–50*. British Columbia: UBC Press.

Kibler, K., Cook, G., Chambers, L., Donnelly, M., Hawthorne, T., Rivera, F., & Walters, L. (2018). Integrating Sense of Place into Ecosystem Restoration: A Novel Approach to Achieve Synergistic Social-Ecological Impact. *Ecology and Society, 23*(4).

King, J. T. (2006). The Value of Water and the Meaning of Water Law for the Native Americans Known as the Haudenosaunee. *Cornell Journal of Law and Public Policy, 16*, 449.

Kukutai, T. (2013). The Structure of Urban Maori Identities. *Indigenous in the City: Contemporary Identities and Cultural Innovation,* 311–333.

Kurashima, N., Jeremiah, J., & Ticktin, T. (2017). I Ka Wā Ma Mua: The Value of a Historical Ecology Approach to Ecological Restoration in Hawai'i. *Pacific Science, 71*(4), 437–456.

Lake, P. S., Bond, N., & Reich, P. (2017). Chapter 5.4 – Restoration Ecology of Intermittent Rivers and Ephemeral Streams. In T. Datry, N. Bonada, & A. Boulton (Eds.), *Intermittent Rivers and Ephemeral Streams* (pp. 509–533). London: Academic.

Langton, M. (2006). *Settling with Indigenous People: Modern Treaty and Agreement-Making.* Annandale, NSW: Federation Press.

Leonard, S., Parsons, M., Olawsky, K., & Kofod, F. (2013). The Role of Culture and Traditional Knowledge in Climate Change Adaptation: Insights from East Kimberley, Australia. *Global Environmental Change, 23*(3), 623–632.

Lidström, S., West, S., Katzschner, T., Pérez-Ramos, M. I., & Twidle, H. (2016). Invasive Narratives and the Inverse of Slow Violence: Alien Species in Science and Society. *Environmental Humanities, 7*(1), 1–40.

Light, A. (1994). I. Hegemony and Democracy: How Politics in Restoration Informs the Politics of Restoration. *Restoration & Management Notes, 12*(2), 140–144.

Long, J. W., Goode, R. W., Gutteriez, R. J., Lackey, J. J., & Anderson, M. K. (2017). Managing California Black Oak for Tribal Ecocultural Restoration. *Journal of Forestry, 115*(5), 426–434.

MacGregor, S. (2014). Only Resist: Feminist Ecological Citizenship and the Post-Politics of Climate Change. *Hypatia, 29*(3), 617–633.

Maniapoto Maori Trust Board. (2017). *Review of the Deed In Relation to the Co-Governance and Co-Management of the Waipa River.* Maniapoto Maori Trust Board.

Māori Business Owner 1. (2019, August 29). Māori Business Owner 1.

Marshall, K., Koseff, C., Roberts, A., Lindsey, A., Kagawa-Viviani, A., Lincoln, N., & Vitousek, P. (2017). Restoring People and Productivity to Puanui: Challenges and Opportunities in the Restoration of an Intensive Rain-Fed Hawaiian Field System. *Ecology and Society, 22*(2).

McCormack, F. (2012). Indigeneity as Process: Māori Claims and Neoliberalism. *Social Identities, 18*(4), 417–434.

McDonald, T., Gann, G. D., Jonson, J., & Dixon, K. W. (2018). *International Standards for the Practice of Ecological Restoration – Including Principles and Key Concepts* (p. 48). Washington, DC: Society for Ecological Restoration.

McGregor, D. (2014). Traditional Knowledge and Water Governance: The Ethic of Responsibility. *AlterNative: An International Journal of Indigenous Peoples, 10*(5), 493–507.

McGregor, D. (2018). Indigenous Environmental Justice, Knowledge, and Law. *Kalfou, 5*(2), 279.

McLeod, N.C. (1964). District Commissioner of Works N.C. McLeod to Walsh, 29 September 1964, R17280156, AATE 5113 A1002, 321/a, 13/125/1, Archives New Zealand, Auckland.

Metge, J., & Kinloch, P. (2014). *Talking Past Each Other: Problems of Cross Cultural Communication.* Wellington: Victoria University Press.

Moran, S. (2010). Cities, Creeks, and Erasure: Stream Restoration and Environmental Justice. *Environmental Justice, 3*(2), 61–69.

Morandi, B., Piégay, H., Lamouroux, N., & Vaudor, L. (2014). How Is Success or Failure in River Restoration Projects Evaluated? Feedback from French Restoration Projects. *Journal of Environmental Management, 137,* 178–188.

NIWA. (2014). *Maniapoto Priorities for the Restoration of the Waipā River Catchment.* Wellington: NIWA.

Norgaard, K. M., Reed, R., & Bacon, J. M. (2018). How Environmental Decline Restructures Indigenous Gender Practices: What Happens to Karuk Masculinity When There Are No Fish? *Sociology of Race and Ethnicity, 4*(1), 98–113.

Nursey-Bray, M. (2016). Cultural Indicators, Country and Culture: The Arabana, Change and Water. *The Rangeland Journal, 37*(6), 555–569.

Nursey-Bray, M., & Palmer, R. (2018). Country, Climate Change Adaptation and Colonisation: Insights from an Indigenous Adaptation Planning Process, Australia. *Heliyon, 4*(3), e00565.

Palmer, M. A., Bernhardt, E. S., Allan, J. D., Lake, P. S., Alexander, G., Brooks, S., et al. (2005). Standards for Ecologically Successful River Restoration: Ecological Success in River Restoration. *Journal of Applied Ecology, 42*(2), 208–217.

Palmer, M. A., Filoso, S., & Fanelli, R. M. (2014). From Ecosystems to Ecosystem Services: Stream Restoration as Ecological Engineering. *Ecological Engineering, 65,* 62–70.

Palmer, M. A., Zedler, J. B., & Falk, D. A. (2016). Ecological Theory and Restoration Ecology. In M. A. Palmer, J. B. Zedler, & D. A. Falk (Eds.), *Foundations of Restoration Ecology* (pp. 3–26). Washington, DC: Island Press/ Center for Resource Economics.

Panelli, R., & Tipa, G. (2007). Placing Well-Being: A Maori Case Study of Cultural and Environmental Specificity. *EcoHealth, 4*(4), 445–460.

Park, G. (2018). *Nga Uruora*. Wellington: Victoria University Press.

Parsons, M. (2010). Defining Disease, Segregating Race: Sir Raphael Cilento, Aboriginal Health and Leprosy Management in Twentieth Century Queensland. *Aboriginal History*, 85–114.

Paterson-Shallard, H., Fisher, K., Parsons, M., & Makey, L. (2020). Holistic Approaches to River Restoration in Aotearoa New Zealand. *Environmental Science & Policy, 106*, 250–259.

Pearce, L. (2019). *Critical Histories for Ecological Restoration*. Thesis, Australian National University, Canberra. Retrieved June 8, 2020, from https://search.proquest.com/docview/2343447929/?pq-origsite=primo.

Perring, M. P., Standish, R. J., Price, J. N., Craig, M. D., Erickson, T. E., Ruthrof, K. X., et al. (2015). Advances in Restoration Ecology: Rising to the Challenges of the Coming Decades. *Ecosphere, 6*(8), 1–5.

Peters, M. A., Hamilton, D., & Eames, C. (2015). Action on the Ground: A Review of Community Environmental Groups' Restoration Objectives, Activities and Partnerships in New Zealand. *New Zealand Journal of Ecology, 39*(2), 179.

Pluymers, K. (2011). Taming the Wilderness in Sixteenth- and Seventeenth-Century Ireland and Virginia. *Environmental History, 16*(4), 610–632.

Pouliot, A. (2016). *A Thousand Days in the Forest: An Ethnography of the Culture of Fungi*.

Ratana, K., Herangi, N., & Murray, T. (2019). Me pēhea te whakarauora i ngā repo o Ngāti Maniapoto? How Do We Go About Restoring the Wetlands of Ngāti Maniapoto? *New Zealand Journal of Ecology, 43*(3), 1–12.

Reyes-García, V., Fernández-Llamazares, Á., McElwee, P., Molnár, Z., Öllerer, K., Wilson, S. J., & Brondizio, E. S. (2019). The Contributions of Indigenous Peoples and Local Communities to Ecological Restoration. *Restoration Ecology, 27*(1), 3–8.

Rose, D. B. (2004). *Reports from a Wild Country: Ethics for Decolonisation*. Sydney: University of New South Wales Press.

Salmond, A. (2014). Tears of Rangi: Water, Power, and People in New Zealand. *HAU: Journal of Ethnographic Theory, 4*(3), 285–309.

Saulters, O. (2014). Undam It? Klamath Tribes, Social Ecological Systems, and Economic Impacts of River Restoration. *American Indian Culture and Research Journal, 38*(3), 25–54.

Schotborgh, H. M. (2005). *An Analysis of Home Ranges, Movements, Foods, and Breeding of Kereru (Hemiphaga novaeseelandiae) in a Rural-Urban Landscape on Banks Peninsula, New Zealand.* Thesis, Lincoln University. Retrieved July 4, 2020, from https://researcharchive.lincoln.ac.nz/handle/10182/2681.

Schuelke, N. (2014). *Urban River Restoration and Environmental Justice: Addressing Flood Risk Along Milwaukee's Kinnickinnic River.* Thesis, The University of Wisconsin – Milwaukee, Milwaukee (Wisconsin). Retrieved July 4, 2020, from http://search.proquest.com/docview/1617457744/abstract/C58A60BB963D422EPQ/2.

Scientist 2. (2019, November 7). Interview with Scientist 2.

Senos, R., Lake, F. K., Turner, N., & Martinez, D. (2006). Traditional Ecological Knowledge and Restoration Practice. In D. Apostol & M. Sinclair (Eds.), *Restoring the Pacific Northwest: The Art and Science of Ecological Restoration in Cascadia* (pp. 393–496). Washington, DC: Island Press.

Statistics New Zealand. (2020). *Livestock Numbers.* Statistics New Zealand. Government. Retrieved August 3, 2020, from https://www.stats.govt.nz/indicators/livestock-numbers.

Steffen, W., Persson, Å., Deutsch, L., Zalasiewicz, J., Williams, M., Richardson, K., et al. (2011). The Anthropocene: From Global Change to Planetary Stewardship. *AMBIO, 40*(7), 739.

Steffen, W., Broadgate, W., Deutsch, L., Gaffney, O., & Ludwig, C. (2015). The Trajectory of the Anthropocene: The Great Acceleration. *The Anthropocene Review, 2*(1), 81–98.

Stein, E. D., Dark, S., Longcore, T., Grossinger, R., Hall, N., & Beland, M. (2010). Historical Ecology as a Tool for Assessing Landscape Change and Informing Wetland Restoration Priorities. *Wetlands, 30*(3), 589–601.

Stokes, E. (2000). *The Legacy of Ngatoroirangi: Maori Customary Use of Geothermal Resources.* Department of Geography, University of Waikato. Retrieved June 20, 2018, from https://researchcommons.waikato.ac.nz/handle/10289/6323.

Te Aho, L. (2010). Attempting to Integrate Indigenous Traditional Knowledge of Waterways with Western Science: To Restore and Protect the Health and Well-Being of an Ancestral River. In *4th International Traditional Knowledge Conference 2010* (p. 328).

Te Aho, L. (2019). Te Mana o te Wai: An Indigenous Perspective on Rivers and River Management. *River Research and Application, 35*(10), 1615–1621.

Tipa, G., & Nelson, K. (2017). Eco-Cultural Restoration Across Multiple Spatial Scales: A New Zealand Case Study. *Water History, 9*(1), 87–106.

Tipa, G., & Teirney, L. (2006). *A Cultural Health Index for Streams and Waterways: A Tool for Nationwide Use.* Wellington: Ministry for the Environment.

Tobias, J. K., & Richmond, C. A. (2014). "That Land Means Everything to Us as Anishinaabe....": Environmental Dispossession and Resilience on the North Shore of Lake Superior. *Health & Place, 29*, 26–33.

Todd, Z. (2014). Fish Pluralities: Human-Animal Relations and Sites of Engagement in Paulatuuq, Arctic Canada. *Études/Inuit/Studies, 38*(1–2), 217–238.

Turner, N. J., & Clifton, H. (2009). "It's So Different Today": Climate Change and Indigenous Lifeways in British Columbia, Canada. *Global Environmental Change, 19*(2), 180–190.

Uprety, Y., Asselin, H., Bergeron, Y., Doyon, F., & Boucher, J. F. (2012). Contribution of Traditional Knowledge to Ecological Restoration: Practices and Applications. *Ecoscience, 19*(3), 225–237.

van Dooren, T. (2014). *Flight Ways: Life and Loss at the Edge of Extinction.* New York: Columbia University Press.

Veracini, L. (2010). *Settler Colonialism: A Theoretical Overview.* New York: Springer.

von Wehrden, H., Hanspach, J., Kaczensky, P., Fischer, J., & Wesche, K. (2012). Global Assessment of the Non-Equilibrium Concept in Rangelands. *Ecological Applications, 22*(2), 393–399.

Waikato Regional Council. (2012). *Waipā Zone Management Plan.* Hamilton: Waikato Regional Council.

Waikato Regional Council. (2014). *Waipā Catchment Plan.* Hamilton: Waikato Regional Council.

Waltham, N. J., Barry, M., McAlister, T., Weber, T., & Groth, D. (2014). Protecting the Green Behind the Gold: Catchment-Wide Restoration Efforts Necessary to Achieve Nutrient and Sediment Load Reduction Targets in Gold Coast City, Australia. *Environmental Management, 54*(4), 840–851.

Wanhalla, A. (2015). *In/Visible Sight: The Mixed-Descent Families of Southern New Zealand.* Wellington: Bridget Williams Books.

Watts, V. (2013). Indigenous Place-Thought and Agency Amongst Humans and Non Humans (First Woman and Sky Woman Go On a European World Tour!). *Decolonization: Indigeneity, Education & Society, 2*(1) Retrieved May 16, 2020, from https://jps.library.utoronto.ca/index.php/des/article/view/19145.

Weber, C., Åberg, U., Buijse, A. D., Hughes, F. M. R., McKie, B. G., Piégay, H., et al. (2018). Goals and Principles for Programmatic River Restoration Monitoring and Evaluation: Collaborative Learning Across Multiple Projects: Programmatic River Restoration Monitoring and Evaluation. *Wiley Interdisciplinary Reviews: Water, 5*(1), e1257.

Wehi, P. M., & Lord, J. M. (2017). Importance of Including Cultural Practices in Ecological Restoration. *Conservation Biology, 31*(5), 1109–1118.

Wehi, P. M., Beggs, J. R., & McAllister, T. G. (2019). Ka mua, ka muri: The Inclusion of Mātauranga Māori in New Zealand Ecology. *New Zealand Journal of Ecology, 43*(3), 1–8.

White, C. A., Perrakis, D. D. B., Kafka, V. G., & Ennis, T. (2011). Burning at the Edge: Integrating Biophysical and Eco-Cultural Fire Processes in Canada's Parks and Protected Areas. *Fire Ecology, 7*(1), 74–106.

Whyte, K. (2017). Indigenous Climate Change Studies: Indigenizing Futures, Decolonizing the Anthropocene. *English Language Notes, 55*(1), 153–162.

Whyte, K. (2018). Settler Colonialism, Ecology, and Environmental Injustice. *Environment and Society, 9*(1), 125–144. https://doi.org/10.3167/ares.2018.090109.

Winter, C. J. (2018). The Paralysis of Intergenerational Justice: Decolonising Entangled Futures. Retrieved January 11, 2020, from https://ses.library.usyd.edu.au/handle/2123/18009.

Winter, C. J. (2019). Does Time Colonise Intergenerational Environmental Justice Theory? *Environmental Politics*, 1–19.

10

Rethinking Freshwater Management in the Context of Climate Change: Planning for Different Times, Climates, and Generations

Climate change is overwhelmingly framed in global narratives as one of (if not the most) pressing issue facing humanity in the twenty-first century. Common depictions declare it as a fundamentally new phenomenon that threatens apocalyptic environmental changes and the survival of both human and nonhuman beings. Climate change (as one of the markers of the Anthropocene epoch) is framed as fundamentally new and unpredicted and therefore requires radical interventions, global actions, and new approaches. Warnings from the 2018 Intergovernmental Panel on Climate Change (IPCC) report *Global Warming of 1.5 C* identified the need for collective actions to lower the amount of greenhouse gas (GHG) emissions entering the atmosphere to avoid dangerous climate change (Hoegh-Guldberg et al. 2018; Roy et al. 2019). If societies fail to reduce GHGs within the next few decades, the global earth system could cross the climate 'tipping point' (of 2°C increase in the global average air temperature). The results of crossing this point would be widespread loss and damage to human and ecological communities, including the extinction of certain species, collapse or serious disruption of ecosystem functioning, ocean acidification, sea-level rise, more intense extreme weather events and increased food and water insecurity (Hoegh-Guldberg et al.

M. Parsons et al., *Decolonising Blue Spaces in the Anthropocene*, Palgrave Studies in Natural Resource Management, https://doi.org/10.1007/978-3-030-61071-5_10

2018). Climate change is framed as an urgent issue that requires immediate and largescale collective actions by individuals, civil society, industries, and governments to address to avoid a litany of environmental and societal disasters, which raises critical questions for freshwater governance and management in Aotearoa.

Indigenous peoples are one of the most important groups of voices in climate change debates, with their discourses referencing environmental and intergenerational justice. Their concerns are two-fold, including both the environment and the continuation of specific cultures linked to particular places and ecosystems. It is these relationships "between the processes of the natural and social worlds" (Schlosberg 2012, p. 451), which are crucial to Indigenous framings of Indigenous Environmental Justice (IEJ). Such a framing does not imply a dogmatic "living in the past" wherein Indigenous Peoples' must choose to live how their ancestors did. Instead, it emphasises how Indigenous societies are dynamic, living, and unique cultures and that a diversity of ways of life (incorporating different knowledges and values) are possible (Durie 1998). Instead, this discourse supports the creation of environments that maintain and enhance dignity (for both current and future generations) in which there are intimate and reciprocal relationships between humans and nonhumans (Alfred 2008, 2015; Watene 2016).

Notwithstanding the urgency of acting and the need to take climate change seriously, Indigenous and decolonial scholars increasingly challenge hegemonic narratives that attribute responsibility for climate change to human actions without also recognising the impacts of colonisation on Indigenous peoples. Such scholars raise critical questions about what climate change means in the context of IEJ and how efforts to address climate change can be (or already are being) interwoven with Indigenous peoples' efforts to reassert their knowledge, values, and practices in the context of freshwater governance, management, and restoration efforts. As we demonstrated earlier, settler-colonialism has violently uprooted Indigenous peoples (including Māori), communities, altered environments, and undermined human-more-than-human relations. Indeed, Indigenous peoples have (over the past two hundred plus years) faced (and continue to face) catastrophic social and environmental changes (dubbed "the end of worlds" by some Indigenous scholars) including radical changes to freshwater systems. This has led some scholars to point out that "colonialism is itself a form of anthropogenic climate change" (Parsons and Nalau 2016; Whyte 2020, p. 5), and to suggest Indigenous peoples are among the first survivors or victims of climate change (Whyte 2017). Whyte asserts that the heightened 'vulnerability' of Indigenous peoples to the negative impacts of climate change (rising sea levels, extreme weather events, biodiversity loss, and so on) are not the result of either "bad luck" or the confluence of two unfortunate events (colonisation and climate change), but are critical components of the structures and practices that created the Anthropocene.

For many Indigenous peoples, the drastic alterations experienced because of colonisation are (in many instances) more extreme than the apocalyptic warnings written by climate change scholars, journalists, and activists if we exceed 2.0°C of warming (Cameron 2012; Carter 2018; McMillen et al. 2017; Rumbach and Foley 2014; Veland et al. 2013; Vinyeta et al. 2016; Whyte et al. 2019). Indigenous scholars, including Whyte (2017) and Todd (2016), argue that achieving climate justice for and by Indigenous people requires addressing the multiple ways in which global environmental changes (including freshwater degradation and the various impacts of climate change) are also inextricably related to, and in reality, predicated on, settler-colonialism (Zahara 2017). This includes drawing attention to the ontological assumptions that underpin understandings of climate change and which shape responses at local, national and international scales. Key among these assumptions is that conceptions of time are universal across cultures and peoples, and that time is linear. For many Indigenous cultures, however, including Māori, time is non-linear and conceptualised as a spiral or temporal loop (Stewart-Harawira 2005, 2018); past/present/future are conceived as holding concurrent status and existing together such that no time is privileged over another. For Indigenous scholars, such as Christine Winter, Indigenous conceptions of time raise important questions about the intergenerational environmental justice (EJ) implications of climate change for Indigenous peoples.

In this chapter, we explore justice as an intergenerational imperative for Indigenous peoples by examining how different conceptions of time shape responses to climate change. We offer insights into how bringing

Māori understandings of time can open new spaces for thinking about and planning for climate change in ways that do not reinforce and rearticulate the multiple environmental injustices (disproportionately experienced by Indigenous peoples because of settler-colonialism). The histories of Māori and other Indigenous cultures, over the last two hundred plus years of colonisation, offer important lessons about what constitutes a life well-lived and how to maintain cultures, identities, a sense of belonging and connectivity in the face of radical social, economic, political, cultural, and environmental changes. We first begin by drawing attention to growing critiques amongst Indigenous scholars to the dominant framing of climate change as an unprecedented environmental crisis for all of humanity, which overlooks Indigenous peoples' experiences of the social and ecological disaster that was (is) colonialism. We then consider developments in EJ theories that attempt to accommodate multiple generations and intergenerational rights, responsibilities, and obligations and focus particularly on the ontological framing of time as a way of overcoming the limitations of Western liberal theories founded on linearity and progress. We take these critiques through into our next section that explores climate change policies in Aotearoa, where we argue that, once again, Euro-Western knowledge and value systems are privileged over Māori ways of knowing. We contrast this by exploring Māori conceptualisations of intergenerational duties of care, specifically kaitiakitanga (environmental guardianship), and time as a way of recuperating Indigenous ways of knowing and being in space and across time to overcome environmental injustices against Māori and their environments. Next, we spiral back to the Waipā River and examine how local governments, Māori, and other actors within the Waipā catchment (and more broadly within the Waikato Region) are framing climate change, and what actions are being planned or taken in response to climate risks. In this section, we critique the dominant framing of climate change as an economic and technical problem and highlight Māori efforts to reassert their knowledge, values, and mana to address the cascade of risks (both climatic and non-climatic) facing their awa (river). We also consider the justice implications of the responses being taken (or not being taken) to address climate change for Māori and their awa by focusing on tuna (freshwater eels) as a way to highlight the ontological conflicts that manifest due to different conceptualisations of time, nonhumans and things that matter in the Anthropocene. Moreover, there remain critical conceptual differences when different cultures imagine what is a dignity-supporting environment for future generations (Watene 2016; Winter 2019).

Indigenous Critiques of Climate Change: Indigenising Intergenerational Climate Justice

Climate change, biodiversity loss, environmental degradation, and pollution bring attention to the longer-term future and raise questions about the temporal dimensions of environmental justice. Decisions of individuals, businesses, and governments that affect lands, waters, biodiversity, and atmosphere have both current and longer-term impacts. The formulation of these decisions within the dominant Western worldview continues to marginalise Indigenous worldviews, modes of living, and cultural norms. Candis Callison writes, in regard to the Indigenous peoples of the North American Arctic, that scholars and decision-makers need to acknowledge what "climate change portends for those who have endured a century of immense cultural, political and environmental changes" (Callison 2015, p. 42).

Settler-colonialism was (is) "an attack" on the capacities of Indigenous peoples' to adapt to variable (climatic, ecological, economic) conditions; colonisation involved deliberate and incidental actions by settlers to erase Indigenous landscapes and waterscapes and supplant and naturalise the values, economies, political structures, and human-environment relations of settler-societies. Many of the anticipated climate change-related losses, damages and shocks that non-Indigenous peoples are now increasing concern and alarm about, were experienced by Indigenous peoples as a consequence of climate change. These include: drastically altered ecosystems; loss of biota; environmental degradation; destruction of economies; decline in livelihood opportunities; forced relocation; political and ontological conflicts; and socio-cultural disintegration (Brännlund and

Axelsson 2011; Cameron 2012; Nursey-Bray 2016; Nursey-Bray et al. 2019; Parsons and Nalau 2016; Veland et al. 2013).

International conventions and agreements, including Brundtland's *Our Common Future* (1987), the United Nations Framework on Climate Change Convention (UNFCCC), the Sustainable Development Goals (2016), and the Report of the Indigenous Peoples' Global Summit on Climate Change (United Nations 2009) all emphasise the need to protect and maintain the climate system, and the need to ensure environmental benefits to future generations. Each document implies that intergenerational justice should be addressed, but at the same time make clear that the most important obligations must be to the living. Such documents (excluding the Report of the Indigenous Peoples) are largely written from a Western liberal worldview, and despite the various reports and agreements, environmental degradation and worsening climate change continue.

Whyte (2020) argues it is already too late to avoid climate injustices against Indigenous peoples as they are already affected by biophysical phenomena as well as socio-economic and political processes. A key reason why it is too late, Whyte argues, is that the urgency and alarm about climate change—being problematically expressed in the media, research, education, advocacy, and political decision-making—could result in actions that contribute to further injustices; chiefly, through strategies undermining Indigenous peoples' worldviews, values, governance and management approaches, and sovereignties (Whyte 2020, pp. 2–3). His argument, based on his personal and academic experiences involving his own Indigenous people (Anishinaabe peoples of the United States and Canada) as well as other North American Indigenous nations, maintains that actions to address climate change are failing to empower Indigenous collective self-determination and advance Indigenous aspirations. Instead, climate change, under the guise of urgency and the prospect of dangerous climate change, can be mobilised to justify interventions that disrespect Indigenous rights, knowledge, and ways of life. Scholars such as Cameron (writing about Canadian Inuit and First Nations) and Veland and Howitt (researching Australian Aboriginal peoples) have drawn similar conclusions (Cameron 2012; Howitt et al. 2012; Veland et al. 2013).

A wealth of scholarship around the world highlights how governments and industries are continuing to disrespect Indigenous values and sovereignties by establishing high carbon developments; a layering of multiple recognition, distributive and procedural social and environmental injustices. These include fossil fuel exploration and extraction operations, mining, and the development of new oil and gas pipelines constructed on and under Indigenous lands and waters without the consent of Indigenous peoples. Even when there is a legal requirement for consultation with Indigenous groups and/or obtaining consent to such developments, it is frequently bypassed by claims of national security and/or urgency, as occurred in relation to the Dakota Access Pipeline and the Standing Rock Sioux Tribe and others (Baum 2019; Gilio-Whitaker 2019; LeQuesne 2019; Whyte 2017). Also, there is continued dispossession of Indigenous peoples from their lands and waters (through new hydropower and forest conservation schemes) under the auspices of climate mitigation and low-carbon energy transition policies (Baldwin 2009; Dressler et al. 2012; Li 2010; Yenneti et al. 2016). Similarly, scholars argue that the majority of government- and NGO-led programmes aimed at encouraging low-carbon energy developments or the resettlements of Indigenous communities deemed highly vulnerable to climate risks in the Arctic violate the justice principles of contest, trust, accountability and reciprocity. This means Indigenous peoples experience further environmental injustices because of climate change policies and actions.

Climate change has focused Western scholars such as Derek Bell, Tim Hayward, Simon Caney, and Henry Shue on the issue of climate justice and intergenerational justice largely because climate change and its impacts are so long-lived (Bell 2011, 2013; Caney 2009, 2014; T. Hayward and Iwaki 2016; Shue 2014). For example, the lag time between when a GHG is emitted and its warming effect on the atmosphere occur is on average 50 years. Likewise, the cumulative and cascading effects of climate change on the environment are predicted to worsen in the future. Invariably, these scholars draw on Western liberal justice traditions and their arguments centre on the claim that dignity and equality, guaranteed under international agreements such as the United Nations Declaration of Human Rights, are not temporally bounded. This means it is insufficient to just guarantee and protect the rights of human beings living now; those same obligations and duties need to be carried forward to include future generations of people (Caney 2008). This entails ensuring the state of the environment remains at the same or similar state (with no worse degradation and similar benefits) from one generation to the next so that the environment passed on to future generations is in no worse state than what was inherited from ancestors.

Western liberal theories of climate justice (which remain dominant within the international scholarship and policymaking domains) claim neutrality, impartiality, and universality and, thus, do not consider the life-and dignity-supporting environments peoples around the globe (including those from non-Western cultures) need in the context of changing climate conditions. Western justice theorists, like politicians and economists, have struggled to conceptualise obligations of justice that include distant futures where the beneficiaries are not clearly identified, and their future circumstances cannot be truly fathomed or predicted with any certainty. In developing policies and taking action to address climate change and other environmental problems, decisions often favour the present over the future. For instance, cost-benefit analysis, which is used frequently by economists and policymakers to evaluate the effectiveness of spending and investment to address climate change, apply uncertain discount rates over a relatively short time frame (100-150 years) that can unfairly distribute benefits to people living today and disadvantage future generations (Caney, 2008; Stern et al., 2007; Winter 2018).

Simon Caney argues that future generations of people possess the same basic rights as current living people (the right to life, the right to health, and the right to subsistence) (Caney 2008, 2009). Each of these rights is threatened by climate change (both now and in the future) and impacts on environments that undermine peoples' rights to life, health and subsistence. This argument could be expanded to include the pollution of waterways, land degradation, biodiversity loss, and so forth (on a more localised level). A limitation of this rights-based justice structure is, however, that it focuses on protecting the environment only insofar as it supports the health, wellbeing, and capabilities of human beings (as individuals). Life, moreover, is a constrained definition. Human beings are conceptualised primarily as individuals and individual human rights are emphasised. Moreover, communities do not extend to include the interests (or people's obligations towards) ancestors (or the past more broadly), ecological communities, or the environment (and more-than-human actors). Although, under the United Nations Declaration on the Rights of Indigenous Peoples (UNDRIP), the declaration states that Indigenous knowledge, values, and relationships with nonhumans must be guaranteed and protected (Articles 1, 8, 12, 25). For Indigenous peoples, these protections are issues of Indigenous rights.

In contrast to intergenerational environmental justice theorists who claim a linear projection in which both the living and generations of people in the near future accumulate the majority of duties and obligations, some justice scholars emphasise obligations to ancestors in their theorising. Duncan Ivison's exploration of the normativity of Western liberal political thought raises questions about how the assumed universality in "understanding and reflecting upon social and political relations" affects Indigenous peoples' rights (and their understandings of rights) as they relate to their land, culture and self-rule (Ivison 2014, p. 1). Ivison argues (Ivison, 2003, p. 336) that accommodating Indigenous rights may require departures from Western liberal norms to conceive Indigenous rights as coexisting with those of the settler state. Edward Page suggests a way to overcome the hopelessness of establishing social contracts with undefinable future individuals is to view intergenerational justice through the lens of reciprocity: "considered as either mutual advantage or fair play" (Page 2007a, p. 226). This theorisation acknowledges the benefits inherited from one's ancestors and the attendant obligations for future generations and positions intergenerational justice as a form of intergenerational custodianship or environmental stewardship.

Page criticises the neglect of ancestors and argues the need to consider ancestors in accounts of the impacts and implications of environmental degradation and change and in determining how the benefits and costs associated with environmental changes are distributed. As custodians, there are constraints on how one generation may use the environment: "existing persons are bound by duties of indirect reciprocity to protect environmental and human resources for posterity in return for the benefits inherited from their ancestors" (Page 2007b, p. 233). He emphasises how people's lives are built on the work of past generations and depend on a stable climate system conducive to human dignity and flourishing, clear waterways and air, stable and productive solids, and abundance of forest, fisheries, and agricultural products needed to sustain humans on earth. However, Page's conceptualisation of time to generations past, present, and future remains distinctly (like other scholars, including Parfit) anthropocentric (Page 2007a; Parfit 2011). Page overlooks the intrinsic need to preserve and protect the nonhuman (Schlosberg 2012; Winter 2018). In his conceptualisation of intergenerational environmental justice, the protection and responsibilities attached to environmental custodianship are for the benefits of future generations of people. Although the result may be the protection and care of nonhumans, the primary focus is still human needs and desires. He does not express any normative obligation to nonhuman actors. In marked contrast, within Indigenous philosophies (including the concept of kaitiakitanga from Māori people of Aotearoa and the concept of kanyini from the Anangu people from central Australia) there are normative responsibilities and duties of people towards nonhumans, and these duties are built upon a non-Western conceptualisation of time (Winter 2019).

Scholars, such as Winter (2018), offer conceptualisations of intergenerational environmental justice that take into account Indigenous ontologies and which expand on theorisations of justice focused on individuals located in the present and which disrupt Western assumptions about time; specifically, the ontological presumption that time always moves forward and that individuals exist only in one discrete time. Linear temporality is, according to Tilley, perhaps the most "taken-for-granted Western social discourse" (Love and Tilley, 2013). In Western imaginaries, time is a progressive forward seeking arrow. Time is a commodity or material to be lost, to be utilised wisely (but not to be squandered), and to be measured by minutes and seconds, by light years, and the moon and sun (Winter, 2019). Time (conceptualised as a time as a spiral) is something that stands still, persists, speeds by, and repeats itself. It is something dark, fleeting, and illuminating.

Winter (2018) refers to her own whakapapa to demonstrate the relatedness and interconnectedness of time and all other things, human and nonhuman, and the multiple temporalities embodied within Māori.

I have ancestors; I live now, and I have children who will (all being equal continue live when I am gone; I will have grandchildren (all things equal, etc. While I am living, I am also a (potential) ancestor, and my living children were once a future generation to me, as are my potential grandchildren, as was I to my ancestors. In time I will be an ancestor as will my children and my grandchildren will be living and thinking of future generations. The generations are coexisting, the past is always in the present, and the future is always in the past.

Kia whakatōmuri te haere whakamua (I walk backwards into the future with my eyes fixed on my past) is a whakataukī (proverb) that speaks to Māori conceptualisations of time, where past, present and future as perceived as interwoven in a temporal loop, and life is an ongoing cosmic process underpinned by whakapapa (genealogical connections). Within this neverending spiralling movement, there are no temporal restrictions (it is both past, present, and future) (Rameka 2016; Winter 2019). For Māori society, like some other Indigenous cultures, time is not framed in a way that privileges the present (over the past and future), and time does not proceed on a linear forwards pathway from past to present then future. In this way, Māori understand the past as a constant reference point. A person's actions in the present day are linked to both past and future generations of people, with the future/present/past linked together in the ongoing spiral. Likewise, Australian Aboriginal and First Nations' peoples' imaginings of time are cyclic (Davis and Todd 2017; Mckay and Walmsley 1969; Stewart-Harawira 2005; Whyte et al. 2019; Winter 2019), and emphasise how the past and future are close companions to the present day (Povinelli 2016; Povinelli et al. 2017). What these diverse Indigenous philosophies hold in common is that the "past and future are intimate bedfellows to present" (Winter 2018, p. 34). How one behaves in the present-day thus is always referenced from the point of view of past and future generations (who are referees of what is happening now).

In contrast, Western perspectives of time tend to view the past as something behind oneself and that people's goals, aspirations and plans should be directed at the future (Patterson 1998; Winter 2018). The fundamentally different way in which Indigenous and Western cultures understand time is also reflected in the different ways in these cultures live with or experience the loss of the past (Stewart-Harawira, 2005; Watson, 2015). Accordingly, Indigenous people living now include obligations to past, present, and future generations in their considerations. Embedded within Indigenous environmental justice, therefore, is always an intergenerational element that embraces communal continuum, inclusive and interconnected communities, and the places responsibilities on the living to connect future generations into their ethical communities.

Framing Climate Change in Aotearoa as an Economic and Technical Problem

Freshwater systems in Aotearoa are under pressure from a diversity of factors, which will likely intensify because of climate change (Ballantine and Davies-Colley 2014; Chapman 1996; Dudley et al. 2020). Small alterations in weather conditions are likely to produce significant effects in places with temperate climates such as Aotearoa, including biodiversity loss and heighten threats to people's livelihoods (Manning et al. 2015). In the Waikato region, climate change is meant to produce a small number of positive impacts such as warmer temperatures (1-3.5 degrees warming by 2090) enabling farmers to grow new crops and pasture as well as extending growing seasons (Pearce 2019). However, a large number of negative impacts are also predicted for the region (New Zealand et al. 2020; Reisinger et al. 2014; West 2007). Warmer air and water temperatures are likely to increase the frequency and spread of invasive species and water- and vector-borne diseases, which will place additional stress on native biota as well as (non-native) livestock (Pearce 2019; Reisinger et al. 2014). The number of hot days is forecast to increase and heatwaves are likely to result in a greater incidence of heat stress amongst people and fauna. The seasonal distribution of rainfall is expected to alter as a consequence of climate change (wetter in winter and autumn and drier in summer and spring) (New Zealand et al. 2020). Heavy rainfall events, whereby large amounts of rain falling in a short amount of time, are more likely to occur in the region under changing climate conditions, resulting in saturated soils and flooding, slips and sedimentation of waterways. Similarly, droughts are predicted to increase (especially during spring and summer), causing reduced river flows, greater pressure of water supplies, and stress of biota (Pearce 2019; Reisinger et al. 2014).

At the national scale, efforts to address climate change focus on both mitigation and adaptation using a mix of regulatory and market-based tools. The New Zealand Emissions Trading Scheme (ETS) was created in 2007 to record GHG emissions and facilitate a new 'carbon trading' market as a way to encourage businesses to reduce their GHG emissions. The carbon-trading scheme ensured the country complied with its international obligations under the Kyoto Protocol; however, the ETS has been consistently criticised, including by Māori commentators, for failing to reduce GHGs (Harawira, 2007). For instance, Māori parliamentarian Hone Harawira (from Northland iwi Ngāpuhi and a member of the Māori Party) spoke forcefully in parliamentary debates opposing the creation of the ETS in 2007. He referenced emails he received from concerned Māori individuals around the country as well as emails from the Indigenous Environmental Network (an international NGO) in his speeches. He asked:

> [I]s this emissions trading scheme really the answer to all our climate change problems, or is it just creating another property rights regime to let the world's biggest polluters continue along their merry, filthy way? Charging people for greenhouse gas emissions was supposed to encourage businesses to come up with alternatives to fossil fuels, but all it is doing is giving them an excuse to continue. Why bother with the expensive, long-term structural changes if we can meet our targets by simply buying pollution rights from operations that can reduce their carbon cheaply?

He warned the inclusion of Māori owned forests (mostly, exotic pine plantations) in the ETS would "not make any difference [to reducing GHG emissions] because all that it would do is let industrialised nations and companies [around the world] buy their way out of emissions reductions", and any financial benefits Māori received would not compensate to the damage done to the environment (Harawira 2007). Harawira sought to draw his fellow parliamentarians attention to Māori principles ("our kaupapa [principles and ideas] of rangatiratanga [authority], manaakitanga [hospitality, respect and showing care for others generosity], and whanaungatanga [the centrality of kinship]") and Māori obligations to care for and preserve the environment as encapsulated in kaitiakitanga.

[It is] our responsibility to care for our world through the reduction of those activities that would harm and, indeed, destroy that world. In the interests of life itself, let alone social, economic, and environmental sustainability, we have a responsibility to reduce our carbon output. Māori have a role to play in the reduction of greenhouse emissions, and we do not resile [sic] from that responsibility, but Māori also have the right to manage what little assets they may have for the betterment of their people. We realise that in order to manage both roles effectively we must—and we do—appreciate that our total wellbeing, our health, our economy, and our sustenance are dependent on the wellbeing and health of our world, just as all indigenous peoples across the globe understand their unique role of caring for and conserving mother Earth.

While Harawira's comments, as an individual and a member of a particular iwi (Ngāpuhi), cannot be taken to mean that all Māori in Aotearoa felt the same way about the ETS; however, it does indicate that Māori were seeking to (once again) get their knowledge, values, and intergenerational generational responsibilities recognised by the settler-nation but were still being disregarded (misrecognition) by those holding power.

The nation's total GHG emissions continued to rise, chiefly from methane generated from cows (with the agricultural industry not part of the ETS) (National Business Review 2018; Unknown Author 2009). Since its creation, the ETS has been subject to review and reform to improve how it functions and to ensure it works as a mitigation strategy. With regard to adaptation, the Climate Change Response (Zero Carbon) Amendment Act 2019 situates adaptation planning and action as primarily the responsibilities of local governments; however, the Minister of Climate Change is required to prepare a national adaptation plan to guide local governments.

Both the ETS and the Climate Change Response (Zero Carbon) Amendment Act 2019 reflect Western ontological and epistemological views that seek to command-and-control nature (with nature kept apart from culture) through technical knowledge (scientific and economic) and maximise economic productivity. For instance, the ETS (and other carbon trading schemes) rearticulates many of the same ideas and approaches used by past generations of settlers (as individuals and governments) to address environmental problems (rendered a scientific, technical, managerial problem) (Baldwin 2009; Driver et al. 2018; Gerrard 2012). However, there remain critical issues with the ETS as a supposed solution to rising GHG emissions, particularly since the same capitalist market forces that drove the industrial revolution, colonialism and imperialism, and high carbon economies and lifestyles, are now meant to be able to solve the problem of climate change. Recent research by Māori scholar Lyn Carter (from iwi Ngāi Tahu also spelt Kāi Tahu) argues that the ETS, from its initial creation through to its implementation, persistently excluded mātauranga Māori (Māori knowledge) and values (Carter 2018). There was (and continues to remain) a persistent tension between the ETS (as an economic-focused climate mitigation approach built on exploitative capitalist models of the endless accumulation of goods) and the principles and practices of mātauranga Māori and tikanga, which centre on notions of kaitiakitanga (guardianship) and whakapapa, and intergenerational justice.

The Zero Carbon Act is framed solely from a Euro-Western (or more specifically, Te Ao Pākehā) perspective of time and decision-making time frames, which overlooks Māori understandings of time and intergenerational justice. In addition, the requirement on governments to recognise and take into account Māori values provides weaker statutory requirements than the Resource Management Act (RMA), 1991 (which we critiqued in Chap. 5). There is an absence of recognition-based environmental (climate) justice considerations and lack of a clear pathway for achieving procedural-based justice (with no specific mechanisms detailed to allow Māori to meaningfully participate—as Treaty partners—in national adaptation planning). Thus, the legislation leaves room for interpretation that could compound environmental injustices for future generations through inequitable planning regimes (which reinforce settler-colonial supremacy, western scientific knowledge, and Pākehā values at the exclusion of Māori and other ways of knowing and being).

The framing of climate change through a Te Ao Pākehā economic lens aligns with ongoing neoliberalism and a techno-managerialist approach in which climate change is a problem to be expertly remedied, avoided or mitigated through market-led mechanisms (most notably the ETS) and new low-carbon economic development schemes. The economic threat posed by climate change is especially strong due in part to the large portion of GHG emissions produced through agriculture (methane from cows) and the dairy industry's ranking as the country's top international export earner (Adler et al. 2013; Driver et al. 2018; Hopkins et al. 2015). Within the current paradigm, climate change is largely rendered an object to be measured (through scientific knowledge) and managed by selected actors (which includes governments, iwi organisations, businesses, share market brokers and other experts), but not something that requires radically different knowledge, values, or practices to be adopted. This means that Māori experiences of disastrous environmental changes, their knowledge of how to live with and adapt to uncertain and changeable environmental conditions, and their visions of what constitutes living well remains largely marginalised within the dominant narratives of climate change.

In the context of the Waipā River, the framing of climate change as an economic problem has been internalised at the local level (including the Waikato Regional Authority, hereafter, WRA) and is evident in institutional responses to climate change. For instance, in December 2019 the WRA held its first workshop on climate change to "dig into what they could see [happen in the Waikato catchment] over the next 20 to 60 years" (Unknown Author 2019). In an official statement, Chief Executive of the WRA Bob Penter stated (in a local newspaper):

It's hard to imagine that climate change isn't going to be a factor if we've got 70 years still to run on achieving the vision of a restored Waikato and Waipā river catchment. … For us [at the WRA] it raises questions around climate change resilience, adaptation and starting to at least consider what we might need to think about when making funding decisions for [river governance and management] projects in the future to ensure they will endure if climate change is going to be a factor in their success. (Unknown Author 2019)

While the workshop included presentations on climate change impacts, resilience and the need for adaptation (Linwood 2019; Pearce 2019), overwhelming focus was on the economic dimension of climate change (Brownsey 2019; Dickie 2019; Ledgard 2019; Waikato River Authority 2019). Indeed, the narrative of climate change as an economic problem (as opposed to social, political, ecological) for institutions, businesses, farmers, iwi, and individuals, including those involved in freshwater governance and management, translates into an emphasis on economic and technical solutions designed to maintain, protect and intensify economic productivity (Brownsey 2019; Dickie 2019; Ledgard 2019; Linwood 2019). Thus, responses to climate change were limited to identifying financial opportunities (co-benefits) to those landowners and businesses in the Waikato willing to invest rather than being recording and addressing the social, economic, cultural, and ecological risks.

Iwi representatives noted that (within Waikato and Rohe Pōtae) local governments (regional councils and district councils) continued to pay limited attention to climate change aside from the potential for economic development (for example, through increased flood risk) as well as limited discussion of technical interventions designed to address individual climate risks. Where mitigation strategies were promoted, these remained underpinned by a language of ongoing economic growth and technological advances, which echo the earlier decades' narratives of settler-colonial land and river 'improvements', development and settlement, and ever-increasing productivity. In many instances, international scholarship demonstrates, such technical and economic 'fixes' provide new footholds for neoliberal capitalism (which remained deeply intertwined with settler-colonial projects) (Nightingale et al. 2019; Pelling et al. 2012).

Flooding of the Waipā River remains an ongoing concern for the Waikato Regional Council (WRC) and the district councils in regards to climate change (which came a feature of the waterscape after drainage works and deforestation as shown in Fig. 10.1 and Fig. 10.2). Previous government-led interventions focused on draining the wetlands, straightening waterways, dredging riverbeds, removing vegetation, supplanting Indigenous biota with exotic, and controlling flooding through levees; all of which took place in the context of Māori dispossession and marginalisation. Such interventions were narrated as vital necessities to ensure

Fig. 10.1 Photograph of flooding in Otorohanga. Published in the *Auckland Weekly News* 17 June 1920. Photograph taken by J. A. Parry. Source: Auckland Libraries Heritage Collections AWNS-19200617-31-4

economic security and societal (more specifically Pākehā) progress. Yet, radically remaking Aotearoa's freshwater systems (as we demonstrate in Chaps. 3 and 4) through the removal of wetlands resulted in more frequent and greater intensity flood events within the Waipā catchment (see Fig. 10.2, Fig. 10.3 and Fig. 10.4).

In the context of the Waipā River, iwi representatives recall the long history of local government-led engineering works along their awa and the ways in which flood levees (stopbanks) caused negative effects on tangata whenua (people of the land) and on aquatic biota. One iwi member recalled her family being made homeless because the local government acquired their land to build the Otorohanga flood control scheme (which was established following the 1958 flooding see Fig. 10.5) (discussed in Chap. 4). Others spoke of how engineered "solutions" to flood controls destroyed flora that were harvested by tangata whenua for food, medicine, and cultural practices. The loss of vegetation also meant there was little habitat for aquatic fauna and no places for metaphysical beings (such as

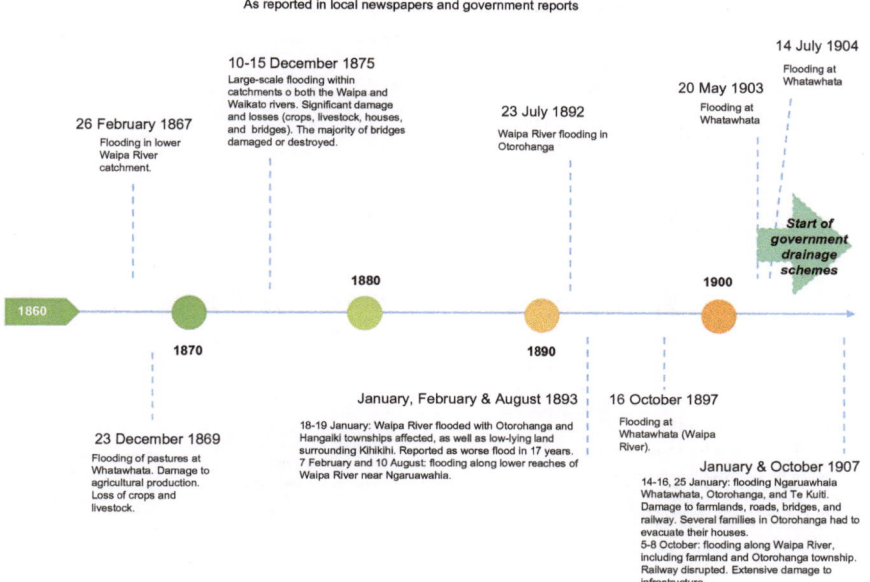

Flooding Waipā River catchment 1867-1907

As reported in local newspapers and government reports

26 February 1867
Flooding in lower
Waipa River
catchment.

10-15 December 1875
Large-scale flooding within
catchments o both the Waipa and
Waikato rivers. Significant damage
and losses (crops, livestock, houses,
and bridges). The majority of bridges
damaged or destroyed.

23 July 1892
Waipa River flooding in
Otorohanga

20 May 1903
Flooding at
Whatawhata

14 July 1904
Flooding at
Whatawhata

**Start of
government
drainage
schemes**

23 December 1869
Flooding of pastures at
Whatawhata. Damage to
agricultural production.
Loss of crops and
livestock.

January, February & August 1893
18-19 January: Waipa River flooded with Otorohanga and
Hangaiki townships affected, as well as low-lying land
surrounding Kihikihi. Reported as worse flood in 17 years.
7 February and 10 August: flooding along lower reaches of
Waipa River near Ngaruawahia.

16 October 1897
Flooding at
Whatawhata (Waipa
River).

January & October 1907
14-16, 25 January: flooding Ngaruawhaia
Whatawhata, Otorohanga, and Te Kuiti.
Damage to farmlands, roads, bridges, and
railway. Several families in Otorohanga had to
evacuate their houses.
5-8 October: flooding along Waipa River,
including farmland and Otorohanga township.
Railway disrupted. Extensive damage to
infrastructure.

Fig. 10.2   Timeline of flooding 1880–1910. Source: Created by Meg Parsons

the taniwha Waiwaia) to rest. Thus, Māori lost more than kai (food) from the imposition of hard adaptations designed to keep water away from people and properties (Iwi Rep 4 2020; Iwi Rep 6 2020). As discussed previously in Chap. 4, Ngāti Maniapoto interviewees spoke of how floods were not perceived as "bad things, [it] would help [with tuna and other fish] migration" (Iwi Rep 6 2020), and emphasised the disaster of colonisation. Hapū-led restoration projects along the Waipā River and its tributaries are deliberately considering how they can address flooding without using hard adaptations and seeking to use their knowledge to facilitate more just outcomes (for themselves, their awa, and their more-than-human kin). As one iwi representative recalled: "we thought of the [incidence] of the … 100-year flood, if we [didn't] put a stopbank [a flood levee] around our marae [tribal meeting complex], how could we slow her [the river] down" and start "planting along the river" (Iwi Rep 4 2020).

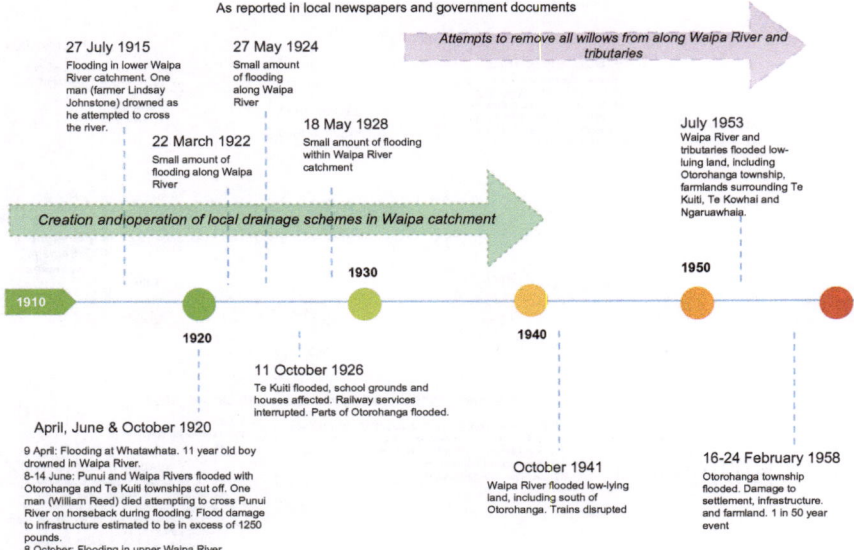

Fig. 10.3 Timeline of flooding 1910–1960. Source: Created by Meg Parsons

The continued use of engineered flood control structures within the Waipā River catchment illustrates clear evidence of path dependency within flood risk management institutions (see Fig. 10.5: Flood control scheme in Otorohanga established after the 1958 flood); it also demonstrates the persistent lack of recognition and procedural inclusion of Māori values, knowledge, and practices (all of which was supposedly guaranteed under various legislation). For more than a century, Māori along the Waipā River have expressed their concerns about the negative impacts of such physical structures on flora and fauna, their capacities to access mahinga kai (food gathering sites), and wahi tapū (sacred sites) sites (as we discussed previously in Chaps. 3 and 4) but they still lack equitable and fair outcomes.

Non-Māori local government officials and staff from NGOs in the Waipā catchment we interviewed situated "flooding issues" as complex matters of science and technology, rather than of importance to Māori.

Flooding Waipā River catchment 1958-2010

As reported in local newspapers and official government reports

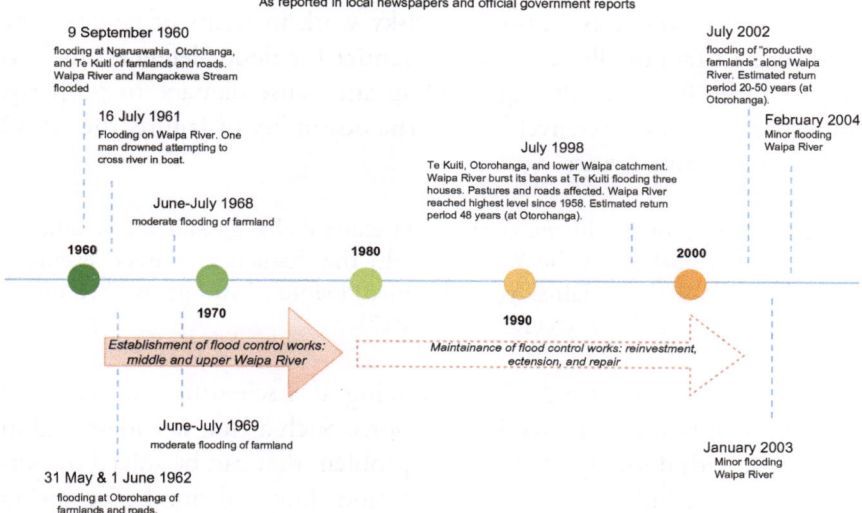

9 September 1960
flooding at Ngaruawahia, Otorohanga, and Te Kuiti of farmlands and roads. Waipa River and Mangaokewa Stream flooded

16 July 1961
Flooding on Waipa River. One man drowned attempting to cross river in boat.

June-July 1968
moderate flooding of farmland

July 2002
flooding of "productive farmlands" along Waipa River. Estimated return period 20-50 years (at Otorohanga).

February 2004
Minor flooding Waipa River

July 1998
Te Kuiti, Otorohanga, and lower Waipa catchment. Waipa River burst its banks at Te Kuiti flooding three houses. Pastures and roads affected. Waipa River reached highest level since 1958. Estimated return period 48 years (at Otorohanga).

1960   **1980**   **2000**

1970   **1990**

Establishment of flood control works: middle and upper Waipa River

Maintenance of flood control works: reinvestment, extension, and repair

June-July 1969
moderate flooding of farmland

January 2003
Minor flooding Waipa River

31 May & 1 June 1962
flooding at Otorohanga of farmlands and roads.

Fig. 10.4 Timeline of flooding 1960–2004. Source: Created by Meg Parsons

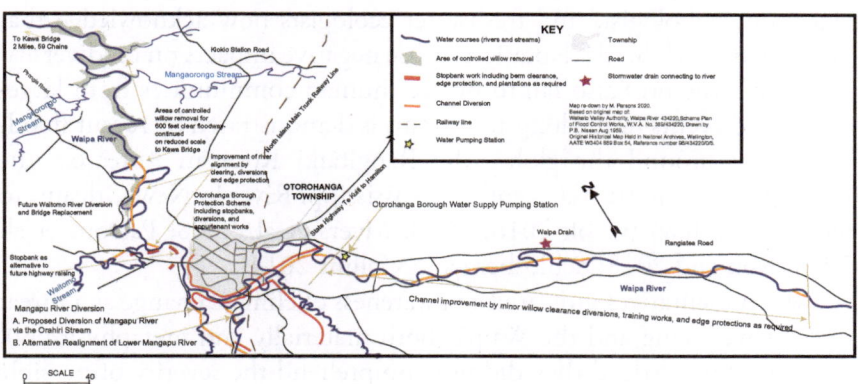

Fig. 10.5 Otorohanga flood control scheme established in 1959 following a major flood event affect Otorohanga township in 1958. Source: Created by Meg Parsons

No mention was made of mātauranga Māori for flood control. Flooding was, one Environmental NGO employer declared, a "technically difficult" problem and was "reasonably risky work in terms of failure"; the reference to failure alludes to the potential for flood defences, such as levees, to be breached during flooding and cause damage to property, infrastructure, and lives/livelihoods. The possibility of failure, the NGO worker maintained, was:

> particularly [notable in the context of] climate change and the weather we've been having [in the Waikato with] the frequency of flood events enabling those things [already] establish[ed] before ... [to] get washed out [was a] bit of an issue. (NGO Rep 1 2017)

One NGO employee depicted flooding as a scientific and technical issue to be solved by engineering solutions. Such a view is widespread in Aotearoa, with flooding narrated as a problem that can be solved by better science (including improved flood modelling and flood monitoring systems) and more extensive flood levees (Parsons et al. 2019). Yet, a wealth of scholarship criticises such a narrow perspective of how to manage flood risk, particularly in the context of climate change. Indeed, fluvial geomorphologists and freshwater ecologists now acknowledge that not only do such hard adaptations cause negative impacts on biodiversity, but such structures also fail to protect human communities from large-scale flood events in the long-term. This is demonstrated by recent flooding events around the globe (the Rangitāiki River in Aotearoa, the Brisbane River in Australia, and the Mississippi River in New Orleans, to name just a few) (Cook 2016, 2018; Myers et al. 2008; Parsons et al. 2019; Rohland 2018; Schlosberg and Collins 2014).

Iwi representatives indicated an awareness of climate change as a threat to their wellbeing and the Waipā (both materially and metaphysically), though acknowledged they did not comprehend the severity of possible climate change impacts on their awa fully. Two iwi representatives, who were involved in hapū-led environmental management and river restoration projects, noted that district councils were "not to our knowledge" considering the risks of climate change in their planning and strategies (Iwi Rep 4 2020; Iwi Rep 6 2020).

[F]or me anyway—[the councils' focus is] about getting more industry in the area to enable the councils to do more. They're [the council] more focused on ... the day to day running [of council services] and how the long term [development] plans and those sorts of things, as opposed to restoring the awa. (Iwi Rep 6 2020)

Iwi representatives were concerned about the focus by local governments on metrics of poor river health (poor water quality) but not on other issues affecting their awa beyond its chemical and physical properties. Ngāti Maniapoto interviewees expressed concerns over water scarcity due to the lack of consideration by WRC of over-allocation and over-extraction, and the risks climate change poses to the health and wellbeing of human and more-than-human actors within the Waipā catchment. Iwi representatives were particularly concerned that WRC continued to grant resource consents to allow extraction (by residential, commercial, and agricultural users), despite claims that water was already over-allocated. One iwi representative argued that the WRC (and district councils) could learn from experiences in Australia with water over-allocation, insecurity, and degradation as a "guide" of what not to do in the Waipā catchment. Specifically, one iwi representative warned that continued extraction could lead to the river system becoming like the systems of South-Eastern Australia (most notably the Murray-Darling Basin) where states allocated water to residential, commercial, and agricultural operations with almost no consideration of water supplies and climate conditions. This led to reduced instream flows and a situation in which rivers could no longer sustain human or ecological communities, with huge costs and damages not just in monetary terms, but also in terms of Aboriginal communities' capacities to access and use resources, maintain their relationships with their tribal lands and waters (their country), and protect their physical and spiritual health and wellbeing. Thus, decision-makers could learn from mātauranga Māori and recent histories of river mismanagement in Aotearoa, as well as other global histories and other knowledges as a way to adapt to the impacts of climate change:

I think water again in the future is going to be a huge issue if we don't sort it out now, because climate change is upon us and really, we're only guessing what's going to happen in the future… [and] the water's getting less and less every year. (Iwi Rep 5 2019)

According to one iwi representative, local governments were short-sightedly exploring economic and technical solutions, including transporting water from one catchment to another (from the Waikato to the Waipā) to offset growing water demand in the Waipā catchment. The iwi representative spoke of writing to the Waipā District Council to complain about proposals to allocate and transfer water outside of the catchment and to draw attention to the interrelatedness of the Waipā River and the land:

why can't you just go through a catchment response, talk to [people up in the headwaters near Waitomo] about using up some of the [water] allocation from up [the Waipā] river … rather than bringing it across from Waikato into the Waipa? I said it doesn't make any sense to me. … It's a bit like a person who needs a blood transfusion because they've been going out partying all their life and I go to you oh, can you give me a blood transfusion? You say yeah, but you carry on your lifestyle, eh? You don't change. Well in time you're going to become unwell giving me all your blood.

The reference to blood reflects Māori understandings of the river that they whakapapa (genealogically connect) to as their lifeblood (Fox et al. 2017; Salmond 2020); in this case, the Waipā River and its tributaries. The consequence of transferring water from one catchment to another place is that it reduces the mauri (life force) of the water in both places and increases the risk of sickness for those who consume it. River restoration, rather than being treated as a specific practice aimed at improving water quality, is narrated by iwi as being interrelated with adjusting the ways in which people engage with and care for (enact kaitiakitanga) the whenua and the awa. It does not just focus on a single dimension of rivers, lands, peoples, or ecosystems (as everything—from a Māori worldview—is interwoven together and bound by ongoing relationships) nor does it focus on short-term decisions. Instead, kaitiakitanga allows for the possibilities of engendering adaptation through local actions.

Climate adaptation actions around the world, including in Aotearoa, remain primarily focused on maintaining existing infrastructure or constructing new physical structures to reduce climate risks. These actions are located within Western ontologies and epistemologies (B. Hayward 2008; Parsons et al. 2019; Reisinger et al. 2011; Nightingale et al. 2019; Barnett et al. 2015; Klein 2011; Manning et al. 2015; Klein et al. 2001; Rangel-Buitrago et al. 2017). However, hard adaptations can be path dependent (simply repeating past decisions and approaches without consideration to new circumstances or past failures) and maladaptive (Barnett et al. 2015; Barnett and O'Neill 2010; Magnan et al. 2016).

Ngāti Maniapoto iwi representatives argue that, in order to address the cascade of environmental crises facing not only Te Waipā Awa (but the world as a whole), attention needs to be directed at holistic understandings of places, peoples, and human and more-than-human relations across generations. Such environmental justice is, thus, not just recognition-, procedural-, and distributive-based but also intergenerational. Only by doing this can people address the root causes of the Anthropocene as well as adapt to the impacts of climate change. Iwi representatives argue, within the context of the Waipā River, attention needs to be given to mātauranga Māori and tikanga in the context of iwi- and hapū-led decision-making processes and practices about how institutions and individuals manage (or rather maintain relationships with) the land (whenua), rivers (awa), and more-than-human actors.

Kaitiakitanga and Climate Justice for the Waipā River

In considering Māori understandings of intergenerational justice and environmental justice, it is critical to recognise justice as place-based, holistic, community-centred, and temporally spiralling. If (neo)colonial ontologies and epistemologies of ignorance (Mills 2015) are premised with existing conceptualisations of intergenerational justice, justice is incorrect, and injustices to Indigenous Peoples distributional. Worse still, as philosophers Dotson and Winter argue, if a theory of justice enacts "epistemological violence against and epistemological exclusion of any group of people, it is unjust" (Dotson 2011, 2014; Winter 2018, p. 36).

Within Māori ontological and cosmological paradigms, it is impossible to conceive the present and future as separate and distinct from the past, for the past is constitutive of the present and, as such, is inherently reconstituted within the future (Stewart-Harawira 2005, p. 42). Winter (2019) raises important questions about the intergenerational environmental justice implications of climate change on Māori duties as kaitiaki (guardians) and the ways in which different perspectives of time are critical considerations. For instance, questions about what time and climate change mean in the contexts of rivers as more-than-human entities (such as the Waipā), and how to formulate plans that address their best interests require careful consideration. A river is more than tens of thousands of years old, a tuna that swims in its waters can be up to one hundred years old (*Anguilla dieffenbachii* or *Anguilla australis*), each exists individually and collectively in multiple realms (material and metaphysical, more-than-human beings and ancestors) that do accord ways of seeing the world.

Within Te Ao Māori (the world of Māori), the principle of kaitiakitanga (environmental guardianship) is founded on the notion that the descendants of Papa-tū-ā-nuku (Earth Mother) need to take care of their environment. Kaitiakitanga emphasises the connections and relationships, and in so doing does not distinguish the "when" of existence thereby, creating a web of intergenerational communitarian obligations founded on the understanding that all life is entangled (human with human, human with nonhuman, nonhuman with nonhuman). It is inconceivable within this way of knowing to locate an individual outside of this entanglement. Indeed, the entanglement provides the power of the whole that must be sustained and encompasses all generations (past, present, future). In essence, the contemporaneous of kaitiakitanga binds generations together so that the mauri, mana, and wairua of all generations of iwi/hapū/whānau (living and not) are given equal consideration within decision-making; thus, when selecting what actions should be taken now (or in the near future) to address climate change, the needs and interests of current generations are not meant to be given priority over those of future generations. One of our Ngāti Maniapoto interview participants described how they understand the temporal and whakapapa entanglements encapsulated in kaitiakitanga and which binds generations:

> kaitiakitanga is an obligation from me as her uri—Papa-tū-ā-nuku's descendant to pay attention to her hā [breath of life] to look after her. To care for her... So kaitiaki—tiaki means to look after, right—kai being I'm responsible for this—kaitiaki has a whakapapa obligation to the relationship with Ranginui [Sky Father] and Papa-tū-ā-nuku ... That's where that kaitiaki comes from. [It's] our whakapapa. Our genealogical relationship to her, to him, to all of ... the cosmos of creation [tracing] right down to us and our kai. (Iwi Rep 9 2019, p. 9)

At present, the majority of decisions about how to address freshwater degradation and climate change are still largely made using Eurocentric framings and tools of assessment that do not take into account Māori conceptualisations of "concurrent past-present-future time" and intergenerational responsibilities of care (embedded in the principle of kaitiakitanga). While Māori, such as Ngāti Maniapoto, continue to practice over their rivers and lands, the privileging of state apparatuses and institutions to determine responses to climate change continue to dominate.

Tuna and Climate Change

At present, there is limited research into the impacts of climate change on Aotearoa's two species of freshwater eels (*Anguilla dieffenbachia* and *Anguilla australis*) (see Fig. 10.6); however, international research indicates that temperate eels around the world are especially vulnerable to changes associated with climate change (Aarestrup et al. 2010; Arai 2014; Drouineau et al. 2018; Jellyman et al. 2009). Freshwater eels' distinct life cycles, which include oceanic spawning grounds and the growth stage in freshwater systems, are impacted by five elements of global environmental change: 1) climate change affects the survival and drift of eel larval; 2) increased pollution contributes to the accumulation of high levels of contamination; 3) greater habitat loss and fragmentation that decrease the

Fig. 10.6 Sketch of a freshwater eel—'Longfin eel (*Anguilla dieffenbachia*)'. Source Alicia Wong (artist)

amount of available habitat radically and induce higher eel mortality; 4) warmer temperatures linked to the appearance of parasitic nematode (roundworms) that reduces the success of eel spawning; and 5) the effect of recreational and commercial fisheries on eel populations (Drouineau et al. 2018, p. 903). In this context, ongoing social and environmental processes of change are likely to surpass the capacity of eels to adapt to changing environmental conditions around the world. Some scientists suggest that the cumulative impacts of global environmental change, including climate change, may lead to the collapse of eel and other of aquatic species, even in species that possess a high adaptive capacity (August and Hicks 2008; Capon et al. 2013; Death et al. 2016; Drouineau et al. 2018; Pacariz et al. 2014). Despite the limited research (mātauranga Māori and/or scientific studies) into the impacts of climate change on Aotearoa's tuna and reducing the vulnerability of tuna (Jellyman et al. 2009; 'Tuna—customary fisheries' 2012), there are indications within the Waipā catchment that climate change (in combination with other environmental stresses) will negatively affect Māori capacities to harvest tuna in the future.

The lack of attention to the health of tuna, in part, highlights the ways in which climate change and other environmental problems are often treated (by those employing Western knowledge) as distinct and entirely separate issues rather than holistic and interconnected issues (markers of the Anthropocene). Moreover, it demonstrates how Te Ao Pākehā (the Pākehā worldview) continues to shape what researchers, decision-makers, and environmental management practitioners consider to be worthy topics of study and what actions should be taken to conserve particular plants and animals. Tuna remains largely overlooked by those looking for evidence of the Anthropocene (as opposed to those interested in water quality) because tuna are not highly valued within Pākehā society (although tuna are commercially fished in small numbers). By and large, Pākehā recreational fishers seek to catch trout and salmon from Aotearoa's rivers and lakes (not eels); whereas, for Māori hapū and iwi, tuna remains a taonga. Tuna, through the settler-colonial gaze, are not viewed as 'iconic' evidence of climate change (instead Aotearoa-based researchers search for signs of rising sea levels, "sinking" islands, and Pacific climate refugees). For Māori, in contrast, the decline and loss of tuna (alongside many other native fauna) are important markers of the interconnected consequences of Anthropocene-settler-colonialism-climate change that requires action.

Our Ngāti Maniapoto research participants report the worsening health of tuna, despite ongoing river restoration projects along the Waipā River, but do not necessarily draw clear linkages between declining tuna numbers, poor-health, and climate change. In the upper catchment of the Waipā River, kaitiaki report that the tuna caught had liver fluke (a name for the group of parasitic trematodes *Platyhelminthes*) that can cause serious health problems in both tuna and other animals (including people) (Iwi Rep 2 2020; Iwi Rep 3 2020). Other kaitiaki reported working with scientists from the National Institute of Water and Atmospheric Research (NIWA—the New Zealand government research institute that focuses on climate, freshwater and ocean science) to try to understand why their tuna are sick. The scientists conducted autopsies on tuna and found parasitic worms (probably nematode) as well as liver fluke (Iwi Rep 4 2020; Iwi Rep 6 2020). Even tuna caught in the headwaters of the Waipā River, where the water quality is far better than downstream, have been found filled with invasive parasites. Tangata whenua are not sure what is causing the infestation of their tuna (which was a food source, a member of their extended family, and a taonga), though some suggest farming run-off might be responsible for parasitic infections (none mentioned climate change). One iwi representative asked:

> what do you do to get rid of it is what I want to know in your scientific world. How do you drench an eel? What sort of chemical … it would be an ideal to find out [what we could use] if we had one eel that did worms and give it a [chemical or] cider vinegar and if it does bring the worms out? (Iwi Rep 2 2020)

The causes of declining tuna health (both in terms of health status and numbers) as well as other aquatic biota in the Waipā River was uncertain (even amongst the freshwater scientists we interviewed), although some raised the issues of increased water allocation, intensification of dairy farming, and climate change. Instead, declining river health was discussed in terms of loss of vegetation and biodiversity, land-use changes, marginalisation of mātauranga Māori and iwi rangatiratanga (tribal authority), and polluting practices (sedimentation, agricultural run-off, effluent discharge) rather than global climate change. Yet, iwi representatives argued they were in dire need of information about climate change, and the topic was not widely discussed amongst government or co-governance bodies. Iwi representatives argued they wanted research and practical explanations of the ways in which they (and others) could protect and enhance tuna habitat in the context of changing environmental conditions. For iwi, being able to care for the tuna both now and in the future was an essential part of their intergenerational duties as kaitiaki (Iwi Rep 2 2020).

In previous chapters, we recount stories of tuna prized, caught, and lost. When once abundant, tuna swam in the peat lakes, flowing rivers and porous wetlands. The destruction of pā tuna, straightening of the river, pollution and presence of unwanted vegetation have led to reduced catches by Ngāti Maniapoto, and also speak to the complexities, multiple histories, and socio-cultural and political vitality of storied freshwater spaces. Tuna, like land, water, and other taonga are figuratively and literally connections between humans and more-than-human worlds that spiral through time and bind everything together through whakapapa (including mātauranga Māori, tikanga, and Māori as members of whānau, hapū and iwi). It is worthwhile for Indigenous and non-Indigenous scholars and freshwater water management practitioners to consider freshwater, and its functions, as more than sites upon which humans enact history or as physical locations in which history (and the sediment washing off cleared land) accumulates. Rather, historic and contemporary manifestations of colonial violence are deeply interwoven with ecological violence (including those linked to climate change).

Scientific indicators and measures are frequently used as the only worthy (accurate and objective) evidence of ecological destruction; at present, tuna are not worthy of much attention, but herein we suggest it should be. Yet, as our research (and the work of other Indigenous scholars) demonstrates, Indigenous experiences, stories, histories, and knowledge(s) provide a rich body of evidence about how colonial invasion, violence, violations, oppression of Indigenous and other sovereign peoples resulted in radical challenges to social and ecological systems, including the impacts of climate change. Scientific reports and academic studies documenting the commencement, speed, and pace of the Anthropocene and the impacts of climate change rarely include such "fleshy stories" (to borrow Métis scholar Zoe Todd's term) that Indigenous elders around the globe tell younger generations about how once (before colonisation, hydro-electric dams, introduced species, pollution, and commercial fisheries) there was an abundance of fish that swam in their waterways. Nor do they tell stories of the fish that were (are) caught, cooked, preserved and fed to family, friends, visitors and kin members. As Todd writes, the evidence used to record the Anthropocene precludes "the flash of a school of minnows in the clear prairie lakes I intimately knew as a child ... the succulent white fish my stepdad caught from us from the Red Deer River when I was growing up" (Davis and Todd 2017, p. 767). The consequences of loss and damage to fish, therefore, extend beyond the physical and metaphysical conditions. Such lessons (evidence of nuanced human and nonhuman relations and environmental changes), she explains, are often overlooked and are:

> deeply erased from dominant (non-Indigenous) public discourse in … [Canada] and I had not recognized the implicit ways fish were woven into my own life as more than food. This is the thing about colonization: it tries to erase the relationships and reciprocal duties we share across boundaries, across stories, across species, across space, and it inserts new logics, new principles, and new ideologies in their place. (Todd 2016)

In Aotearoa, similarly, such fleshy stories of the interconnected, inter-generational, and reciprocal relationships between living entities (waters, flora and fauna, supernatural beings and peoples) are continually margin-alised within how freshwater is governed and managed. In the Waipā River, river restoration is primarily directed at improving water quality (specifically through sedimentation reduction) rather than addressing other climate and non-climate risks (including flood control and loss of biodiversity) (Local Government Rep 1 2018; Local Government Rep 2 2019; Local Government Rep 3 2017). As local one government official informed us:

> Restoration practices, such as fencing and planting of riparian zones, are important tools to improve waterway health and play a significant part in [regional] council achieving [its] goals and vision. This is reflected by the catchment management funding incentive available, not just in the Waipā catchment, but across the region to encourage landowners to undertake measures on their properties to improve the ecological health of waterways. (Local Government Rep 3 2017)

Many tangata whenua critique this narrow framing of river restora-tion, which only seeks to draw linkages between water quality and ecosys-tem integrity, but not to the health of nonhuman beings nor the reciprocal (kinship-based) relations between people, plants, insects, animals and water. Healthy waters equal healthy biota and healthy people (Iwi Rep 5 2019; Iwi Rep 6 2020; Kaitiaki 1 2020; Kaitiaki 5 2020).

In contrast, mātauranga Māori emphasises interconnectivity between all beings (living and non-living) across time, and encompasses all generations (ancestors, people living now, and those who will live in the future). Within this holistic ontology, restoration efforts are inseparable from current and future efforts to reduce the pollution of air, water, land, and actions to mitigate and adapt to the impacts of climate change. All are focused on the same goal: the protection and enhancement of the mauri of all beings, and these goals and duties of care are intergenerational. Such a view is steeped in Māori perspectives of time as a temporal spiral wherein the present (such as the current state of the environmental degradation within the Waipā River) is not seen as more important than the past or future: it is one of equals (Rameka 2016; Winter 2019). Despite the emergence of co-governance and co-management arrangements for the Waipā River, Western ontologies and epistemologies continue to dominate river restoration (and the freshwater management more generally) and frame the past/present/future as discreetly separate domains. Although there is formal recognition in legislation that past policies and actions contributed to current environmental degradation in the Waipā River, for Pākehā that past is behind them and they do not carry the past (including their ancestors and their histories) with them in the present and future times. The goal of river management, therefore, continues to focus on the present and immediate future needs and priorities of people alive together, while other framings of time (including the distant past and future) and different generations (of human and more-than-human actors) remain excluded from consideration.

Conclusion

With ontological underpinnings premised on communitarianism and holism, Ngāti Maniapoto articulations of IEJ emphasise the ways in which everything is bound together across the neverending temporal spiral that interweaves past/present/future/past as coexisting times. Just as an injustice against one member of one's kin group (a river) is an injustice to all (metaphysical and social realms), similarly an injustice against a tūpuna (an ancestor), a person living today, or one living in the future is an injustice for all generations. As the previous chapters on wetland drainage and water pollution attest too, IEJ, in the context of Ngāti Maniapoto relationships with their awa, involves the lassoing and braiding together of different dimensions of justice, wherein pluralism (in ontologies, laws, management approaches) is an essential pathway by which just (tika) interactions with water, land, biota, time, and generations can be (re)established. We argue, following on from McGregor (2014) and Winter (2018), the ontological clashes between Te Ao Pākehā and Te Ao Māori rest in part on the former's anthropocentrism and individualism and latter's holism and communitarianism. Western intellectual frameworks underpinning EJ is framed in terms of justice for human beings, specifically individual rights and entitlements, including people's rights to clean water, air, food, dignity, participatory parity, and recognition of their cultural differences. In contrast, Ngāti Maniapoto EJ is about justice for both human and more-than-human entities within their rohe, with threats to the existence of beings that live within their awa threatening the health and wellbeing of mana whenua (past/present/future generations). Moreover, from a Te Ao Māori perspective, environmental justice lassos human and more-than-human entities together reciprocal relationships premised on kinship ties. In this way the Waipā River, a non-human actor with its own agency, mana, and mauri, also provides waters, habitat, and connections that are life-affirming for diverse entities, including providing the lifeblood for the Ngāti Maniapoto people. In the worldview of Ngāti Maniapoto, all beings within their landscapes and waterscapes possess reciprocal responsibilities to each other which stretch across generations spiralling backwards and forwards through time. EJ in the context of freshwater governance and management is much wider than the 'impacts' of freshwater degradation on people's health; it also extends to include duties that ensure the continuation of more-than-human entities and processes linked to the continuation of metaphysical realms and the current and future impacts of climate change on material and metaphysical worlds. Climate injustices, as we demonstrate in regard to the example of tuna (eels), can and are occurring for more-than-human beings not only when their habitats are degraded and destroyed through drainage, forest clearance, water pollution, and climate change, but also because of misrecognition of the mana (power and status) of tuna, its mauri, and the interconnection of tuna to the health and wellbeing of tangata whenua.

References

Aarestrup, K., Thorstad, E. B., Koed, A., Svendsen, J. C., Jepsen, N., Pedersen, M. I., & Økland, F. (2010). Survival and Progression Rates of Large European Silver Eel Anguilla Anguilla in Late Freshwater and Early Marine Phases. *Aquatic Biology, 9*(3), 263–270.

Adler, A. A., Doole, G. J., Romera, A. J., & Beukes, P. C. (2013). Cost-effective Mitigation of Greenhouse Gas Emissions from Different Dairy Systems in the Waikato Region of New Zealand. *Journal of Environmental Management, 131*, 33.

Alfred, T. (2008). *Peace, Power, Righteousness: An Indigenous Manifesto* (2nd ed.). New York: Oxford University Press Canada.

Alfred, T. (2015). Cultural Strength: Restoring the Place of Indigenous Knowledge in Practice and Policy. *Australian Aboriginal Studies, 1*, 3.

Arai, T. (2014). Do We Protect Freshwater Eels or Do We Drive Them to Extinction? *SpringerPlus, 3*(1), 534.

August, S. M., & Hicks, B. J. (2008). Water Temperature and upstream migration of glass eels in New Zealand: implications of climate change. *Environmental Biology of Fishes, 81*(2), 195–205.

Baldwin, A. (2009). Carbon Nullius and Racial Rule: Race, Nature and the Cultural Politics of Forest Carbon in Canada. *Antipode, 41*(2), 231–255.

Ballantine, D. J., & Davies-Colley, R. J. (2014). Water Quality Trends in New Zealand Rivers: 1989–2009. *Environmental Monitoring and Assessment, 186*(3), 1939–1950.

Barnett, J., Evans, L., Gross, C., Kiem, A., Kingsford, R., Palutikof, J., et al. (2015). From Barriers to Limits to Climate Change Adaptation: Path Dependency and the Speed of Change. *Ecology and Society, 20*(3).

Barnett, J., & O'Neill, S. (2010). Maladaptation. *Global Environmental Change, 20*(2), 211–213.

Baum, A. (2019). Mni Wiconi (Water is Life): Knowledge, Power and Resistance at Standing Rock. *Ideas from IDS: Graduate Papers from 2017/18*, 9.

Bell, D. (2011). Global Climate Justice, Historic Emissions, and Excusable Ignorance. *The Monist, 94*(3), 391–411.

Bell, D. (2013). Climate Change and Human Rights. *WIREs Climate Change,* *4*(3), 159–170.

Brännlund, I., & Axelsson, P. (2011). Reindeer Management During the Colonization of Sami lands: A Long-term Perspective of Vulnerability and Adaptation Strategies. *Global Environmental Change, 21*(3), 1095–1105.

Brownsey, P. (2019, December 12). *Addressing Climate Change Through Investment in Global Markets, Paul Brownsey, CIO Pathfinder Asset Management.* Workshop Presentation presented at the The Waikato River Authority, Responding to Climate Change for the Waikato and Waipā rivers—implications for Te Ture Whaimana/Vision & Strategy, Waikato-Tainui Endowed College, Ngaruawahia. Retrieved April 29, 2020, from https://waikatoriver.org.nz/climate-change-workshop/.

Callison, C. (2015). *How Climate Change Comes to Matter: The Communal Life of Facts.* Durham: Duke University Press.

Cameron, E. S. (2012). Securing Indigenous Politics: A Critique of the Vulnerability and Adaptation Approach to the Human Dimensions of Climate Change in the Canadian Arctic. *Global Environmental Change, 22*(1), 103–114.

Caney, S. (2008). Global Distributive Justice and the State. *Political Studies, 56*(3), 487–518.

Caney, S. (2009). Justice and the Distribution of Greenhouse Gas Emissions. *Journal of Global Ethics, 5*(2), 125–146.

Caney, S. (2014). Two Kinds of Climate Justice: Avoiding Harm and Sharing Burdens. *Journal of Political Philosophy, 22*(2), 125–149.

Capon, S. J., Chambers, L. E., Mac Nally, R., Naiman, R. J., Davies, P., Marshall, N., et al. (2013). Riparian Ecosystems in the 21st Century: Hotspots for Climate Change Adaptation? *Ecosystems, 16*(3), 359–381.

Carter, L. (2018). *Indigenous Pacific Approaches to Climate Change: Aotearoa/New Zealand.* Cham: Palgrave Pivot.

Chapman, M. A. (1996). Human Impacts on the Waikato River System, New Zealand. *GeoJournal, 40*(1), 85–99.

Cook, M. (2016). Damming the 'Flood Evil' on the Brisbane River. *History Australia, 13*(4), 540–556.

Cook, M. (2018). "It Will Never Happen Again": The Myth of Flood Immunity in Brisbane. *Journal of Australian Studies, 42*(3), 328–342.

Davis, H., & Todd, Z. (2017). On the Importance of a Date, or Decolonizing the Anthropocene. *ACME: An International E-Journal for Critical Geographies, 16*(4), 1.

Death, R., Bowie, S., & O'Donnell, C. (2016). 3. Vulnerability of freshwater ecosystems due to climate change. *Freshwater conservation under a changing climate*, 14.

Dickie, B. (2019, December 12). *The challenge of climate change adaptation and mitigation in the Waikato Region, Blair Dickie, Waikato Regional Authority*. Workshop Presentation presented at the The Waikato River Authority, Responding to Climate Change for the Waikato and Waipā rivers—implications for Te Ture Whaimana/Vision & Strategy, Waikato-Tainui Endowed College, Ngaruawahia. Retrieved June 8, 2020, from https://waikatoriver.org.nz/climate-change-workshop/.

Dotson, K. (2011). Tracking Epistemic Violence, Tracking Practices of Silencing. *Hypatia, 26*(2), 236–257.

Dotson, K. (2014). Conceptualizing Epistemic Oppression. *Social Epistemology, 28*(2), 115–138.

Dressler, W., McDermott, M., Smith, W., & Pulhin, J. (2012). REDD Policy Impacts on Indigenous Property Rights Regimes on Palawan Island, The Philippines. *Human Ecology, 40*(5), 679–691.

Driver, E., Parsons, M., & Fisher, K. (2018). Technically Political: The Post-politics(?) of the New Zealand Emissions Trading Scheme. *Geoforum, 97*, 253–267.

Drouineau, H., Durif, C., Castonguay, M., Mateo, M., Rochard, E., Verreault, G., et al. (2018). Freshwater Eels: A Symbol of the Effects of Global Change. *Fish and Fisheries, 19*(5), 903–930.

Dudley, B. D., Burge, O., Plew, D. R., & Zeldis, J. (2020). Effects of Agricultural and Urban Land Cover on New Zealand's Estuarine Water Quality. *New Zealand Journal of Marine and Freshwater Research, 54*, 1–21.

Durie, M. (1998). *Te mana, te kāwanatanga : The Politics of Māori self-determination*. Auckland: Oxford University Press.

Fox, C. A., Reo, N. J., Turner, D. A., Cook, J., Dituri, F., Fessell, B., et al. (2017). "The River Is Us; The River Is in Our Veins": Re-defining River Restoration in Three Indigenous Communities. *Sustainability Science, 1*, 1–13.

Gerrard, E. (2012). Towards a Carbon Constrained Future: Climate Change, Emissions Trading and Indigenous Peoples' Rights in Australia. In J. K. Weir (Ed.), *Country, Native Title and Ecology* (1st ed., pp. 135–174). Canberra: ANU Press.

Gilio-Whitaker, D. (2019). *As Long as Grass Grows: The Indigenous Fight for Environmental Justice from Colonization to Standing Rock*. Boston: Beacon Press.

Harawira, H. (2007, December 11). Harawira, Hone: Climate Change (Emissions Trading and Renewable Preference) Bill—First Reading—New Zealand Parliament. *Hansard (Parliamentary Debates)*. New Zealand Parliament. Retrieved June 10, 2020, from https://www.parliament.nz/en/pb/hansard-debates/rhr/document/48HansS_20071212_00000875/harawira-hone-climate-change-emissions-trading-and-renewable.

Hayward, B. (2008). 'Nowhere Far from the Sea': Political Challenges of Coastal Adaptation to Climate Change in New Zealand. *Political Science, 60*(1), 47–59.

Hayward, T., & Iwaki, Y. (2016). Had We But World Enough, and Time: Integrating the Dimensions of Global Justice. *Critical Review of International Social and Political Philosophy, 19*(4), 383–399.

Hoegh-Guldberg, O., Jacob, D., Bindi, M., Brown, S., Camilloni, I., Diedhiou, A., et al. (2018). Impacts of 1.5°C Global Warming on Natural and Human Systems. In *Global warming of 1.5°C.: An IPCC Special Report* (pp. 175–311). IPCC Secretariat. Retrieved May 1, 2020, from https://researchportal.helsinki.fi/en/publications/impacts-of-15c-global-warming-on-natural-and-human-systems.

Hopkins, D., Campbell-Hunt, C., Carter, L., Higham, J. E. S., & Rosin, C. (2015). Climate Change and Aotearoa New Zealand: Climate Change and Aotearoa New Zealand. *Wiley Interdisciplinary Reviews: Climate Change, 6*(6), 559–583.

Howitt, R., Havnen, O., & Veland, S. (2012). Natural and Unnatural Disasters: Responding with Respect for Indigenous Rights and Knowledges. *Geographical Research, 50*(1), 47–59.

Ivison, D. (2003). The Logic of Aboriginal Rights. *Ethnicities, 3*(3), 321–344. https://doi.org/10.1177/14687968030033003.

Ivison, D. (2014). Indigenous Peoples' Rights. In *The Encyclopedia of Political Thought* (pp. 1815–1817). American Cancer Society. Retrieved June 28, 2020, from https://doi.org/10.1002/9781118474396.wbept0503.

Iwi Rep 2. (2020, February 13). Interview with Iwi Representative 2.

Iwi Rep 3. (2020, February 13). Interview with iwi Representative 3.

Iwi Rep 4. (2020, February 14). Interview with Iwi Representative 4.

Iwi Rep 5. (2019, March 25). Interview with Interview Iwi Representative 5.

Iwi Rep 6. (2020, February 14). Interview with Iwi Representative 6.

Iwi Rep 9. (2019, October 9). Interview with Iwi Representative 9.

Jellyman, D. J., Booker, D. J., & Watene, E. (2009). Recruitment of Anguilla spp. glass eels in the Waikato River, New Zealand. *Evidence of declining migrations? Journal of Fish Biology, 74*(9), 2014–2033.

Kaitiaki 1. (2020, February 4). Interview with Kaitiaki 1.

Kaitiaki 5. (2020, February 5). Interview with Kaitiaki 5.

Klein, R. J., Nicholls, R. J., Ragoonaden, S., Capobianco, M., Aston, J., & Buckley, E. N. (2001). Technological Options for Adaptation to Climate Change in Coastal Zones. *Journal of Coastal Research, 20*, 531–543.

Klein, R. J. (2011). Adaptation to Climate Change. In I. Linkov & T. S. Bridges (Eds.), *Climate* (pp. 157–168). Dordrecht: Springer Netherlands.

Ledgard, S. (2019, December 12). *Addressing Biological Emissions: Mitigation and Co-benefits on Farms, Dr Stewrad Ledgard, Principal Scientist AgResearch.* Workshop Presentation presented at the The Waikato River Authority, Responding to Climate Change for the Waikato and Waipā rivers—implications for Te Ture Whaimana/Vision & Strategy, Waikato-Tainui Endowed College, Ngaruawahia. Retrieved April 29, 2020, from https://waikatoriver.org.nz/climate-change-workshop/.

LeQuesne, T. (2019). Petro-hegemony and the Matrix of Resistance: What Can Standing Rock's Water Protectors Teach Us About Organizing for Climate Justice in the United States? *Environmental Sociology, 5*(2), 188–206.

Li, T. M. (2010). Indigeneity, Capitalism, and the Management of Dispossession. *Current Anthropology, 51*(3), 385–414.

Linwood, R. (2019, December 12). *Freshwater Policy and Climate Change: Synergy and Co-benefits, Rachelle Linwood, Manager Freshwater Policy, Ministry of Primary Industries.* Workshop Presentation presented at the The Waikato River Authority, Responding to Climate Change for the Waikato and Waipā rivers—implications for Te Ture Whaimana/Vision & Strategy, Waikato-Tainui Endowed College, Ngaruawahia. Retrieved April 29, 2020, from https://waikatoriver.org.nz/climate-change-workshop/.

Local Government Rep 1. (2018, October 4). Interview with Local Government Representative 1.

Local Government Rep 2. (2019, March 25). Interview with Local Government Representative 2.

Local Government Rep 3. (2017, September 29). Interview with Local Government Representative 3.

Love, T., & Tilley, E. (2013). Temporal Discourse and the News Media Representation of Indigenous-Non-Indigenous Relations: A Case Study from Aotearoa New Zealand. *Media International Australia, 149*(1), 174–188. https://doi.org/10.1177/1329878X1314900118.

Magnan, A. K., Schipper, E. I. F., Burkett, M., Bharwani, S., Burton, I., Eriksen, S., et al. (2016). Addressing the Risk of Maladaptation to Climate Change. *Wiley Interdisciplinary Reviews: Climate Change, 7*(5), 646–665.

Manning, M., Lawrence, J., King, D. N., & Chapman, R. (2015). Dealing with Changing Risks: A New Zealand Perspective on Climate Change Adaptation. *Regional Environmental Change, 15*(4), 581–594.

McGregor, D. (2014). Traditional Knowledge and Water Governance: The Ethic of Responsibility. *AlterNative: An International Journal of Indigenous Peoples, 10*(5), 493–507.

Mckay, B., & Walmsley, A. (1969). Maori Time: Notions of Space, Time and Building Form in the South Pacific. *Idea Journal, 20*, 85–95.

McMillen, H., Ticktin, T., & Springer, H. K. (2017). The Future Is Behind Us: Traditional Ecological Knowledge and Resilience Over Time on Hawai'i Island. *Regional Environmental Change, 17*(2), 579–592.

Mills, C. W. (2015). Race and Global Justice. In *Domination and Global Political Justice* (pp. 193–217). New York: Routledge.

Myers, C. A., Slack, T., & Singelmann, J. (2008). Social Vulnerability and Migration in the Wake of Disaster: The Case of Hurricanes Katrina and Rita. *Population and Environment, 29*(6), 271–291.

National Business Review. (2018, July 11). Iwi Leaders Complain Weak ETS Destroying Maori Asset Value. *NBR*. Retrieved June 10, 2020, from https://www.nbr.co.nz/article/iwi-leaders-complain-weak-ets-destroying-maori-asset-values-bd-130795.

New Zealand, Ministry for the Environment, New Zealand, & Stats NZ. (2020). *Our freshwater 2020*. Retrieved April 28, 2020, from https://www.mfe.govt.nz/sites/default/files/media/Environmental%20reporting/our-freshwater-2020.pdf.

NGO Rep 1. (2017, September 28). NGO Representative 1.

Nightingale, A. J., Eriksen, S., Taylor, M., Forsyth, T., Pelling, M., Newsham, A., et al. (2019). Beyond Technical Fixes: Climate Solutions and the Great Derangement. *Climate and Development, 12*(4), 343–352.

Nursey-Bray, M. (2016). Cultural Indicators, Country and Culture: The Arabana, Change and Water. *The Rangeland Journal, 37*(6), 555–569.

Nursey-Bray, M., Palmer, R., Smith, T. F., & Rist, P. (2019). Old Ways for New Days: Australian Indigenous Peoples and Climate Change. *Local Environment, 1*, 1–14.

Pacariz, S., Westerberg, H., & Björk, G. (2014). Climate Change and Passive Transport of European Eel Larvae. *Ecology of Freshwater Fish, 23*(1), 86–94.

Page, E. A. (2007a). Intergenerational Justice of What: Welfare, Resources or Capabilities? *Environmental Politics, 16*(3), 453–469.

Page, E. A. (2007b). *Climate Change, Justice and Future Generations*. Cheltenham; Northampton, MA: Edward Elgar Publishing.

Parfit, D. (2011). *On What Matters*. Oxford: Oxford University Press.

Parsons, M., & Nalau, J. (2016). Historical Analogies as Tools in Understanding Transformation. *Global Environmental Change, 38*, 82–96.

Parsons, M., Nalau, J., Fisher, K., & Brown, C. (2019). Disrupting Path Dependency: Making Room for Indigenous Knowledge in River Management. *Global Environmental Change, 56*, 95–113.

Patterson, J. (1998). Respecting Nature: A Maori Perspective. *Worldviews: Global Religions, Culture, and Ecology, 2*(1), 69–78.

Pearce, P. (2019, December 12). *Climate Change—What Does It Mean for Waikato? Petra Pearce, Manager, Climate, Atmosphere and Hazards, NIWA*. Workshop Presentation presented at the The Waikato River Authority, Responding to Climate Change for the Waikato and Waipā rivers—implications for Te Ture Whaimana/Vision & Strategy, Waikato-Tainui Endowed College, Ngaruawahia. Retrieved April 29, 2020, from https://waikatoriver. org.nz/climate-change-workshop/.

Pelling, M., Manuel-Navarrete, D., & Redclift, M. (2012). *Climate Change and the Crisis of Capitalism: A Chance to Reclaim, Self, Society and Nature*. Routledge. Retrieved April 27, 2017, from https://books.google.co.nz/books ?hl=en&lr=&id=Wy7zD66p9ngC&oi=fnd&pg=PP2&dq=Pelling+Navarret te&ots=R2gdQaucD7&sig=899ymbdnBRY7apxRN6NmLE5yLmE.

Povinelli, E. A. (2016). *Geontologies: A Requiem to Late Liberalism*. Durham: Duke University Press.

Povinelli, E. A., Coleman, M., & Yusoff, K. (2017). An Interview with Elizabeth Povinelli: Geontopower, Biopolitics and the Anthropocene. *Theory, Culture & Society, 34*(2–3), 169–185.

Rameka, L. (2016). Kia whakatōmuri te haere whakamua: 'I Walk Backwards into the Future with My Eyes Fixed on My Past': *Contemporary Issues in Early Childhood*.

Rangel-Buitrago, N., Williams, A., & Anfuso, G. (2017). Hard Protection Structures as a Principal Coastal Erosion Management Strategy Along the Caribbean Coast of Colombia. A Chronicle of Pitfalls. *Ocean & Coastal Management, 156*(58–75), 1.

Reisinger, A., Kitching, R. L., Chiew, F., Hughes, L., Newton, P. C., Schuster, S. S., et al. (2014). Australasia. In V. R. Barros, C. B. Field, M. D. Dokken, P. R. Mastrandrea, & L. L. White (Eds.), *Climate Change 2014: Impacts, Adaptation, and Vulnerability. Part B: Regional Aspects. Contribution of Working Group II to the Fifth Assessment Report of the Intergovernmental Panel on Climate Change* (pp. 1371–1438). Cambridge and New York: Cambridge University Press. Retrieved April 21, 2017 from https://www.researchonline. mq.edu.au/vital/access/manager/Repository/mq:60146.

Reisinger, A., Wratt, D., Allan, S., & Larsen, H. (2011). The Role of Local Government in Adapting to Climate Change: Lessons from New Zealand. In J. D. Ford & L. Berrang-Ford (Eds.), *Climate Change Adaptation in Developed Nations: From Theory to Practice* (pp. 303–319). Dordrecht: Springer Netherlands.

Rohland, E. (2018). Adapting to Hurricanes. A Historical Perspective on New Orleans from Its Foundation to Hurricane Katrina, 1718–2005. *Wiley Interdisciplinary Reviews: Climate Change, 9*(1), e488.

Roy, J., Tschakert, P., Waisman, H., Halim, S. A., Antwi-Agyei, P., Dasgupta, P., et al. (2019). Sustainable Development, Poverty Eradication and Reducing Inequalities. In V. Masson-Delmotte, P. Zhai, H. O. Pörtner, D. Roberts, J. Skea, P. R. Shukla, et al. (Eds.), *Global Warming of 1.5 °C. An IPCC Special Report on the impacts of global warming of 1.5 °C above pre-industrial levels and related global greenhouse gas emission pathways, in the context of strengthening the global response to the threat of climate change, sustainable development, and efforts to eradicate poverty* (pp. 445–538). Intergovernmental Panel on Climate Change. Retrieved June 9, 2020, from https://www.ipcc.ch/site/assets/uploads/sites/2/2019/05/SR15_Chapter5_Low_Res.pdf.

Rumbach, A., & Foley, D. (2014). Indigenous Institutions and Their Role in Disaster Risk Reduction and Resilience: Evidence from the 2009 Tsunami in American Samoa. *Ecology and Society, 19*(1), 1–9.

Salmond, D. A. (2020). Afterword: 'I Am the River, and the River Is Me'. *The Contemporary Pacific, 32*(1), 164–171.

Schlosberg, D. (2012). Climate Justice and Capabilities: A Framework for Adaptation Policy. *Ethics & International Affairs, 26*(4), 445–461.

Schlosberg, D., & Collins, L. B. (2014). From Environmental to Climate Justice: Climate Change and the Discourse of Environmental Justice. *Wiley Interdisciplinary Reviews: Climate Change, 5*(3), 359–374.

Shue, H. (2014). *Climate Justice: Vulnerability and Protection.* Oxford: Oxford University Press.

Stern, N., Stern, N. H., & Treasury, G. B. (2007). *The Economics of Climate Change: The Stern Review.* Cambridge: Cambridge University Press.

Stewart-Harawira, M. (2005). Cultural Studies, Indigenous Knowledge and Pedagogies of Hope. *Policy Futures in Education, 3*(2), 153–163.

Stewart-Harawira, M. (2018). *Indigenous Resilience and Pedagogies of Resistance: Responding to the Crisis of Our Age (May 27, 2018).* Available at SSRN: https://ssrn.com/abstract=3185625 or https://doi.org/10.2139/ssrn.3185625

Todd, Z. (2016). *'You Never Go Hungry' : Fish Pluralities, Human-fish Relationships, Indigenous Legal Orders and Colonialism in Paulatuuq, Canada*

(Thesis). University of Aberdeen, Aberdeen. Retrieved June 8, 2019, from http://digitool.abdn.ac.uk:80/webclient/DeliveryManager?pid=231448.

Tuna—customary fisheries. (2012, January 26). *NIWA*. Retrieved August 13, 2017, from https://www.niwa.co.nz/te-k%C5%ABwaha/tuna-information-resource/pressures-on-new-zealand-populations/customary-tuna-fisheries.

Unknown Author. (2009, November 16). National's ETS to Include Special Treatment for Maori. *News Hub*. Retrieved June 10, 2020, from https://www.newshub.co.nz/general/nationals-ets-to-include-special-treatment-for-maori-2009111617.

Unknown Author. (2019, November 19). Waikato River Authority Workshop to Tackle Climate Change | Stuff.co.nz. *Waikato Times*. Hamilton. Retrieved April 28, 2020, from https://www.stuff.co.nz/waikato-times/news/117357753/waikato-river-authority-workshop-to-tackle-climate-change.

Veland, S., Howitt, R., Dominey-Howes, D., Thomalla, F., & Houston, D. (2013). Procedural Vulnerability: Understanding Environmental Change in a Remote Indigenous Community. *Global Environmental Change, 23*(1), 314–326.

Vinyeta, K., Whyte, K., & Lynn, K. (2016). *Climate Change Through an Intersectional Lens: Gendered Vulnerability and Resilience in Indigenous Communities in the United States* (SSRN Scholarly Paper No. ID 2770089). Rochester, NY: Social Science Research Network. Retrieved February 8, 2018, from https://papers.ssrn.com/abstract=2770089.

Waikato River Authority. (2019). Responding to Climate Change for the Waikato and Waipā rivers—implications for Te Ture Whaimana/Vision & Strategy, Hui summary and notes, Waikato-Tainui Endowed College, Ngaruawahia, 12 December 2019. Unpublished. Retrieved April 29, 2020, from https://waikatoriver.org.nz/climate-change-workshop/.

Watene, K. (2016). Valuing Nature: Māori Philosophy and the Capability Approach. *Oxford Development Studies, 44*(3), 287–296.

Watson, I. (2015). *Aboriginal Peoples, Colonialism and International Law*. Abingdon, Oxon: Routledge.

West, D. W. (2007). *Responses of Wild Freshwater Fish to Anthropogenic Stressors in the Waikato River of New Zealand* (Thesis). The University of Waikato. Retrieved August 15, 2019, from https://researchcommons.waikato.ac.nz/handle/10289/2601.

Whyte, K. (2020). Too Late for Indigenous Climate Justice: Ecological and Relational Tipping Points. *WIREs Climate Change, 11*(1), e603. https://doi.org/10.1002/wcc.603.

Whyte, K. L., Talley, J., & Gibson, J. (2019). Indigenous Mobility Traditions, Colonialism, and the Anthropocene. *Mobilities, 14*(3), 319–335. https://doi.org/10.1080/17450101.2019.1611015.

Whyte, K. (2017). The Dakota Access Pipeline, Environmental Injustice, and U.S. Colonialism. *Red Ink: An International Journal of Indigenous Literature, Arts, & Humanities,* (19.1). Retrieved June 8, 2019, from https://ssrn.com/abstract=2925513.

Winter, C. J. (2018). The Paralysis of Intergenerational Justice: Decolonising Entangled Futures. Retrieved January 11, 2020, from https://ses.library.usyd.edu.au/handle/2123/18009.

Winter, C. J. (2019). Does Time Colonise Intergenerational Environmental Justice Theory? *Environmental Politics, 1,* 1–19.

Yenneti, K., Day, R., & Golubchikov, O. (2016). Spatial Justice and the Land Politics of Renewables: Dispossessing Vulnerable Communities Through Solar Energy Mega-projects. *Geoforum, 76,* 90–99.

Zahara, A. (2017, March 14). Difference in the Anthropocene: Indigenous Environmentalism in the Face of Settler Colonialism. *Discard Studies.* Retrieved June 10, 2020, from https://discardstudies.com/2017/03/14/difference-in-the-anthropocene-indigenous-environmentalism-in-the-face-of-settler-colonialism/.

11

Conclusion: Spiralling Forwards, Backwards, and Together to Decolonise Freshwater

Within Te Ao Māori (Māori worldviews), wai (water) is at the heart of identity and life itself. The interconnections between wai and humans abound within the Māori language (Te Reo Māori) as one common whakataukī (proverb) states: "Kei te ora te wai, kei te ora te whenua, kei te ora te tangata. When the water is healthy, the land and the people are nourished". The word for water—wai—also means who and memory. Thus, when a Māori person meets someone new, they ask "Koi wai koe?" which translates as "Who are you" or more specifically "Who are your waters?". To answer that question necessitates that a person possesses the knowledge of their genealogical connections (whakapapa) to their tribe (iwi), sub-tribe (hapū) and ancestral river (Ruru 2012, p. 110). Fisher, for instance, belongs to Ngāti Maniapoto and her awa is the Waipā River. All iwi throughout Aotearoa use rivers as their ancestral identity markers (alongside mountains/maunga). Throughout this book, we sought to articulate how Indigenous interests in and rights to water go beyond access to freshwater (for drinking, sanitation, development) and extend to encompass identity, wellbeing, and authority.

In this concluding chapter, we seek to bring together our earlier analyses of the historical and contemporary waterscapes of the Waipā River

© The Author(s) 2021
M. Parsons et al., *Decolonising Blue Spaces in the Anthropocene*, Palgrave Studies in Natural Resource Management, https://doi.org/10.1007/978-3-030-61071-5_11

(Te Awa o Waipā), its interwoven histories, geographies, meanings, and physical and metaphysical entities. We braid together the examples outlined in previous chapters of this book to consider, once again, the theory and practice of Indigenous environmental justice (IEJ). We adopt the view that decolonisation is a process and that are numerous possibilities to decolonise freshwater governance and management approaches through recognising, procedurally including, and providing for Indigenous peoples' to express and enact their ontologies and epistemologies. Such work provides for expanding theorising about environmental justice (EJ) and providing empirical evidence of what practical (context-specific) efforts to achieve IEJ can consist of. While the works of scholars, including Schlosberg and Fraser, made significant contributions towards more pluralistic conceptualisations of justice, still such EJ frameworks continue to be rooted within Western intellectual thinking that takes universality and time-as-linear as givens of justice. In doing so, such theorising forecloses different understandings of what justice is and how it should be delivered. In the context of water justice, moreover, the discourse of 'rights' (water rights, Indigenous rights to water, human rights) are based on the Western legal 'rights' discourse, which does little to account for Indigenous conceptualisations of water responsibilities to and for water as a living being (or multiple beings). Indeed, by deliberately situating our book within the scholarship of EJ, rather than water justice (Perreault et al. 2018; Robison et al. 2018), we draw on Māori ontological thinking wherein all things are connected by reciprocal relationships (based on whakapapa), and that water cannot be separated from land (whenua), people (tangata), and all other parts of the cosmos (see Fig. 11.1). Everything is related and interwoven together through whakapapa (first discussed in Chap. 3). Each component (a plant, a river, a person, a mountain) both depend on and possesses responsibilities to care for one another. The goal is to ensure balance within the totality of Te Ao Māori, which involves the protection and enhancement of the life force (mauri) of all beings.

In this book, we sought to consider EJ in terms of our perspectives as scholars who identify as Māori/Pākehā/Hybrid Others as well as members of particular iwi (tribes), hapū (sub-tribes), and whānau (family) examining freshwater degradation of Te Awa o Waipā (which is the

Fig. 11.1 Photograph of Lake Ngāroto showing algae bloom in 2019. Source: Meg Parsons, 2019

ancestral river for two of our authors). We discuss how settler-colonialism resulted in violence (against people and ecosystems) and the dispossession of Māori iwi from their land and awa. Colonisation in Aotearoa, like in other settler-colonial societies, involved settlers physically inscribing their values, imagined geographies, and collective continuance through (what we now realise) unsustainable methods: deforestation, removal of endemic biodiversity, drainage of wetlands, productivist agriculture, air and water pollution, and so on. These means are underpinned on the settler-colonial narrative of a homeland (Aotearoa New Zealand as the 'Britain of the South Seas') but also frequently concealed in plain sight by stories of Māori 'wastelands' and untamed (unproductive) wilderness that mask histories of violence and dispossession of Māori (Hursthouse 1861; Whyte 2016, 2018). The inscriptions of the settler-colonial spaces provided the foundational conditions needed for settler collective continuance within Māori rohe (traditional lands and waters), while negatively impeding the capacities of Māori to maintain their cultural continuation. Settler-colonialism is, Whyte argues, a "structure of oppression based on one society's interference with and erasure of another society" which is both a driver and an outcome of the Anthropocene (Whyte 2016).

Given that settler-colonialism is an ongoing process that is "deeply ecological", it is always related to environmental injustices. In the catchment of the Waipā River, the settler-state and settlers, as we document in-depth in Chaps. 4 and 5, sought to "establish their collective continuance" over that of other societies (Ngāti Maniapoto and other iwi) (Whyte 2016). The Pākehā-dominated settler society imposed preventable harms on Māori communities to facilitate the former's process of making a new home, a place of belongingness, and security. The inscription process replaced Māori knowledge, values, laws, institutions, and ecologies with those of settler political institutions, social norms, environments and relationships. The foundation of EJ is centred on how people see, exist, and interact with the world. We demonstrate, throughout this book, that the environmental changes that took place within the Waipā catchment were (and are still) unjust because those changes (directed by one society for its benefit) robbed local Māori iwi, hapū and whānau of their capacities to experience their landscapes and waterscapes (their worlds) on their terms; which included their subsistence and flourishing as well as their abilities to maintain their systems of responsibilities. Although some scholarship on EJ only emphasise the distribution of environmental burdens or risks, our examples (drawn predominately from the Waipā) illustrate the challenges of negotiating relationships between different cultures, systems of law, governance and management, and what these negotiations mean in the context of addressing worsening freshwater crises in the Anthropocene. Therefore, let us briefly return to the three components of EJ discussed throughout the book as a means to tease out some of our thinking, but also to emphasise the ways in which these three categories blur together and are interwoven within IEJ.

Distributive Justice

In the previous chapters, we demonstrate how settler-colonial-led acts to transform Māori waterscapes into drained and canalled farmlands, straightened rivers, and flood levee-protected townships negatively impacted the health and wellbeing of iwi, hapū and whānau. Māori could no longer depend on their different mahinga kai (food gathering sites) to provide them with an abundance of foods (shellfish, fish, birds, plants) due to deforestation, drainage and flood control works, pollutants (human waste, livestock effluent, fertilisers and agri-chemicals), invasive introduced species, as well as the imposition of private property rights that restricted access. At the same time, as Māori faced ongoing losses of environmental "goods", Pākehā derived greater and greater material benefits from their newly created landscapes and waterscapes. Farms, factories, townships, piped and treated water supplies, as well as newly introduced exotic plants and animals (including cows, sheep, pigs, deer and trout) all ensured Pākehā communities' free access to environmental benefits (clean water, food) while Māori communities encountered more restricted access. When placed through the lens of distributional justice, there is clear evidence that the distribution of environmental risks and goods was inequitable in the Waipā catchment. Indeed, as we demonstrated in Chap. 4, government officials took deliberate actions to prevent Māori access to parts of the freshwater system of the Waipā (such as wetlands) which underpinned the health and wellbeing of local iwi, hapū, and whānau. Likewise, water pollution, a result of the ongoing discharges of effluent, chemicals and fertilisers (from towns, factories, and farms) onto lands and into waterways, disproportionately impacted Māori due to their use of waterways for swimming, bathing, drinking, and food sources. The distribution of environmental risks was (and are still) not equitable, we argue, because of the different ways in which Māori and Pākehā use and relate to the river. For instance, in 2020 many Māori (unlike Pākehā) continue to swim in and harvest foods from waters of the Waipā River and its tributaries (despite the waterways remaining highly unhealthy due to incredibly high counts of bacteria (E. Coli), nitrogen and phosphorus levels. They make the decision to use the polluted water and foods from the river (despite their awareness of potential health risks) because they consider such practices critical to retaining their cultural identity and continuance, mātauranga (knowledge), and spiritual integrity (wairua), with bathing and eating food collected from their ancestral waters critical to maintaining connections between tūpuna (ancestors which includes the river itself) and living people (discussed in Chap. 10). Yet, the distribution of environmental goods and harms are only a segment of the story of how Māori were (and still are) negatively impacted by changes wrought by settler-colonialism on their relationships with awa.

One of the pitfalls of much of the EJ scholarship relates to the employment of distributive justice (environmental equity) as the solution to address environmental injustices. Environmental equity is often simplistically conceived of in terms of the equitable distribution of society's environmental risks and benefits. To be sure, over that last two decades, more and more scholars' critiques within the EJ literature calls for more in-depth attention to examining the reasons underpinning such maldistribution (Agyeman et al. 2016; Schlosberg 2003; Swyngedouw and Heynen 2003). However, as we have demonstrated throughout this book's previous chapters, the distributive justice approach often renders invisible Indigenous peoples' experiences of environmental harms and benefits. Indeed, when many scholars discuss distributive equity, there is an underlying assumption that nature or environments can be exploited and turned into a distributable good (a resource, commodity). However, this conception is challenged by Indigenous relational ways of thinking and modes of life.

The struggles of Ngāti Maniapoto against settler colonialism, dispossession, and the destruction of their rohe were historically (throughout the last one hundred years) and now, in the present-day (2020), not just a challenge or fight against the "distribution of risks and impacts" (to borrow the words of Coulthard); but also about the right to live "concerning one another and the natural world in non-dominating and nonexploitative terms" (Coulthard 2014, p. 13). In consideration of this, it is difficult to see how the distributional equity of environmental resources amongst different populations would address many of the environmental injustices at hand. As demonstrated throughout our book, the narrow conceptualisation of distributional EJ, underpinned with Western intellectual framings (materialism, anthropocentrism, individualism, land/water/biota as property), is incompatible with Māori conceptualisations of the taiao (environment), awa (river) and whenua (land) as being part of one's extended family. These genealogical relationships (whakapapa), that stretch back to the creation of the cosmos (starting with Io/the supreme being, and continuing to Ranginui/Sky Father and Papa-tū-ā-nuku/Earth Mother, and the creation of the living beings), are centred on the idea that everything is connected and that everything possesses dignity which must be respected. The reciprocal relations between human and more-than-human entities, in the Waipā freshwater system, challenge Western views of water as a resource that can be commanded and controlled or quantified and allocated. This highlights the critical need for recognitional justice, wherein Indigenous ontologies and epistemologies are not only recognised by settler-states but also allowed to be enacted in ways that Indigenous peoples' themselves define as appropriate within contemporary settings. Our descriptions of Ngāti Maniapoto perspectives of freshwater management and governance offer an important entry point for others, in Aotearoa and around the globe, to think about what it means to move beyond dominant Euro-Western framings of water as a resource to measure, allocate, and control. We highlight how such challenges to the colonial capitalist order of things (which continues to the norm around the world) require the embrace of ontological pluralism wherein the existence of multiple worlds (not just Te Ao Māori and Te Ao

Pākehā) are not only acknowledged but also that practical actions are taken to ensure that these worlds (and worldviews) be allowed to flourish (Escobar 2016; Grosfoguel 2006; Rojas 2016; Salmond 2017). In light of these different ideas, ways of knowing, and modes of life, current framings of distributional EJ appears incompatible with many Indigenous ontologies and epistemologies. Moreover, efforts to address environmental injustices through distributive means (equitable distribution of environmental goods and burdens) do little to address many of the environmental injustices faced by Indigenous peoples.

Procedural Justice

Procedural injustice, as documented in Chap. 2, occurs when people possess "no voice or capacity to exercise self-determination in decision-making processes that affect their lives" and there are no acceptable reasons why those persons lack information or a voice (Whyte 2017, p. 117). As Shrader-Frechette writes, procedural justice premised on the principle of "participative justice" that aims to "ensure that there are institutional and procedural norms that guarantee that all people [have] equal opportunity for consideration in decision-making" (Shrader-Frechette 2002, p. 28). Since institutions and procedures (established by the settler-state) did not allow Ngāti Maniapoto to participate in decision-making processes about freshwater (and environmental management more generally) during the nineteenth century and the majority of the twentieth century, procedural EJ did occur. However, since the 1990s, procedural justice has been pursued through many different strategies that aim to redress the historical social and environmental injustices experienced by Māori. The passage of the Resource Management Act (1991) (RMA), outlined in Chaps. 5 and 6, required that local governments consider Māori interests in their responsibilities to create and implement regional and district plans, and consult with iwi. The RMA, for instance, directs local authorities to acknowledge Māori connections with rivers, including their wāhi tapu (sacred sites), harvesting practices, and understandings of rivers as more-than-human beings that possess agency and mauri (Ruru 2012). The RMA does provide some possibilities for Māori to be procedurally included within freshwater governance and management by providing Māori with a platform to speak about their concerns (in consultation meetings, public hearings and submissions to local authorities). However, Māori are still only participants in public consultative planning processes that are designed and administered by the settler-colonial-state rather than Māori and do not necessarily accord to mātauranga Māori (knowledge), tikanga (laws), and kawa (ceremonies) (Bell 2018; Ruru 2012). Local governments (and if the decision is appealed, the Environmental Court) are expected to consider Māori interests and concerns (often focused on the need to prevent and/or mitigate environmental degradation) over those of other interest groups (district councils, local developers, large energy corporations, farmers) who seek to maintain existing or create new infrastructure development opportunities (Ruru 2012). Ultimately, when local authorities and the courts make decisions about resource consents, both government and the judiciary frequently favour groups seeking to maintain the settler-colonial status quo (centred on the endless expansion of development and accumulation of material assets), and the priorities of Māori (most notably the need to enact the practices of kaitiakitanga—environmental guardianship) continued to be marginalised. The RMA does not give Māori the decision-making power to veto resource consent applications that breach their tikanga (Māori laws), such as those that threaten to diminish the mauri of their ancestral river and whenua.

Recently, new legislation created co-governance and co-management arrangements over the Waipā River that provided positions for iwi to negotiate with governments and seek to address their EJ issues, which provides for far greater inclusion of iwi interests than previous legislations. In particular, the creation of the joint management agreement (JMA) (signed between Ngāti Maniapoto and different territorial authorities) provides for some of local governments' functions and duties over freshwater management to be transferred to the iwi authority (Maniapoto Māori Trust Board). Ngāti Maniapoto considers the JMA an effective new tool to emerge from the Treaty settlement process which offers the iwi an opportunity to re-assert their rangatiratanga (authority) over their rohe (traditional lands and waters), actively monitor water quality, and seek to design and enact kaitiakitanga-based management processes to restore the health of their awa and their entire iwi. However, Ngāti Maniapoto iwi, hapū, and whānau still, despite the passage of new legislation as well as co-governance and co-management arrangements, encounter substantive challenges when trying to make meaningful actions to address freshwater problems (as we outlined in Chaps. 5, 7, 8, and 9). They do not necessarily possess the required resources (highlighting the distributive inequities) and capacities to support their restoration efforts as well as mount legal actions to hold government authorities, private landholders, and industries that fail to comply with regulations and continue to pollute and degrade their rivers. Despite new legislative and institutional arrangements, the Waikato Regional Council (WRC) and Environmental Court, as we outline in Chap. 5, consistently deny iwi complaints, submissions and court cases seeking to prevent local authorities discharging wastewater (human waste) into waterbodies because Māori claimants' cite only mātauranga Māori, cultural practices and tikanga not scientific knowledge to support their claims. Local governments and the courts argue that Māori experts speaking on mātauranga and tikanga are not equivalent to scientific experts. Furthermore, the courts regularly uphold the view that there is no legal basis (because Māori knowledge and laws are just 'cultural views' rather than legitimate evidence) to support their opposition to the discharge of human waste into water bodies. Thus, though there are now mechanisms that allow for iwi members to participate in decision-making processes, and iwi members take the time to attend council meetings, public hearings, and write and orally submit their views at such official local government forums (or launch legal cases and give evidence before the courts), there is no guarantee that Māori experiences, knowledge, and values will be taken seriously by decision-makers, who operate within Te Ao Pākehā and governance frameworks designed to secure and maintain the sovereignty and authority of the settler-state.

Recognition as Justice

Environmental injustice also occurs when laws, institutions, and practices are organised and enacted in ways that fail to recognise or respect the identities, knowledges, and values of certain populations. In this book, we discuss the numerous ways in which the settler-colonial state persistently did not recognise (or misrecognition) the ontologies and epistemologies of Māori groups. Recognition is a critical element of EJ (Barnhill-Dilling et al. 2020; Whyte 2017). Indeed, increased recognition by the settler-state, since the 1980s onwards, of the specific relationships that Māori iwi have with their rohe, provide an important entry point for addressing environmental injustices. As we outline in Chap. 6, the acknowledgement that iwi (including Ngāti Maniapoto) are not stakeholders but Treaty partners with the New Zealand Government (the Crown), and iwi hold particular interests in and decision-making authority in their traditional lands and waters are now incorporated in various statutes, Treaty settlements, local government policies, as well as co-governance and co-management arrangements. Although Tiriti o Waitangi (Treaty of Waitangi) is not recognised under domestic law, which means Māori do not possess any "general constitutional rights … [to] heard within the court setting" (Ruru 2012, p. 111), informal recognition of the Treaty and Māori interests has shown through in other legislation (such as the New Zealand Bill of Rights Act 1990 and the Constitution Act 1986). A range of statutes to explicitly acknowledge Māori knowledge (mātauranga) and laws (tikanga), including the principles of kaitiakitanga (environmental guardianship) and mauri (life force) as well as specific relationships iwi possess with their ancestral lands and waters (as mana whenua).

Recognition of Māori relational ontologies, which encompass the more-than-human, through new Treaty settlement legislation are presenting the possibility of disrupting Western worldviews and environmental management practices premised on anthropocentrism (and the nature-culture binary). A range of new legislation, imbued with ontological and legal pluralism, adopts different ways to acknowledge the more-than-human. In Aotearoa, this includes the awarding of legal personhood to geo-entities (the Whanganui River and Te Urewera range) and the specific naming of supernatural beings (the taniwha Waiwaia who lives in the Waipā River). Whereas, in South America, Mother Nature is now included in the constitutions of Bolivia and Ecuador. Yet, there remain substantive challenges in theories, laws, and practices that attempt to dissolve the "ontological divide" between Western and Indigenous intellectual traditions (first noted in Chap. 2) and multiple tensions come forth in translating Indigenous ethics of care (which includes Ngāti Maniapoto) mātauranga and tikanga into legislative and institutional frameworks, as we detail in Chap. 7 in regard to co-governance arrangements.

While we demonstrate how 'having a voice' or being recognised within settler-state apparatus is a significant step for Māori, it is not necessarily enough to overcome existing or emerging injustices (such as those associated with climate change). To begin with, as we outline throughout this book, state-based recognition of Indigenous interests can serve to (re) produce environmental injustices and of colonised subjectivities (Álvarez and Coolsaet 2018). A closer examination of existing processes demonstrates how different sorts of settler-colonial mechanisms contribute to influence decision-making processes that do not necessarily benefit Indigenous peoples and instead emphasis the settler colonial status quo. Indeed, in Chap. 6 we note how the practice of implementing Treaty settlement legislation remains problematic because the government decision-makers responsible for interpreting the legislation are frequently ignorant about what the purpose of Treaty settlements are, are unfamiliar with Māori concepts, and do not take the learn about Te Ao Māori to avoid misinterpreting the meaning and functions of new statutes. There is thus a considerable interpretative risk associated with state-based recognition through legislation as those in positions of authority (who are interpreting and implementing the laws) continue to enact the world as if it is only one (Te Ao Pākehā) instead of many. Along similar lines, international decolonial and Indigenous scholarship warn of the potential dangers of integrating "Indigenous philosophies into hegemonic institutions" can lead to "distortion, erasure and co-optation" and the emergence of a "new epistemic extractivism and violence" (Hunt 2014, pp. 24, 29; Sundberg 2014; Watts 2013; Widenhorn 2013). As Temper notes, there is a gap between theory and practice, and the tensions between the needs to make marginalised knowledges, ways of life, and practices politically relevant, which can erase, co-opt and distort Indigenous "knowledge systems through the very act of doing so" (Temper 2019, p. 14).

Interweaving and Layering of Justice: Pluralistic Accounts of IEJ

Herein lies the overlaps (interweaving) between recognitional, procedural, and distributive dimensions of EJ. Recognition of Māori cultural identity, knowledges, values, and practices (by the settler-state) through legislation and co-governance and co-management arrangements is not enough if it is not also accompanied by procedures that ensure the knowledge, meanings, and practices of Te Ao Māori are given equal weighting within legal and institutional regimes (that are responsible for how freshwater (land, sea, air) is governed and managed). For instance, Māori are increasingly recognised as mana whenua (tribal authority holders within their ancestral lands and waters) under legislation and new institutional arrangements, however, settler-state designed procedures (designed for collaborative planning) still inadequately accompany Māori practices (centred on collective discussions and consensus-based decision-making). Likewise, Māori groups (both formal institutions as well as whānau and hāpu) inadequate resources (both in terms of time and money) to allow them to participate in state-based planning processes, and often face the burdens of intergenerational deprivation (poverty, poor health, lack of education).

Ngāti Maniapoto iwi members desire to exercise their rangatiratanga and modes of governance within their rohe through practising ethics of care (kaitiakitanga) towards their waterscapes and landscapes; in doing so, they seek to heal and restore the mauri and wairua (spiritual integrity) of their awa and whenua (their more-than-human kin) as well as themselves as a collective group (as members of Ngāti Maniapoto).

Beyond Recognition to Encompass Indigenous Ontologies and Responsibilities

As an ordering principle, the colonial and capitalist modernity claims of itself "the right to be 'the world', subjecting all other worlds to its own terms, or worse, to non-existence" (Escobar 2016, p. 3). Yet, as our studies and countless others demonstrate, different ontologies and epistemologies (ways of knowing and being) persist around the world that provide possibilities to address social and environmental injustices through a plurality of approaches. A plethora of new research highlights the diversity of Indigenous peoples' relationships with waters and demonstrates that there are multiple ways of seeing the world(s) and responding to changing environmental conditions (Bischoff-Mattson et al. 2018; Castleden et al. 2017; Diver et al. 2019; Jackson 2018; McGregor 2015; Parsons and Fisher 2020; Wilson 2019). Such scholarship provides clear evidence of why freshwater management should not be solely framed through the gaze of scientific knowledge and modernising development. Yet, freshwater management scholars, decision-makers, and practitioners remain far too often situated within the universalising lens of Western ontologies (premised on water/land as property, materialism, individualism, anthropocentrism), wherein matters of water pollution, river management, flooding, and restoration are only situated and assessed through Western ontological and epistemological frameworks (McLean 2014; Parsons et al. 2019; Sarna-Wojcicki et al. 2019). In doing so, freshwater governance and management simply become new exercises in colonial modernity (itself the foundation of the multiple ecological crises of the Anthropocene).

That being said, the movement to embrace legal and ontological pluralism (discussed in Chap. 6) may be useful for destabilising settler-colonial and capitalist modernity, thereby, opening spaces for ontologies that do not fit within the settler-colonial command-and-control approaches to freshwater governance and management. In this newly emergent space (post-Treaty settlements) of co-governance and co-management agreements, Ngāti Maniapoto are seeking to re-assert their own legal order and to live their concept of EJ, through practices that disrupt the socio-economic and ecological logic and production of settler-colonial power. In identifying the contradictions that are central in the contemporary realities of freshwater management, in Aotearoa (specifically in the context of the Waipā River), and the ways in which such colonial orderings rest at the heart of many modern freshwater crises, we elucidate, is far more than the fracturing effects of the settler-colonial imposition of territorial boundaries and binaries (nature/culture and land/water). We demonstrate how the colonial logics of the universal (the unanimous applicability of scientific knowledge and technology, as well as the economy, development, and so forth) function(ed) to undermine and devalue Māori ways of being and their relationships with each other and their environments. Our descriptions of Indigenous perspectives of freshwater management, and socio-cultural and political life-worlds (which stand in contrast to the ordering principles of colonial, scientific, and capitalist modernity) offer an important entry point to integrate one of the core components of modernity (Escobar 2016; Grosfoguel 2006; Rojas 2016). In particular, the emphasis on intergenerational EJ within Te Ao Māori, as discussed in Chap. 10, is premised on the need to ensure that the mauri of the tangata, whenua, awa, and all other entities within their rohe are respected, maintained, and enhanced across temporal scales (including in the context of the impacts of climate change) (Winter 2018, p. 216). The movement towards decolonising freshwater governance and management, in the Anthropocene, requires such new (or some would say old) ways of thinking about and enacting respectful inter-being relationality. Respectful intergenerational and inter-being relationality goes beyond the concepts of 'environmental management' and 'sustainability'. Duties and responsibilities can and do move between times, spaces, realms, and forms and allow for flexibility in usage but always with the emphasis being on maintaining balance and life across generations and beings (maintaining the mauri of the river, the people, plants, animals, and others). As we document through our exploration of the changing waterscapes of Te Awa o Waipā throughout this book and the ways in which Māori experiences of environmental injustices are bound up in iwi webs of multiple interactions, wherein the sense of self is always interwoven in ongoing reciprocal relationships with their kin, biota, land and water, ancestors and future generations. Actions to achieve justice, similarly, reside in an interwoven or multifaceted approach.

Our case study suggests that moving towards more pluralistic forms of EJ requires expansions of the boundaries of justice away from its basis within Western liberal philosophies and theories to encompass different philosophies and modes of thought. Thus, when scholars and activists advocate for the inclusion of inter-species and non-humans (including rivers and "Mother Nature") to be given legal recognition and provided for in accounts of EJ, we do not only cite the work of Western theorists and those who draw on Western intellectual traditions (Rawls, Fraser, and Latour) (Fraser 1995; Latour 1996; Rawls 2009). In this book, we use examples from our small corner of the world (which is obviously not without its limitations and an example of strategic localism rather than universalism) to attest to how Western knowledge and the political, economic and social structures of settler-colonialism created (and still create) ontological and epistemological divides that sought to keep (and treat) nature as separate from society, and land as divided from water. Yet, since the commencement of the settler-colonial acts (military and eco-violence, dispossession and marginalisation), Māori groups have persistently sought to challenge and resist the establishment of ontological and epistemological divisions between tangata whenua, whenua, and awa. Māori consistently advocate for their holistic understandings to be recognised by others but also sought to ensure that they enact their own possibilities to transcend and restrict colonial/Western liberal constructs. In doing so, iwi demonstrate how, despite the depth of colonising lines that were and are still drawn across the land and waters (be it through land surveys and the construction of drainage canals, or social norms and political institutions), iwi are seeking to create and (re)assert their own ways of governing and managing and enacting the practices of kaitiakitanga through using their own mātauranga, language, and ways of seeing the world as well as drawing on Western scientific knowledge, approaches and practices (on their own terms). Rather than a politics of refusal and resistance, noted by some scholars, the process involves healing themselves and their awa and sharing with others as a means to transform their awa and the capacities of others to heal. Most importantly, it goes beyond the theory of decolonising freshwater and involves actions and lived practices of kaitiakitanga (guardianship) and intergenerational EJ based on responsibilities to ancestors (past, present, future) both human and more-than-human.

There is a world of difference between perceiving the river as an object (fluvial geomorphological, H20, commodity) and as a subject (wahine/female Waipā, ancestor, kin). The first viewpoint denotes something that can be measured, quantified, divided and controlled, and something situated at the margin of humanity/distance for the political economy/culture/values (a wet surface that cut across lands). The second viewpoint is a perspective that centres of the totality of connections and inter-relationships. The first perspective frames the Waipā River from the settler (Western/European) viewpoint. Settlers drew both real and imaginary lines and physical structures across landscapes, waterscapes, and seascapes, and in doing so imposed colonial boundaries that sought to define, divide, and confine spaces, biota, and peoples to restricted for the first time. The dividing lines continue in the present-day, and seek to separate: human health from river health; taonga (treasures) from wai (water); material from metaphysical; economic security from water security; kaitiaki (guardians) from practices of kaitiakitanga (environmental guardianship); wairua (spiritual integrity) from mauri (life force); and the expression of rangatiratanga (chiefly authority and rights of self-determination) from that of mana (power, prestige and sovereignty).

References

Agyeman, J., Schlosberg, D., Craven, L., & Matthews, C. (2016). Trends and Directions in Environmental Justice: From Inequity to Everyday Life, Community, and Just Sustainabilities. *Annual Review of Environment and Resources, 41*, 1.

Álvarez, L., & Coolsaet, B. (2018). Decolonizing Environmental Justice Studies: A Latin American Perspective. *Capitalism Nature Socialism, 1*, 1–20.

Barnhill-Dilling, S. K., Rivers, L., & Delborne, J. A. (2020). Rooted in Recognition: Indigenous Environmental Justice and the Genetically Engineered American Chestnut Tree. *Society & Natural Resources, 33*(1), 83–100.

Bell, A. (2018). A Flawed Treaty Partner: The New Zealand State, Local Government and the Politics of Recognition. In D. Howard-Wagner,

M. Bargh, & I. Altamirano-Jimenez (Eds.), *The Neoliberal State, Recognition and Indigenous Rights: New Paternalism to New Imaginings* (pp. 77–92). Canberra: ANU Press.

Bischoff-Mattson, Z., Lynch, A. H., & Joachim, L. (2018). Justice, Science, or Collaboration: Divergent Perspectives on Indigenous Cultural Water in Australia's Murray–Darling Basin. *Water Policy, 20*(2), 235–251.

Castleden, H., Hart, C., Cunsolo, A., Harper, S., & Martin, D. (2017). Reconciliation and Relationality in Water Research and Management in Canada: Implementing Indigenous Ontologies, Epistemologies, and Methodologies. In S. Renzetti & D. P. Dupont (Eds.), *Water Policy and Governance in Canada* (pp. 69–95). Cham: Springer.

Coulthard, G. S. (2014). *Red Skin, White Masks: Rejecting the Colonial Politics of Recognition*. University of Minnesota Press. Retrieved May 19, 2019, from https://muse.jhu.edu/book/35470.

Diver, S., Ahrens, D., Arbit, T., & Bakker, K. (2019). Engaging Colonial Entanglements: "Treatment as a State" Policy for Indigenous Water Co-Governance. *Global Environmental Politics, 19*(3), 33–56.

Escobar, A. (2016). Thinking-feeling with the Earth: Territorial Struggles and the Ontological Dimension of the Epistemologies of the South. *AIBR, Revista de Antropología Iberoamericana, 11*(1), 11–32.

Fraser, N. (1995). Recognition or Redistribution? A Critical Reading of Iris Young's Justice and the Politics of Difference*. *Journal of Political Philosophy, 3*(2), 166–180.

Grosfoguel, R. (2006). World-Systems Analysis in the Context of Transmodernity, Border Thinking, and Global Coloniality. *Review (Fernand Braudel Center), 29*(2), 167–187.

Hunt, S. E. (2014, March 3). *Witnessing the Colonialscape: Lighting the Intimate Fires of Indigenous Legal Pluralism* (Thesis). Environment: Department of Geography. Retrieved from http://summit.sfu.ca/item/14145%23310.

Hursthouse, C. F. (1861). *New Zealand: The "Britain of the South"*. Stanford: Stanford University.

Jackson, S. (2018). Indigenous Peoples and Water Justice in a Globalizing World. In K. Conca & E. Weinthal (Eds.), *The Oxford Handbook of Water Politics and Policy* (p. 120). Oxford: Oxford University Press.

Latour, B. (1996). On Actor-network Theory: A Few Clarifications. *Soziale Welt, 47*(4), 369–381.

McGregor, D. (2015). Indigenous Women, Water Justice and Zaagidowin (love). *Canadian Woman Studies, 30*(2–3), 1.

McLean, J. (2014). Still Colonising the Ord River, Northern Australia: A Postcolonial Geography of the Spaces Between Indigenous People's and Settlers' Interests. *The Geographical Journal, 180*(3), 198–210. https://doi.org/10.1111/geoj.12025.

Parsons, M., & Fisher, K. (2020). Indigenous Peoples and Transformations in Freshwater Governance and Management. *Current Opinion in Environmental Sustainability, 43*, 8.

Parsons, M., Nalau, J., Fisher, K., & Brown, C. (2019). Disrupting Path Dependency: Making Room for Indigenous Knowledge in River Management. *Global Environmental Change, 56*, 95–113.

Perreault, T. A., Boelens, R., & Vos, J. (2018). Conclusions: Struggles for Justice in a Changing Water World. *Water Justice, 1*, 346–360.

Rawls, J. (2009). *A Theory of Justice*. Boston: Harvard University Press.

Robison, J., Cosens, B., Jackson, S., Leonard, K., & McCool, D. (2018). Indigenous Water Justice. *Lewis & Clark L. Rev., 22*, 841.

Rojas, C. (2016). Contesting the Colonial Logics of the International: Toward a Relational Politics for the Pluriverse. *International Political Sociology, 10*(4), 369–382.

Ruru, J. (2012). The Right to Water as the Right to Identity: Legal Struggles of Indigenous Peoples of Aotearoa New Zealand. In F. Sultana & A. Loftus (Eds.), *The Right to Water: Politics, Governance and Social Struggle* (pp. 110–122). New York: Routledge.

Salmond, A. (2017). *Tears of Rangi: Experiments Across Worlds*. Auckland, New Zealand: Auckland University Press.

Sarna-Wojcicki, D., Sowerwine, J., & Hillman, L. (2019). Decentring Watersheds and Decolonising Watershed Governance: Towards an Ecocultural Politics of Scale in the Klamath Basin. *Water Alternatives, 12*(1), 26.

Schlosberg, D. (2003). The Justice of Environmental Justice: Reconciling Equity, Recognition, and Participation in a Political Movement. *Moral and Political Reasoning in Environmental Practice, 77*, 106.

Shrader-Frechette, K. (2002). *Environmental Justice: Creating Equality, Reclaiming Democracy*. Oxford: Oxford University Press.

Sundberg, J. (2014). Decolonizing Posthumanist Geographies. *Cultural Geographies, 21*(1), 33–47.

Swyngedouw, E., & Heynen, N. C. (2003). Urban Political Ecology, Justice and the Politics of Scale. *Antipode, 35*(5), 898–918.

Temper, L. (2019). Blocking Pipelines, Unsettling Environmental Justice: From Rights of Nature to Responsibility to Territory. *Local Environment, 24*(2), 94–112.

Watts, V. (2013). Indigenous Place-Thought and Agency Amongst Humans and Non Humans (First Woman and Sky Woman Go On a European World Tour!). *Decolonization: Indigeneity, Education & Society, 2*(1). Retrieved May 16, 2020, from https://jps.library.utoronto.ca/index.php/des/article/view/19145.

Whyte, K.P. (2016). Indigenous Experience, Environmental Justice and Settler Colonialism. In B. Bannon (Ed.), *Nature and Experience: Phenomenology and the Environment* (pp. 157–174). Lanham: Rowman & Littlefield. Retrieved January 30, 2020, from http://www.ssrn.com/abstract=2770058.

Whyte, K. P. (2017). The Recognition Paradigm of Environmental Injustice. In R. B. Holifield, J. Chakraborty, & G. P. Walker (Eds.), *The Routledge Handbook of Environmental Justice* (pp. 113–123). London; New York: Routledge.

Whyte, K. P. (2018). Settler Colonialism, Ecology, and Environmental Injustice. *Environment and Society, 9*(1), 125–144.

Widenhorn, S. (2013). Towards Epistemic Justice with Indigenous Peoples' Knowledge? Exploring the Potentials of the Convention on Biological Diversity and the Philosophy of Buen Vivir. *Development, 56*(3), 378–386.

Wilson, N. J. (2019). "Seeing Water Like a State?": Indigenous Water Governance Through Yukon First Nation Self-Government Agreements. *Geoforum, 1*, 101–113.

Winter, C. J. (2018). The Paralysis of Intergenerational Justice: Decolonising Entangled Futures. Retrieved January 11, 2020, from https://ses.library.usyd.edu.au/handle/2123/18009.

Correction to: Decolonising Blue Spaces in the Anthropocene

Correction to:

M. Parsons et al., *Decolonising Blue Spaces in the Anthropocene*, Palgrave Studies in Natural Resource Management, https://doi.org/10.1007/978-3-030-61071-5

This book was inadvertently published with incorrect information with reference to The Waikato River Authority on pages 306, 307, 369 and 370. This text has now been revised and updated.

The updated version of these chapters can be found at
https://doi.org/10.1007/978-3-030-61071-5_7
https://doi.org/10.1007/978-3-030-61071-5_9

Appendix: Table of Interview Participants

	Designation	Date of interview
1	Local Government Rep 1	4 October 2018
2	Local Government Rep 2	25 March 2019
3	Local Government Rep 3	29 September 2017
4	Scientist 1	4 September 2017
5	Scientist 2	7 November 2019
6	Iwi Rep 1	29 September 2017
7	Iwi Rep 2	13 February 2020
8	Iwi Rep 3	13 February 2020
9	Iwi Rep 4	14 February 2020
10	Iwi Rep 5	25 March 2019
11	Iwi Rep 6	14 February 2020
12	Iwi Rep 7	16 May 2019; 13 June 2019
13	Iwi Rep 8	9 October 2019
14	NGO Rep 1	28 September 2017
15	Māori business owner 1	29 August 2019
16	Māori business operator 2	10 April 2019
17	Kaitiaki 1	4 February 2020
18	Kaitiaki 2	4 February 2020
19	Kaitiaki 3	5 February 2020
20	Kaitiaki 4	5 February 2020
21	Kaitiaki 5	5 February 2020

© The Author(s) 2021
M. Parsons et al., *Decolonising Blue Spaces in the Anthropocene*, Palgrave Studies in
Natural Resource Management, https://doi.org/10.1007/978-3-030-61071-5

Glossary of Te Reo Māori Terms

ariki paramount chief

atua god

awa river, stream, tributary, riverine

hani the male essence

hapū sub-tribes

hāpua lagoon, pool of water, pond

harakeke *Phormium tenax*—New Zealand flax

hauora health and wellbeing

he Wakaputanga o te Rangatiratanga o Nu Tireni The Declaration of Independence of New Zealand signed by a number of Māori rangatira (chiefs) in 1835

heke migrations

hīnaki woven basket used to catch tuna

hui meeting

inanga *galazias spp.*—whitebait

io-matua-te-kore the supreme being (translates as the parentless one) and original energy

iwi tribes

kainga villages

kaitiaki environmental guardians

© The Author(s) 2021
M. Parsons et al., *Decolonising Blue Spaces in the Anthropocene*, Palgrave Studies in Natural Resource Management, https://doi.org/10.1007/978-3-030-61071-5

kaitiakitanga environmental guardianship, obligations and duties to protect, nurture and care for one's tribal lands, waters, and seas; passed on from one's ancestors to present-day tangata whenua; the basis of māori environmental governance and management practices

kākahi Echyridella spp.—freshwater mussels

karanga part of the cultural protocol of a powhiri (welcome ceremony). Karanga involve the exchange of calls between senior women; either to welcome (if hosts) their visitors or to acknowledge their hosts (if visitors) onto a marae (the meeting place of a hapū)

kaumatua elders

kaupapa principles and ideas

kawa ceremonies

kāwanatanga governorship

kereru New Zealand Pigeon—*Hemiphaga novaeseelandiae*

Kīngitanga the King movement

Kingite Māori who supported Kīngitanga

kōhanga nursery (Te Kōhanga Reo are preschools where children are immersed in the Māori language and values)

Korero language, conversation

kotahitanga collective action

kōuka cabbage trees—Cordyline australis

koura freshwater crayfish—*Paranephrops planifrons*

kuia grandmother

kūmara sweet potato—*Ipomoea batatas*

kura school

kuta *Elochiris sphacelata*—bamboo spike sedge

mahinga kai food gathering sites

mana whenua tribal group with authority over land or territory, power from the land, jurisdiction over land or territory (power associated with occupation and possession and occupation of tribal territories)

mana power, sovereignty, political authority, social standing, and prestige

manaakitanga hospitality, generosity, kindness, support, the process of demonstrating generosity, respect, and care for others

manga creek, river, riverine

marae the open area or courtyard in front of a tribal house (wharenui), where formal greetings take place. The term marae is also used to refer to the complex of buildings surrounding the marae

mātauranga Māori knowledge

matua parent
maunga mountain
mauri life force
Mihinare Christian
moana sea, lacustrine
muriwai estuarine
noa normal, ordinary, safe, not subject to restrictions
o Nu Irene the declaration of independence
pā tuna eel weirs
pā whawhai Fortified settlements used during conflicts
Pākehā non-Māori, Europen, foreigner (modifier) and New Zealand European
 ethnic group (noun)
pakiwaitara Legend or story
Papa-tū-ā-nuku Earth Mother
Paru dirty, filthy
pātaka kai Food store
Patupaiarehe Fairy people (some believe this is the origin of the term Pākehā)
pepeha Recitations linking people to place
piharau *Geotria australis.*—lamprey
poukai ceremonial gathering held 28 days a year at different marae or ceremonial centres supporting the Kīngitanga movement
powhiri welcome ceremony
pūhā *Asteraceae.*—stow thistles
puna the female essence
pūrākau traditions and stories
rāhui temporary restriction placed by a chief on people accessing and using a
 certain area
rangatahi younger generations
rangatira Chief or high rank
rangatiratanga chieftainship, chiefly autonomy and authority, leadership of a
 social group, attributes of a chief, right to exercise authority, sovereignty, self-
 determination, self-management
Ranginui Sky Father
raupatu confiscated lands
rāupo reeds
repo wetlands
rohe region, territory

Rongomātāne First-born child and god of peace and cultivated food; traditional lands and waters

roto lake

Rūaumoko God of seasons

rūnganga tribal councils

taiao environment

Tāmaki Makaurau modern-day Auckland

Tāmaki Makaurau modern-day Auckland

Tāne-mahuta God of the Forest

tangaroa God of sea, rivers lakes and aquatic life

tangata whenua people of the land

taniwha supernatural beings that live in waterbodies

taonga treasures, treasured possessions

tapu When used as a modifier tapu refers to prohibited, sacred, restricted, set apart, forbidden, and under the protection of the gods (modifier). When used as noun it refers to something (person, place, thing) restriction, prohibition, and in a supernatural condition. When person, place or thing is tapu it is removed from ordinary use (noa) placed in the sacred sphere where it is untouchable.

taro Root crop—*Colocasia esculenta*

Tāwhirimātea God of winds

Te Ao Māori the world of Māori

Te Ao Pākehā the world of Pākehā aka the settler world

Te Ika-a-Māui North Island of New Zealand

Te Kore The Void

Te Rohe Pōtae The King Country

Te Tiriti o Waitangi The Treaty of Waitangi

tikanga Māori customary laws and principles

tino rangatiratanga sovereignty, self-government, domination, rule, control, power, authority and autonomy

Tuku to grant or gift, to release or transfer

Tūmatauenga God of war

tuna *Anguilla spp.*—short and long-tailed freshwater eels

Tupuna Ancestors

ture law

Uenuku God of rainbows; god of food gathering

Urukehu Fair-haired people

urupā cemetery

utu revenge, retaliation and retribution to maintain balance and harmony in relationships between individuals and groups within Māori society

wahapū estuarine

wāhi tapu sacred sites

wahine Māori woman

wai mā clear and pure waters

wai mangu dark waters

wai water

waiata song

wairua spiritual integrity or spirit of a person. Some iwi and hapū consider that all human and more-than-human beings possess both a whakapapa and a wairua. Others believe that atua Māori (or Io-matua-kore) can give wairua into something (animate or inanimate).

waka canoe

whakapapa ancestral lineage, ancestral connections, genealogical relationships

whakataukī Māori proverbs

whānau family, extended family

whanaungatanga extended family, responsibilities, relationships, the centrality of kinship, whakapapa that binds the Māori world together

whāngai customary practice of adoption or fostering where a child is raised by a person/family other than their biological parents. Usually the adoptive parent(s) are relatives of the child. Common forms of whāngai include grandparents raising a grandchild and educated in mātauranga and tikanga, or an orphan or child who is not able to be raised by their birth parents.

whare komiti committee meeting

whare house

whenua land

Index[1]

---

[1] Note: Page numbers followed by 'n' refer to notes.

© The Author(s) 2021   **491**
M. Parsons et al., *Decolonising Blue Spaces in the Anthropocene*, Palgrave Studies in
Natural Resource Management, https://doi.org/10.1007/978-3-030-61071-5